VORTEX METHODS

VORTEX METHODS

Selected Papers of the First International Conference on Vortex Methods

Kobe, Japan 4–5 November 1999

Edited by

Kyoji Kamemoto
Yokohama National University

Michihisa Tsutahara
Kobe University

World Scientific
Singapore • New Jersey • London • Hong Kong

Published by

World Scientific Publishing Co. Pte. Ltd.
P O Box 128, Farrer Road, Singapore 912805
USA office: Suite 1B, 1060 Main Street, River Edge, NJ 07661
UK office: 57 Shelton Street, Covent Garden, London WC2H 9HE

British Library Cataloguing-in-Publication Data
A catalogue record for this book is available from the British Library.

VORTEX METHODS
Selected Papers from the First International Conference on Vortex Methods

ISBN 981-02-4277-8

Printed in Singapore by FuIsland Offset Printing

Preface

The vortex methods have been developed mathematically and physically, and applied to many kinds of flows related to various problems in wide engineering and scientific fields. Recently, the JSME (Japan Society of Mechanical Engineers) research committee on vortex methods has been closed after many fruitful contributions to development of the vortex methods in related engineering fields during the five years from April 1994 to March 1999. On the occasion of the closing of the research committee, the Organizing Committee consisting of the members, i.e. Prof. Kyoji Kamemoto, Prof. Masaru Kiya and Prof. Teruhiko Kida, Prof. Michihisa Tsutahara, was set up for the purpose of providing an opportunity for engineers and scientists to present their achievements, exchange ideas and discuss new developments in mathematical and physical modeling techniques and engineering applications of vortex methods. Then, the *First International Conference on Vortex Methods* was held at Maiko Villa, Kobe, Japan, during November 4-5, in 1999, under the sponsorship of Kobe University.

In this conference, six keynote lectures by prominent researchers and twenty-three presentations of contributed papers on the following topics were included: theoretical investigations on vortex methods and numerical simulations on vortex structures, flow around solid bodies, proposals for solution procedures of vortex methods, flows with vibration, vortex methods on heat transfer and two phase flows, and application for turbomachinery and engineering problems.

This book deals with the more recent vortex method research being carried out at different institutions in the world, and the material presented is based on the edited versions of excellent papers selected from those presented at the conference.

We thank all the members of the Organizing Committee for their valuable advice and their assistance in selecting these papers. Acknowledgements are also due to Dr. Kazuhiko Ogawa and Miss Shoko Inoue of Kobe University for their helpful and efficient administrative support. Last but certainly not the least, we thank the authors for providing interesting papers.

January 2000 Kyoji Kamemoto and Michihisa Tsutahara

Contents

VORTEX ELEMENT METHODS, THE MOST NATURAL APPROACH TO FLOW SIMULATION - A REVIEW OF METHODOLOGY WITH APPLICATIONS

R. Ivan Lewis
Department of M.M.M.Eng,, Newcastle University,
Room 2-16 Bruce Building, Newcastle upon Tyne , NE1 7RU, UK/ Email R.I.Lewis@ncl.ac.uk

ABSTRACT

The basis of vortex element methods will be reviewed including the two categories of (i) surface vorticity modelling of inviscid potential flows and (ii) extension of this to full vortex cloud simulation of real flows of a viscous fluid. Examples will be given of the surface vorticity method for aerofoils, cascades related to axial, mixed-flow and radial turbomachines and axisymmetric flow through ducts and turbomachine annuli. Extensions of this to full vortex cloud simulation will also be summarised.

1. VORTEX ELEMENT METHODS

Vortex element methods represent the most natural approach to flow modelling because they provide a direct simulation of the real vortical fluid motion in the most economical manner. Thus it can be argued that all fluid motions are driven and controlled by vorticity created at the body or duct surfaces and shed freely into the surrounding fluid where it undergoes convection and diffusion. The main aim of this paper is to support this statement by reviewing the basic methodology underlying vortex methods and to illustrate its application to several important engineering flow problems. The author[1] undertook a major research review along these lines in 1991 to which the reader is referred for greater detail including a number of computer codes given in the appendix. As shown there, vortex methods have developed along two directions, namely

(i) surface vorticity modelling of potential flows,
(ii) vortex cloud (or vortex dynamics) simulation of real viscous fluid flows,

the fundamentals of which will be given in sections 1.1 and 1.2 respectively of this introduction. Although these have often been persued numerically as separate almost disconnected types of flow, the present paper will show that item(ii) is a natural progression from item(i) and a common numerical methodology embracing both will be presented. In section 2 vortex cloud modelling will be extended to flow past multiple bodies in relative motion. In section 3 treatments will be given for turbomachinery cascade flows and blade rows, including a summary of recent attempts to simulate turbine stator/rotor interactions. In section 4 consideration will be given to vortex element modelling of axisymmetric flows including a summary of a recent attempt to find a random walk simulation for diffusion of a ring vortex for vortex dynamics simulation of axisymmetric flows.

1.1 The Surface Vorticity Method for Analysis of Potential Flows

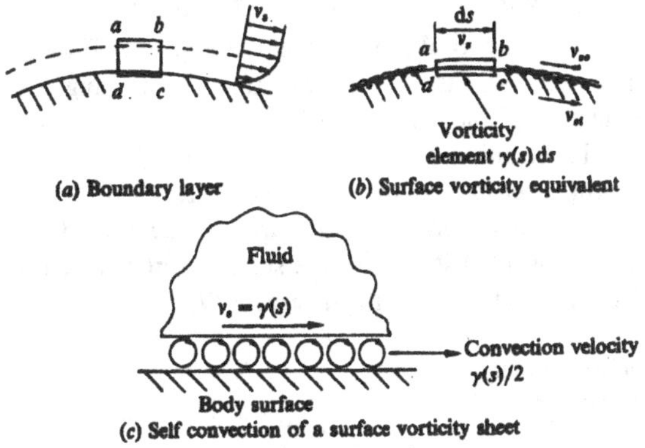

Figure 1 Boundary layer and surface vorticity equivalent in potential flow.

Although the title "Vortex Methods" tends to be linked today with the relatively recent developments of vortex dynamics, the real story began with the modelling of inviscid potential flows for which the key seed-corn paper in this field was presented by E. Martensen[2] in 1959. As illustrated in Figure 1, the flow naturally occurring in real flows at a body surface consists of the familiar boundary layer, comprising a shear layer through which the velocity changes progressively from zero on the wall surface to the main-stream value v_s at the outer edge of the layer. Thus in the real fluid flow vorticity is continuously created at the wall in the laminar sub-layer and once created is free to undergo convection under all surrounding influences and diffusion under the action of viscosity.

At increasing Reynolds numbers the layer will shrink in thickness to the limiting case of an infinitely thin vortex sheet of strength $\gamma(s)$ across which the velocity increases from zero beneath the sheet to the value v_s just outside the sheet, Figure 1(b), where

$$v_s = \gamma(s) \qquad (1)$$

The infinite Reynolds number flow of a real fluid thus approaches the potential flow situation in which the body surface is clothed in an infinitely thin vorticity sheet. We observe however that the vorticity sheet $\gamma(s)$ is not bound to the body surface but convects freely along the surface in potential flow with self-induced velocity of strength.

$$v_c = \tfrac{1}{2}\gamma(s) \qquad (2)$$

From these observations Martensen[2] derived the well known boundary integral equation which states the surface boundary condition applicable just beneath the vortex sheet, namely that the velocity parallel to the surface is zero at any point s on the body. Thus for point s_m on an aerofoil surface, the Dirichlet boundary condition of zero parallel velocity may be stated

$$\tfrac{1}{2}\gamma(s_m) + \oint k(s_m, s_n)\gamma(s_n)ds_n + W_\infty(\cos\alpha_\infty \cos\beta_m + \sin\alpha_\infty \sin\beta_m) = 0 \qquad (3)$$

where β_m is the local profile slope as defined in Figure 2 below and W_∞ is a uniform stream with angle of incidence α_∞ into which the body is placed. The first term in equation(3) states the velocity discontinuity stepping from the centre of the vorticity sheet onto its inside in contact with the body surface. The second term provides the contribution of the surface vorticity and the third term that of the uniform stream W_∞.

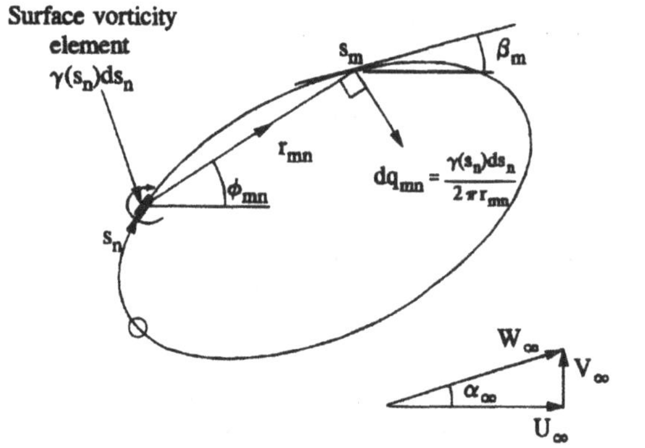

Figure 2 Surface vorticity model.

The coupling coefficient $k(s_m, s_n)$ linking points **m** and **n** for plane two-dimensional flow then follows from Figure(2), namely

$$k(s_m, s_n) = \frac{1}{2\pi}\left\{\frac{(y_m - y_n)\cos\beta_m - (x_m - x_n)\sin\beta_m}{(x_m - x_n)^2 + (y_m - y_n)^2}\right\} \qquad (4)$$

In order to solve this integral equation the standard procedure is to discretise the body surface into M small elements of length Δs_n as illustrated by Figure 3 below

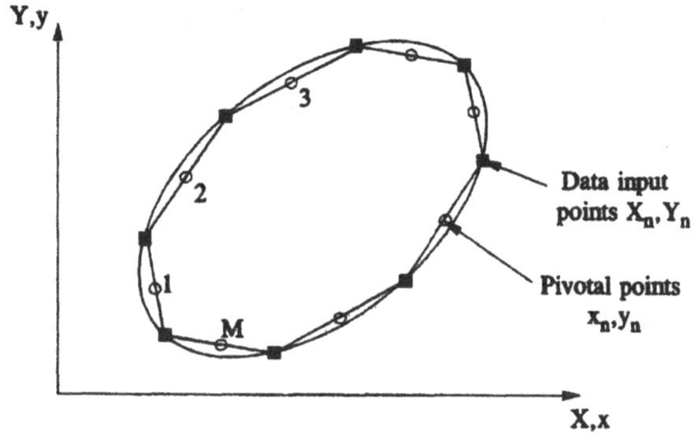

Figure 3 Numerical discretisation of the body surface.

whereupon the boundary integral equation reduces to a set of M linear equations of the form

$$\sum_{1}^{M} K(s_m, s_n)\gamma(s_n) = -U_\infty \cos\beta_m - V_\infty \sin\beta_m \qquad (5)$$

with the modified coupling coefficient $K(s_m, s_n)$ linking elements m and n given by

$$K(s_m, s_n) = \frac{\Delta s_n}{2\pi}\left\{\frac{(y_m - y_n)\cos\beta_m - (x_m - x_n)\sin\beta_m}{(x_m - x_n)^2 + (y_m - y_n)^2}\right\} \qquad (6)$$

To interpret this, we observe that the quantity $K(s_m, s_n)/\Delta s_n$ is equal to the velocity parallel to the body surface at element m induced by a unit strength vortex located at the center of element n. The coupling coefficients $K(s_m, s_n)$ may be evaluated once and for all from the body geometry thus determining the left hand side matrix of equation(5). Fortunately the matrix has a dominant back-diagonal and is thus well disposed to matrix inversion leading to a quick and simple solution for the surface vorticity elements $\gamma(s_n)$ and hence the surface velocity v_{sn} from equation(2). As shown by Wilkinson[3], the back diagonal or "self-inducing" coupling coefficients reduce to the form

$$K(s_m, s_m) = -\frac{1}{2} + \frac{\Delta s_m}{4\pi}\left\{\frac{-\dfrac{d^2 y_m}{dx_m^2}}{\left\{1 + \left(\dfrac{dy_m}{dx_m}\right)^2\right\}}\right\} = -\frac{1}{2} + \frac{\Delta s_m}{4\pi r_m}$$

$$\approx -\frac{1}{2} - \frac{\Delta\beta_m}{4\pi} \approx -\frac{1}{2} - \frac{1}{8\pi}(\beta_{m+1} - \beta_{m-1}) \qquad (7)$$

where the first term -½ represents the step from the centre of the vorticity sheet at point m onto the inner surface and the second term accounts for the effect of local surface curvature of the element m. Although the first term is dominant, the second term must be included to deal with the additional self-induced velocity of finite element m due to the local surface curvature.

1.2 Extension of Surface Vorticity Method to Full Vortex Cloud Modelling

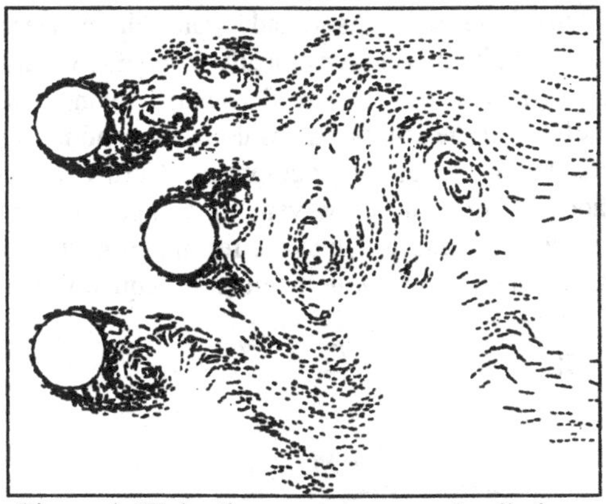

Figure 4 Wake flow past a group of cylinders.

Up to now we have considered the case of irrotational potential (or infinite Reynolds number) flow only. As already pointed out however, even in such ideal fluid cases the surface vorticity is not bound but in fact freely convecting along the body surface. It seems therefore but a small and natural step to model the flow of a real fluid in which the vortex elements at time t created over a small time step Δt are simply shed and allowed to be free thereafter to convect and diffuse. In actual fact the author's first publication[4] on vortex dynamics simulation of bluff body flows in 1981 did not include a model for viscous diffusion and assumed separation of the vortex sheet at prescribed body locations. In parallel with this however Porthouse[5] was developing a random walk model for viscous diffusion combined with more advanced vortex dynamics schemes involving vortex shedding from the entire body surface in the manner illustrated by Figure 5 below and was the subject of a second closely related publication[6]. After many studies and applications, to bluff body flows in particular, the methodology is now well established and was reviewed in depth by the author[1], including reference to a range of other parallel published work.

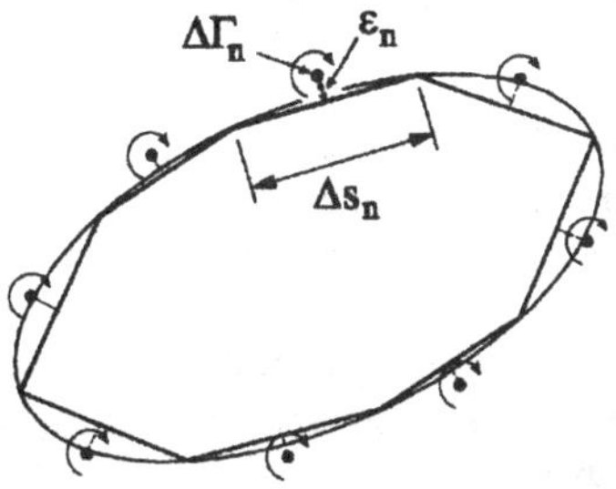

Figure 5 Shedding of discrete vortex elements $\Delta\Gamma_n=\gamma(s_n)\Delta s_n$ from the body surface.

Thus following a potential flow calculation each new surface vorticity sheet element may be discretised into a concentrated vortex of the same strength $\Delta\Gamma_n=\gamma(s_n)\Delta s_n$ and placed a small distance ε_n normal to the centre of the element as illustrated in Figure 5. This is the now well established technique of vortex cloud analysis or vortex dynamics resulting in such simulations as that illustrated in Figure 4. The aim here is to solve the two-dimensional Navier-Stokes equations which may be expressed in vector form through

$$\frac{\partial \hat{q}}{\partial t} + \hat{q}\cdot\nabla\hat{q} = -\frac{\nabla p}{\rho} + \nu\nabla^2\hat{q} \tag{8}$$

(A) (B) (C) (D)

Reading from left to right we are reminded that unsteady (A) and convective (B) fluid motions are related to pressure gradients, i.e. normal stresses (C) and viscous shear stresses (D).

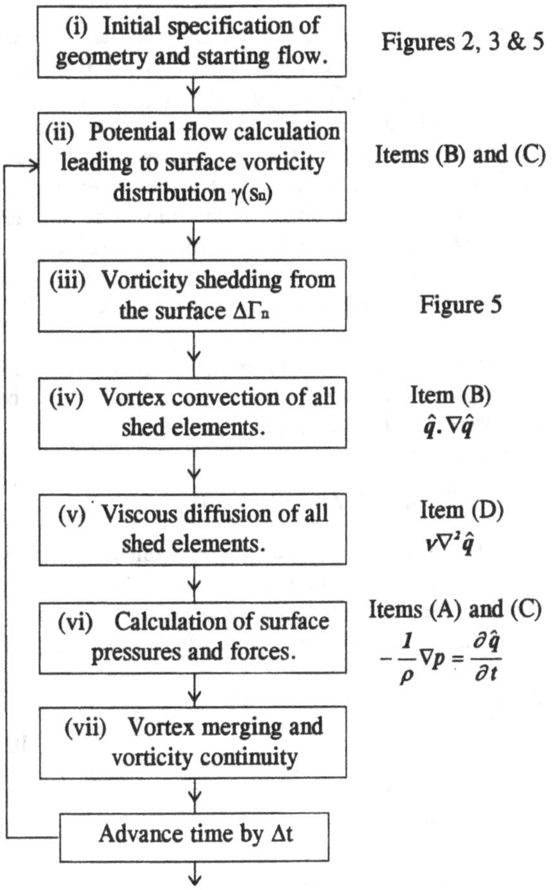

Figure 6 Flow diagram for vortex cloud analysis.

A typical vortex cloud simulation scheme to take account of all these effects will follow an iterative numerical scheme such as that shown in Figure 6. Obviously this is considerably simplified and covers only the principal numerical procedures. In practice, as with all numerical methods, many checks and corrections are built into such schemes both to minimise any errors introduced by discretisation and to ensure that fluid dynamic laws such as the circulation and conservation theorems are satisfied at each time step. A few qualifying comments are required as follows:

1. The author's practice is use straight line elements Δs_n with ends specified by raw input data coordinates X_n, Y_n as illustrated in Figure 3 and to select their mid points x_n, y_n as the pivotal points of the computation.
2. The full Navier-Stokes equation(8) is solved at each time step but in sequential steps of Potential flow, Vortex shedding, Vorticity convection and finally Vorticity diffusion.
3. The vorticity convection process must be reversible as will be explained in section 1.2.1.
4. The vorticity diffusion process is of course thermodynamically irreversible and is modelled using random walks, see section 1.2.2.
5. By trial and error the author has found that it is best to shift vorticies a distance of $\varepsilon_n/\Delta s_n \approx 0.15\text{-}0.25$ normal to the surface before releasing them.
6. Following the random some of the free vorticies will become so close that they can be merged, item (vii). In addition to this some of the recently shed vorticies will be diffused into the body profile. These must be eliminated and a condition of vorticity conservation imposed to compensate their loss. These matters are dealt with in section 1.2.3.

The governing boundary integral equation for the potential flow calculation, item (ii) of Figure 6, may be developed from Equation(3) as follows:

$$-\tfrac{1}{2}\gamma(s_m)+\oint k(s_m,s_n)\gamma(s_n)ds_n + \hat{W}_\infty \cdot d\hat{s}_n + \sum_{j=1}^{Z} L(m,j)\Delta\Gamma_j = 0 \qquad (9)$$

where L(m,j) is a coupling coefficient similar in form to $k(s_m,s_n)$ recording the influence of the Z free vorticies $\Delta\Gamma_j$ upon the boundary condition at pivotal point m. In numerical format for M discrete body elements this becomes

$$\sum_{n=1}^{M} K(s_m,s_n)\gamma(s_n) = -(U_\infty \cos\beta_m + V_\infty \sin\beta_m) - \sum_{j=1}^{Z}\Delta\Gamma_j(U_{mj}\cos\beta_m + V_{mj}\sin\beta_m) \qquad (10)$$

where the unit velocities U_{mj}, V_{mj} are given by

$$U_{mj} = \frac{1}{2\pi}\left(\frac{y_m - y_j}{r_{mj}^2}\right),\quad V_{mj} = -\frac{1}{2\pi}\left(\frac{x_m - x_j}{r_{mj}^2}\right) \qquad (11)$$

with $$r_{mj} = \sqrt{(x_m - x_j)^2 + (y_m - y_j)^2} \qquad (12)$$

1.2.1 *Reversible convection process*

Once the vorticity has been created by the slip flow at the body surface (ii) and then shed freely into the fluid (iii), the flow regime consists of any specified uniform streams W_∞ and the cloud of elementary discrete vortices $\Delta\Gamma_j$. Having completed steps (ii) and (iii) we may now proceed to calculate the convection of this vortex cloud ignoring for the time being the presence of the body. As already stated the convection process must be thermodynamically reversible which for a fluid dynamic flow means physically (geometrically) reversible and the author has paid close attention to this in chapter 8 of his monograph on Vortex Element Methods[1]. The convection velocity components of the free vortex element m due to the remainder of the cloud are given in terms of the unit velocities, equation(11) by

$$u_m = \sum_{\substack{j=1\\ j\neq m}}^{Z}\Delta\Gamma_j U_{mj}\ ,\quad v_m = \sum_{\substack{j=1\\ j\neq m}}^{Z}\Delta\Gamma_j V_{mj} \qquad (13)$$

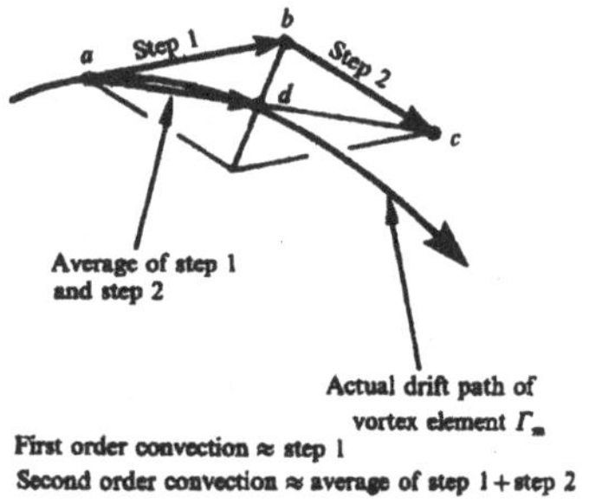

Figure 7 Central difference vortex convection

If we simply apply a forward difference approach to this over the elementary time step Δt, Step 1 of Figure 7, the typical vortex $\Delta\Gamma_m$ will convect forward from ***a*** to ***b*** according to

$$\left.\begin{aligned} x_{mb} &= x_{ma} + u_m\Delta t \\ y_{mb} &= y_{ma} + v_m\Delta t \end{aligned}\right\} \qquad (14)$$

However, if we then repeat the same convection calculation from ***b*** to ***c***, Step 2, and take the average of the two steps, ***a*** to ***d***, we are able to obtain a much improved prediction of the actual curved drift path of each vortex element. In fact if this process is repeated many times to a point of convergence, the final predicted convection may actually be put into reverse by switching to the negative time step $-\Delta t$ and repeating the process backwards in time. The importance of exploring such matters when developing numerical schemes must be emphasised since their neglect can lead to serious numerical error equivalent to the diffusive effect of viscosity, the so-called "numerical viscosity". Such errors obviously decrease for smaller the time steps Δt and in general the author has found the central difference approximation as illustrated by Figure 7 to be adequate.

1.2.2 *The random walk simulation of viscous diffusion*

The random walk model was originally attributable to Chorin[7], but independently of that a similar model was developed by Porthouse[5][6] designed especially for incorporation into vortex cloud modelling. This method is based on the well known solution to the vorticity diffusion equation expressed in polar coordinates,

$$\frac{\partial \omega}{\partial t} = \nu \left\{ \frac{\partial^2 \omega}{\partial r^2} + \frac{1}{r}\frac{\partial \omega}{\partial r} \right\} \qquad (15)$$

Thus for a vortex of initial strength Γ at time t=0, the vorticity at radius r at time t is given by

$$\omega(r,t) = \frac{\Gamma}{4\pi \nu t} e^{-r^2/4\nu t} \qquad (16)$$

where ν is the kinematic viscosity. As shown by Porthouse and in full detail by Lewis[1], if the point vortex is replaced by a number N of equal elements Γ/N, all initially located at r=0 at time t=0, the above diffusion process is well matched if these are scattered according to the following r,θ displacements.

$$r_i = \{4\nu t \ln(1/P_i)\}^{1/2}, \quad \theta_i = 2\pi Q_i \qquad i = 1 \ldots N \qquad (17)$$

This solution of the Lagrangian model we have adopted follows naturally from considerations of probability using statistical theory and in some senses reflects the type of diffusion activity we would expect at molecular level. A full investigation of its accuracy has been given by Lewis[1] and comparisons with the exact solution equation(16) are shown below in Figure 8 for either a single time step Δt=1.0 or ten successive time steps Δt=0.1, both in good agreement.

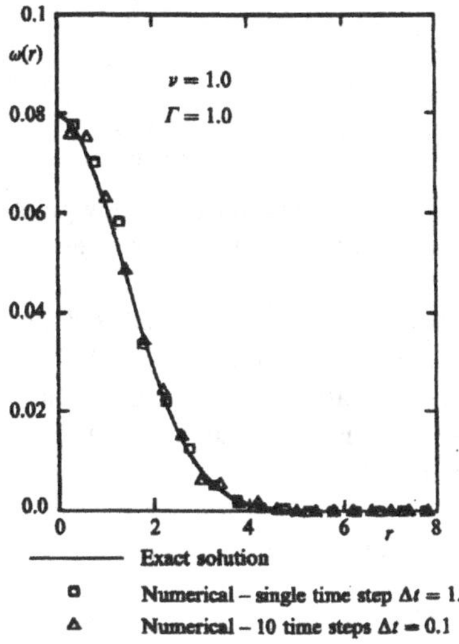

Figure 8 comparison of the random walk with exact solution for a diffusing vortex.

1.2.3 *Merging of vorticies and vorticity conservation*

As mentioned previously, during the random walk due to the statistical scattering of the free vortex elements $\Delta\Gamma_j$, some will be placed so close to each other that their mutual convection velocities become excessive and unrealistic. The author's practice is to merge the vorticies in such cases into a single vortex element equal to the joint strength and mid way between if their convection velocities exceed that of the uniform stream W_∞. This has the beneficial effect of reducing the total number of vortex elements in the cloud and hence the volume of data with some loss of resolution.

A second problem arises during the random walk due to the scatter of some newly shed vorticies into the body interior region. The boundary integral equation demands that there be zero circulation inside the body region since the surface velocity parallel to the body surface is set to zero. There are two options available to handle this as follows:

1. To bounce any such vortex element back out into the mainstream during the random walk by replacing it at its mirror image location as if reflected in the body surface.
2. To snuff the vortex out but to impose an additional circulation correction onto the governing equations to ensure the the same amount of vorticity will be re-shed from the body surface at the next time step. This method has been found to work extremely well and offers the practical advantage of a reduction in the number of shed vorticies and hence the volume of data to be held in arrays.

Thus in response to item 2, a statement may be made of the circulation theorem applied to the whole system when solving the potential flow, that the total vorticity including the surface vorticity + the shed vorticity must retain its initial value of zero, namely

$$\sum_{n=1}^{M} \gamma(s_n)\Delta s_n + \sum_{j=1}^{Z} \Delta\Gamma_j - \Gamma_{circ} = 0 \qquad (18)$$

where Γ_{circ} is the cumulative strength of all of the free vortex elements that have been snuffed out during the computation. To satisfy this condition the governing equation must be modified accordingly to yield

$$\begin{aligned} \sum_{n=1}^{M} (K(s_m, s_n) + \Delta s_n)\gamma(s_n) &= -(U_\infty \cos\beta_m + V_\infty \sin\beta_m) \\ &\quad - \sum_{j=1}^{Z} \Delta\Gamma_j (U_{mj} \cos\beta_m + V_{mj} \sin\beta_m + 1) + \Gamma_{circ} \end{aligned} \qquad (19)$$

2. MULTIPLE BODIES WITH RELATIVE MOTION

Although examples have already been given of flow past groups of mutually interacting bodies such as the three cylinders in Figure 4, no related equations have been presented. As shown by Lewis and Chunjun[8][9], the above analysis for a single body may be extended to multiple bodies where all or some of them are in relative motion. Thus for a total of P bodies the governing equation may be updated to read

$$\begin{aligned} \sum_{q=1}^{P} \sum_{n=1}^{M_q} K_{mn}^{pq} \gamma(s_{qn}) &= -[(U_\infty + U_{RM})\cos\beta_{pm} + (V_\infty + V_{RM})\sin\beta_{pm}] \\ &\quad - \sum_{j=1}^{Z} \Delta\Gamma_j (U_{mj} \cos\beta_m + V_{mj} \sin\beta_m) \end{aligned} \qquad (20)$$

where also $p = 1,2...P$ and $m = 1,2...M_P$. The modified coupling coefficient K_{mn}^{pq} representing the induced velocity at pivotal point m of body p due to element n of body q is then given by

$$K_{mn}^{pq} = \frac{\Delta s_{qn}}{2\pi}\left\{\frac{(y_{pm}-y_{qn})\cos\beta_{pm}-(x_{pm}-x_{qn})\sin\beta_{pm}}{(x_{pm}-x_{qn})^2+(y_{pm}-y_{qn})^2}\right\} \quad (21)$$

For a three body example of this, equations(20) may be stated in matrix form as follows

$$\begin{pmatrix} A^{11} & A^{12} & A^{13} \\ A^{21} & A^{22} & A^{23} \\ A^{31} & A^{32} & A^{33} \end{pmatrix}\begin{pmatrix}\gamma_1\\ \gamma_2\\ \gamma_3\end{pmatrix} = \begin{pmatrix} rhs_1\\ rhs_2\\ rhs_3\end{pmatrix} \quad (22)$$

where we note that the left hand side involves nine sub-matrices of two types, namely

i. A^{11}, A^{22} and A^{33} provide the potential flow influence of each body upon itself.

ii. A^{12}, A^{13} etc. provide the influence of bodies upon one another. Thus A^{12} provides the influence of body 2 upon body 1.

In fact one item has been left out of equation(20), namely the circulation correction Γ_{circ}. If the user decided to apply an overall circulation correction such as that illustrated in section 1.2.3, equation(19), that would certainly satisfy the circulation theorem. On the other hand it would be fluid-dynamically more correct instead to apply the correction individually to each body for vorticies snuffed out by entering that particular body. This is difficult to express clearly in equation(20) but much easier to include when composing the final numerical computer code.

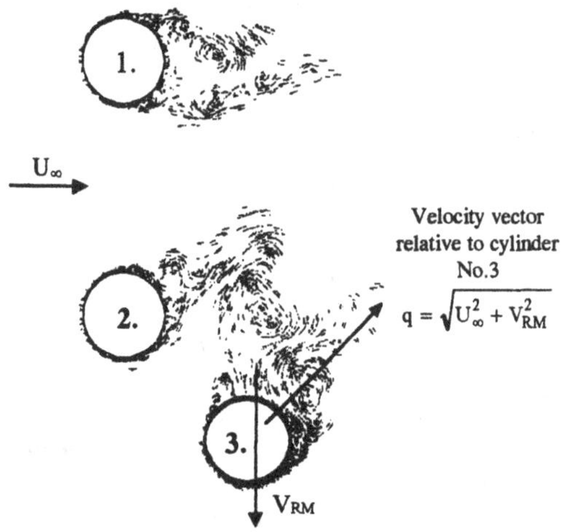

Figure 9 Flow past three cylinders, one in motion, after 200 time steps of $\Delta t=0.015$.

This formulation will also permit one or more of the bodies, say body p, to move independently by applying its relative velocity components U_{RMp}, V_{RMp} when calculating the right hand sides of the M_p equations(21) applicable to that particular body. Thus let us consider the previous example of three cylinders in close proximity in a uniform stream U_∞, $=1.0$m/s parallel to the x axis, one of them traversing the wakes of the other two with equal vertical velocity $V_{Rm}=1.0$m/s. The consequent predicted fluid motion after time t=3.0sec. is shown in Figure 9.

However, caution is needed regarding the geometrical data when calculating the coupling coefficients. Obviously, as there are continual relative displacements between the stationary and moving bodies, the overall geometry must be revised after each time step. A sensible procedure is to use the central position of the translating surface elements as representative of the average geometrical location of the body surface for each time step for computation of the average coupling coefficient for that time step as illustrated in Figure 10. Before moving on to the next time step on the other hand the actual geometry must then be shifted from position ***a*** to position ***b***.

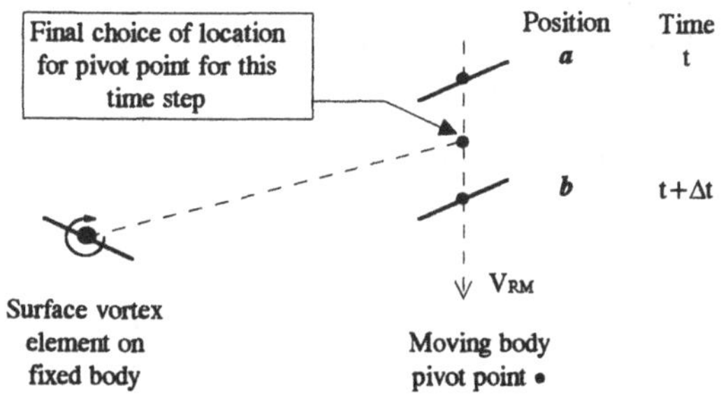

Figure 10 Relocation of a moving body pivot point at central point after one time step.

To conclude this section the predicted flow pattern some time later is shown in Figure 11 below, with cylinder No.3 oscillating behind cylinder No. 2 and on its second downward traverse. The following observations may be made:

- The beginnings of the development of a Von Karman vortex street may be seen to the rear of cylinder No.1, Figure 11(a), although this has been pushed upwards slightly due to interference from cylinder No.3. By the end of the traverse however the vortex street is almost completely dispersed due to the influence and mixing of vorticity shed from the moving cylinder.
- As the traverse begins, the wake from cylinder No.2 is deflected downward due to the blockage effect imposed by the moving cylinder. As the traverse finishes however, only a modest new wake remains behind cylinder No. 2, as might be expected due to entrainment of the previous wake

into the heavily vortical wake of the moving cylinder as it passes in close proximity.

- It can be seen that the wake downstream of the moving cylinder was never able to establish a clear vortex pattern but moved in direction very considerably as might well be expected under the interference of the upstream bodies and wakes.

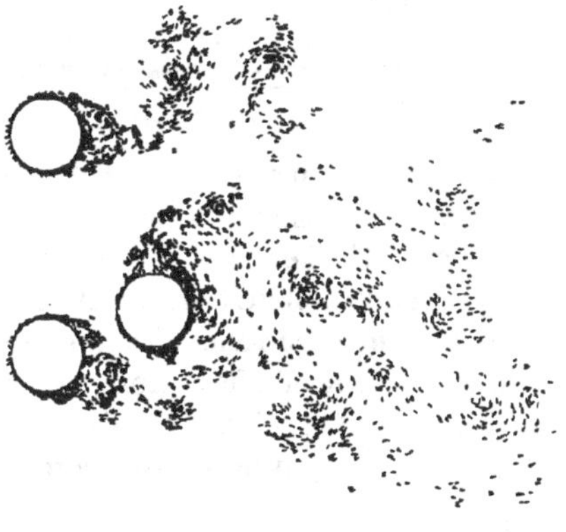

(a) Flow pattern after 460 time steps. Entry into traverse of wake shed by cylinder No.2

(b) Flow pattern after 560 time steps. Approaching finish of wake traverse of cylinder No. 2.

Figure 11 The interaction of cylinder wake flows when in relative motion.

To conclude, it is generally the case that vortex cloud modelling as illustrated here is extremely powerful and revealing for modelling bluff body flows and indeed predicts such unsteady phenomena as Strouhal number remarkably well. This success is probably attributable to the dominating influence in such flows of vorticity convection. The main area of weakness of vortex cloud analysis lies in flows where boundary layer development and stability is a primary determining factor such as aerofoil and turbomachinery cascade flows, particularly when body surface diffusions are present such as in compressor or fan cascades. Two numerical problem arises here, namely:

1) The need for sufficient resolution within the vortex cloud to represent flow within a boundary layer with sufficient accuracy.
2) The selection of a suitable time step size to meet the needs of both convection and diffusion. This matter will be discussed in the next sub-section.

2.1 Selection of Suitable Time Step.

For the problem just considered a reasonable approach to the selection of an appropriate time step Δt might be to focus on the average displacements of the vorticies due to convection and diffusion and attempt to make them equal. The average convective displacement for the cylinder flow might be approximated crudely by

$$\delta_c = \tfrac{1}{2} U_\infty \Delta t \tag{23}$$

whereas, from equation(17), the average random walk displacement (i.e. when the random number $P_i=0.5$) is

$$\delta_d = \sqrt{(4 \nu \Delta t \ln 2)} \tag{24}$$

Equating δ_c and δ_d and rearranging, we have

$$\Delta t = 16 \ln 2 \times D / (U_\infty Re) \tag{25}$$

where $Re=U_\infty D/\nu$ is the Reynolds number based on cylinder diameter D. The dimensionless time step becomes

$$\frac{\Delta t}{D / U_\infty} = 16 \ln 2 \frac{1}{Re} \tag{26}$$

However, except at very low Reynolds numbers this demands extremely small time steps to accept the convection resolution, equation(23) while retaining equal diffusion resolution. Thus with $D=U_\infty=1.0$ and a Re value of 10^5, the appropriate time step would have to be $\Delta t=0.000111$ which is so small that it would lead to ridiculously excessive computational times. The demand here has been imposed by viscous diffusion rather than convection because the Reynolds number is so high. It might be argued that this is precisely why the traditional approach of calculation mainstream and boundary layer flows separately for aerofoils has always been necessary, implying also that all CFD methods may be suspect unless equal discretisation is provided for the convective and diffusive parts of the flow.

A way round this which may help to solve the dilemma is to decide first a suitable time step from convective considerations. For each convective time step the diffusive process may then be conducted over a number of sub time steps to improve accuracy. Thus from equation(23) a suitable estimate of convective time step Δt_c is given by average acceptable drift over the average surface element size Δs through

$$\Delta t_c = \frac{1}{2} \frac{\Delta s}{U_\infty} \tag{27}$$

Dividing this into the previous estimate for the ideal time step, equation(26), the number of sub-iterations N_{sub} for the random walk would need to be

$$N_{sub} \approx \frac{1}{32\ln 2}\frac{\Delta s\, Re}{D} \approx \frac{\pi}{32\ln 2}\frac{Re}{M} \qquad (28)$$

However this proves to be excessive (e.g. about 350 in the present problem) and while unnecessary out in the main flow where viscous stresses are very small could be implemented by improved modelling close to the body surface, for example by the use of a local grid structure. Frequently this problem is ignored for bluff body flows in view of the dominating influence of convection, but further research is needed for flows where the surface boundary layer plays a crucial role such as aerofoil flows. We proceed now to consider the application of vortex modelling to the important problem of turbomachine cascades.

3. TURBOMACHINERY CASCADES

Ever since the early work of Wilkinson[3] there has been a considerable application of the Martensen method for the design and analysis of flow through cascades including its extension to mixed-flow and radial turbomachines. Some of the underlying methodology will now be given in section 3.1 before considering the application of vortex cloud analysis for simulating real cascade flows in section 3.2.

3.1 Surface Vorticity Method for Cascade Analysis

The geometry of a typical compressor cascade is shown in Figure 12 form which we observe that it normally comprises an infinite number of identical blade profiles of chord l pitched t apart between $y=\pm\infty$.

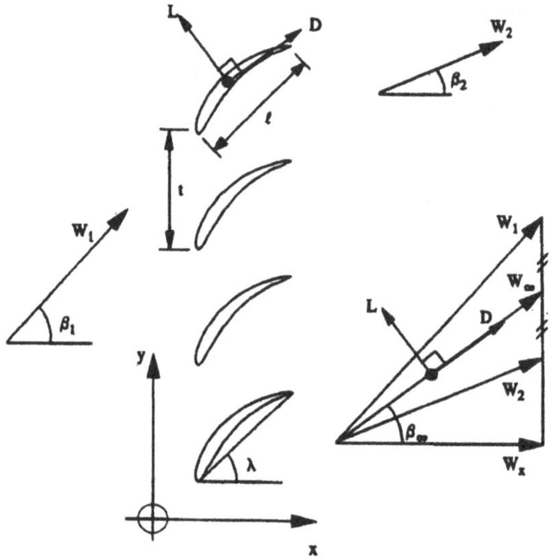

Figure 12 Cascade geometry and velocity triangles

Clearly, the cascade is a multiple body problem but with an infinite number of bodies. Although the previous method could be applied with say 20 or so blades in the hope that the centre blades would approach the desired solution, an alternative approach is to adopt the same governing equations (3) and (5) as for single bodied problems but to introduce a new coupling coefficient[1][3][10] given by

$$K_{mn} = \frac{\Delta s_n}{2t}\left\{\frac{\sin\frac{2\pi}{t}(y_m-y_n)\cos\beta_m - \sinh\frac{2\pi}{t}(x_m-x_n)\cos\beta_m}{\cosh\frac{2\pi}{t}(x_m-x_n)-\cos\frac{2\pi}{t}(y_m-y_n)}\right\} \qquad (29)$$

The assumption here is that the potential flow past each blade is identical and the flow repeats itself in the pitchwise direction. Unlike the bodies we have considered so far, turbomachines are in effect lifting aerofoils and the trailing edge Kutta condition needs to be imposed to ensure smooth flow from the trailing edges of the blade profiles. This best done by first stating the strength of the equivalent bound vorticity Γ

$$\Gamma = \sum_{n=1}^{M}\gamma(s_n)\Delta s_n \qquad (30)$$

and then adding this to every equation giving the following result for element m.

$$\sum_{1}^{M}\left[K(s_m,s_n)+\Delta s_n\right]\gamma(s_n) = -U_\infty\cos\beta_m - V_\infty\sin\beta_m + \Gamma \qquad (31)$$

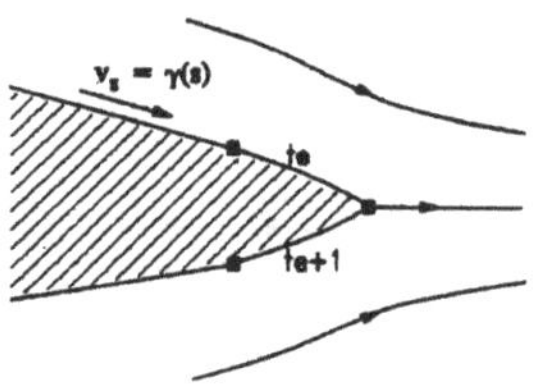

Figure 13 Trailing edge elements.

The next step is to settle upon a suitable statement of the trailing edge Kutta condition and the simplest approach to this is to enforce equal downstream directed surface velocities v_s at the two trailing edge elements te and te+1, Figure 13. With the definition of surface vorticity as clockwise positive we then have

$$\gamma(s_{te}) = -\gamma(s_{te+1}) \qquad (32)$$

The bound vorticity Γ is of course unknown but must be such that equation 32 is satisfied. To achieve this let us break down the surface vorticity into three components as follows

$$v_{sn} = \gamma(s_n) = U_\infty\left[\gamma_U(s_n)+\Gamma_U\gamma_\Gamma(s_n)\right] + V_\infty\left[\gamma_V(s_n)+\Gamma_V\gamma_\Gamma(s_n)\right] \qquad (33)$$

where $\gamma_U(s_n)$ and $\gamma_V(s_n)$ account for the unit uniform streams $U_\infty=1$, $V_\infty=1$ and $\gamma_\Gamma(s_n)$ for unit bound vorticity $\Gamma=1$ and independently satisfy the following equations.

$$\left.\begin{aligned}\sum_{1}^{M}\left[K(s_m,s_n)+\Delta s_n\right]\gamma_U(s_n) &= -\cos\beta_m\\ \sum_{1}^{M}\left[K(s_m,s_n)+\Delta s_n\right]\gamma_V(s_n) &= -\sin\beta_m\\ \sum_{1}^{M}\left[K(s_m,s_n)+\Delta s_n\right]\gamma_\Gamma(s_n) &= 1\end{aligned}\right\}\quad(34)$$

We note that these three equations for unit values of U_∞, V_∞ and Γ may be solved once and for all at the outset of a computation. Final solutions are then readily available from equation (33) for any chosen values of U_∞, V_∞ and Γ. To obtain the latter, equation (33) may be introduced into equation (32), to obtain values of Γ_U and Γ_V for the bound circulations attributable to the mainstream velocity components U_∞ and V_∞ which independently satisfy the trailing edge Kutta condition, namely

$$\left.\begin{aligned}\Gamma_U &= -\frac{\gamma_U(s_{te})+\gamma_U(s_{te+1})}{\gamma_\Gamma(s_{te})+\gamma_\Gamma(s_{te+1})}\\ \Gamma_V &= -\frac{\gamma_V(s_{te})+\gamma_V(s_{te+1})}{\gamma_\Gamma(s_{te})+\gamma_\Gamma(s_{te+1})}\end{aligned}\right\}\quad(35)$$

The total bound circulation is then given by

$$\Gamma = U_\infty\Gamma_U + V_\infty\Gamma_V \quad(36)$$

and the problem is fully stated. Solutions of these equations[1] with 50 elements compare very closely with exact solutions by conformal transformation given by Gostelow[11].

3.2 Vortex cloud Simulation of Cascades

Ji[9] developed vortex cloud procedures for the simulation of both compressor and turbine cascades some of which studies were recently published by Lewis and Ji[8]. Samples of these will be given here for rotating stall in compressor blade row, section 3.2.1 and a full simulation of a turbine blade row with stator and rotor in relative motion, section 3.2.2.

3.2.1 *Rotating stall in a compressor blade row.*

In potential flow modelling of aerofoils and cascades it is clearly necessary to enforce the trailing edge Kutta condition as just explained. Vortex cloud modelling on the other hand attempts solution of the Navier-Stokes equations including viscous effects and thus there is no need to introduce bound vorticity Γ into the analysis. The smooth flow from the trailing edge will look after itself automatically. However, in the case of cascades it is no longer likely that the flow will be identical for each blade profile and therefore it is generally necessary to implement a combination of the multi-body and cascade strategies. Thus for example if we were to define a compressor cascade as a composite of 8 blade profiles, Figure 14, the entire assembly could be converted into a cascade array with pitch $8\times t$ in the y direction. The pattern would then repeat itself every eight blade pitches. Application of the same computer code as that used for the cylinder group in section 3, but with the modified coupling coefficient equation (29), would then lead to the results shown in Figure14 simulating rotating stall within a compressor blade row.

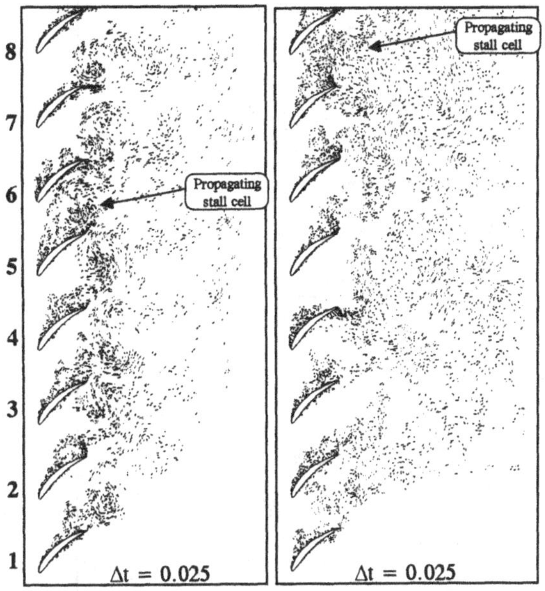

(a) Time step 250 (b) Time step 460

Figure 14 Simulation of rotating stall in an axial compressor cascade.

It will be observed that after 250 time steps $\Delta t=0.025$, Figure 14(a), blade No. 5 has just stalled due to the propagation of the stall cell vertically upwards. Thus blockage due to the presence of the stall cell tends to increase the angle off attack for blade No. 6 while reducing it for blade No. 4, causing cell propagation. Figure 14(b) shows predicted results after 460 time steps, from which it is clear that the stall cell has propagated along the cascade reaching blade row No. 7. It can also be seen that blade row No. 5 is now unstalled and stable again with some evidence of the related anti-clockwise starting vortex just downstream. Blade No.6 is in the process of regaining stability while the opposite is true for blade No.1 which is about to succomb to the influence of the propagating stall cell.

As shown by Lewis and Ji[8] the predicted speed of propagation for this stall cell is almost exactly half of the blade speed for this compressor which was designed for a stage duty of $\phi=0.5, \psi=0.4066$ and with a design diffusion factor of 0.6. The diffusion factor and work coefficient ψ are thus fairly high, which was fully intentional in order to trigger the rotating stall.

3.2.2 *Simulation of flow through a 50% reacion turbine stage.*

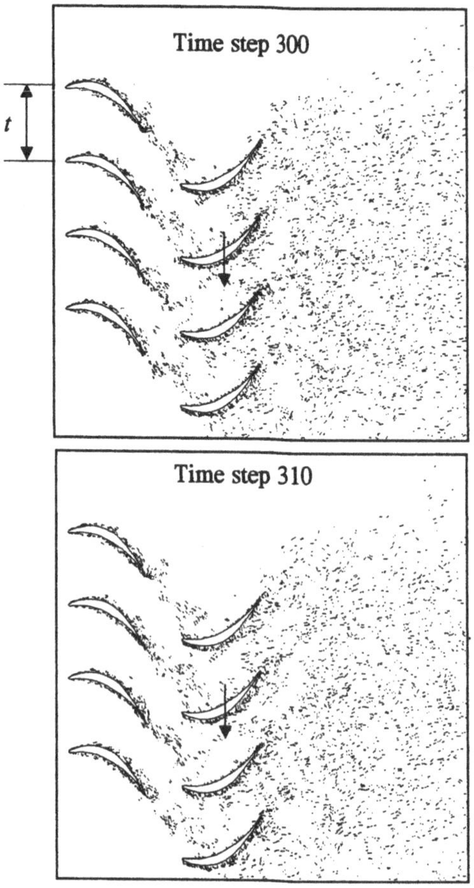

Figure 15 Vortex cloud simulation of the flow through a 50% reaction turbine blade row.

To conclude this review of vortex dynamics simulation of cascades, results are shown in Figure 15 for flow through a 50% reaction turbine stage for which the stator and rotor have identical cascade geometry defined as follows.

Table 1 Cascade geometry for 50% reaction turbine blade row

Base profile T4	Stagger $\lambda = 27.8°$
Parabolic camber line	$t/l = 0.8$
Camber $\theta = 74.635°$	$\beta_1 = 3.577°$
x/l of max camber =0.4	

This turbine stage was designed by means of the programs FIPSI and CASCADE accompanying ref.(10), settling for a conservative stage duty of $\phi = 0.8$ and $\psi = 1.1$, with an expected total-to-total efficiency in the region of 93.3%. As can be seen from Figure 15, both stator and rotor flows are stable and there is evidence of the stator wakes sweeping the passages of the downstream rotor. For this simulation eight blades were used, namely four each for the stator and rotor which were of identical profile and pitch blade pitch *t*. But each package of four blades was converted into an infinite cascade by means of the cascade coupling coefficient equation (29) using the pitch value $4t$. The predicted vortex cloud patterns thus repeat themselves every four blade pitches. Vortex dynamics is clearly the most natural technique for obtaining simulations of this type and given more PC memory and CPU speed will be capable of detailed prediction of blade row interference in turbomachines including the sweeping effects of wakes shed from upstream blade rows. There is one serious limitation however which should be pointed out, namely that the two blade rows must have equal pitch for the blade groups (i.e. both stator and rotor groups of four blade here must have equal pitch 4t). There could of course be different numbers of blades in the stator and rotor groups. Thus we could use three rotor blades instead of four here provided the individual rotor blade pitch was increased to $\frac{4}{3}t$.

To conclude this section we now go on to consider the basic requirements for extension of cascade analysis to radial and mixed-flow turbomachines. So far the author has not attempted to apply vortex dynamics to these situations and the treatment which follows is restricted to potential flow analysis by the surface vorticity method.

3.3 Mixed-flow and Radial Cascades

As shown by Young[12] and fully developed by Lewis[10][1], the "conical" meridional blade-to-blade section of mixed-flow fan of turbine may be transformed conformally into an equivalent infinite straight cascade in the ς-plane as shown by Figure 16.

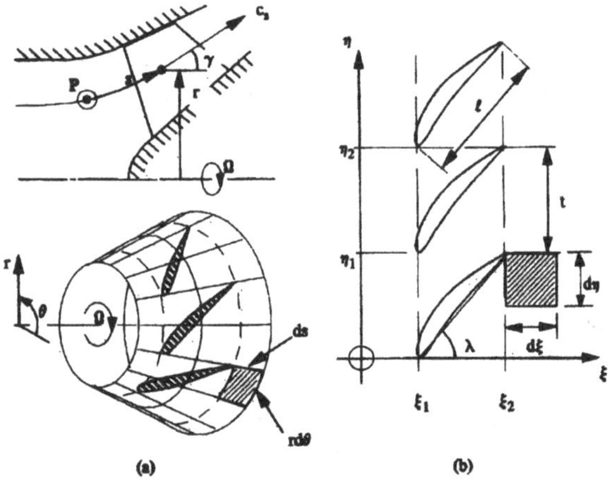

Figure 16 Transformation of a blade meridional flow intersection into a straight infinite cascade.

For conformality the two equivalent elements of area $rd\theta ds$ and $d\xi d\eta$ must have similar shape and proportionate sides and must thus obey the transformation relationships

$$\frac{d\xi}{d\eta}=\frac{ds}{rd\theta}=\frac{1}{r\sin\gamma}\frac{dr}{r\theta} \tag{37}$$

where γ is the local cone angle and s is measured along the meridional surface. This may be achieved by the separate coordinate transformations

$$d\xi=\frac{ds}{r}=\frac{1}{r\sin\gamma}dr, \quad d\eta=d\theta \tag{38}$$

For numerical purposes these equations may be integrated to yield the direct coordinate relationships linking the actual blade profile on the surface of revolution to its equivalent straight cascade.

$$\left.\begin{aligned}\xi-\xi_1&=\int_{s1}^{s}\frac{1}{r}ds=\int_{s1}^{s}\frac{1}{r\sin\gamma}dr\\ \eta-\eta_1&=\theta-\theta_1\end{aligned}\right\} \tag{39}$$

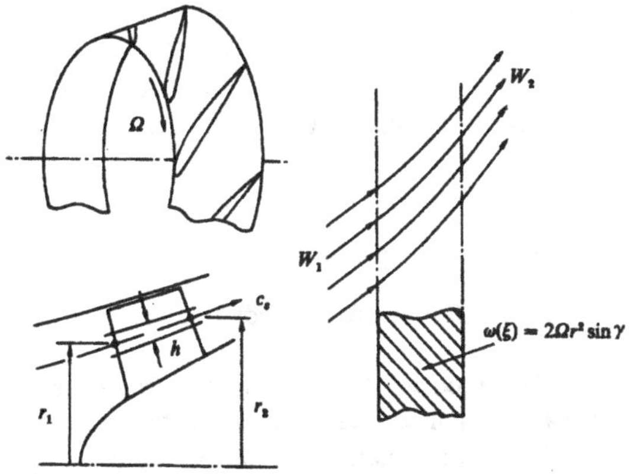

Figure 17 Transformation of relative eddy across for a mixed-flow fan rotor section to its equivalent cascade.

If we consider now the flow relative to the blade section on its surface of revolution, we observe a relative eddy or distributed vorticity of strength 2Ω thoughout the annulus which has components $2\Omega\cos\gamma$ and $2\Omega\sin\gamma$ along and normal to the meridional surface of revolution. The first of these is the throughflow eddy which causes the flow in fact to depart from forming axisymmetric surfaces of revolution. Here we are concerned only with the second component which strongly affects the blade-to-blade flow, introducing Coriolis forces and also the phenomenon known as slip. This vorticity $2\Omega\sin\gamma$ must be transformed across to equivalent vorticity $\omega(\xi)$ in the cascade plane. To achieve this we must apply the circulation theorem to the equivalent elements of area namely, $\omega(\xi)d\xi d\eta=2\Omega\sin\gamma dsrd\theta$. From equations (38) we then have the transformed vorticity strength

$$\omega(\xi)=2\Omega r^2\sin\gamma \tag{40}$$

This vorticity transformation process is illustrated in Figure 17 from which we may make the following observation. In the absence of the blades, the displacement flow into which they are immersed is no longer a uniform stream W_∞ but is a curved stream U,V due to the influence of the distributed vorticity $\omega(\xi)$ and where we may define

$$U=U_\infty\,,\quad V=V_\infty+v_\Omega \tag{41}$$

The vorticity induced component of this v_Ω may be found from the vorticity definition

$$\omega(\xi)=\frac{dv_e}{d\xi}=\frac{dv_e}{dr}r\sin\gamma$$

which may be integrated to yield finally[1]

$$V=V_\infty+\Omega\left\{r^2-\tfrac{1}{2}\left(r_1^2+r_2^2\right)\right\} \tag{42}$$

The governing equation for mixed-flow cascades may now be expressed by upgrading equ. 31 to read:

$$\begin{aligned}\sum_{1}^{M}\left[K(s_m,s_n)+\Delta s_n\right]\gamma(s_n)=&-U_\infty\cos\beta_m\\ &-\left(V_\infty+\Omega\left\{r^2-\tfrac{1}{2}\left(r_1^2+r_2^2\right)\right\}\right)\sin\beta_m+\Gamma\end{aligned} \tag{43}$$

Space prevents a fuller treatment but mention should be made of two other important items fully dealt with elsewhere.

1. The above formulation as it stands implies that the relative eddy vorticity $\omega(\xi)$ also pervades the blade profile inner area. The boundary integral analysis method on the other hand demands zero velocity and thus zero circulation around the inside of the surface vorticity sheet. Thus arrangements must be made to remove the interior vorticity and the relevant technique developed by Wilkinson[13] and checked precisely by Fisher[14] is outlined in reference (1) and implemented in the author's software suit MIXFLO[15].

2. As illustrated in Figure 17, the meridional stream sheet may vary in thickness *h* from leading edge to trailing edge. It is well known from axial cascade studies that this will result in significant changes to its aerodynamic properties[16] (sometimes referred to as AVR or axial velocity ratio effects) and thus must be accounted for. As shown by Lewis[1] including a full analysis, this can be achieved by introducing an equivalent source distribution $\sigma(\xi)$ throughout the fluid which will of course also introduce a second disturbance velocity u_σ to be added to the displacement flow, equation (41)a and will also require corrections to purge the interior region of fluid divergence.

The predicted streamline pattern is compared in Figure 18 with finite difference analysis by Stanitz[17] for a radially bladed centrifugal pump rotor with 20 blades and with prewhirl equal to the inlet blade speed $c_{\theta1}=r_1\Omega$. Agreement is good between the two numerical methods, both of them revealing the slip flow at exit due to the relative eddy.

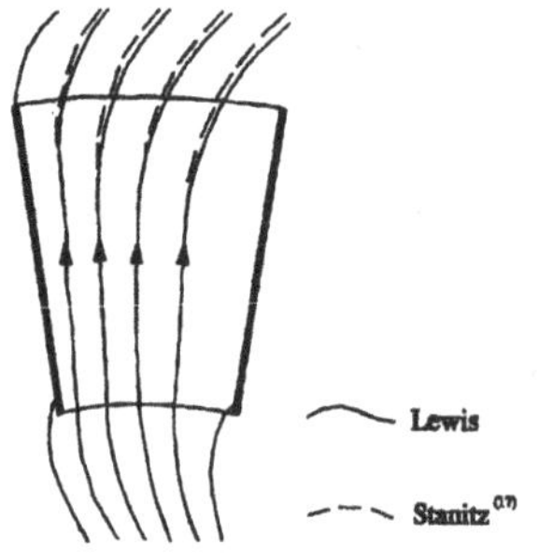

Figure 18 Predicted streamlines for a radial bladed centrifugal pump rotor.

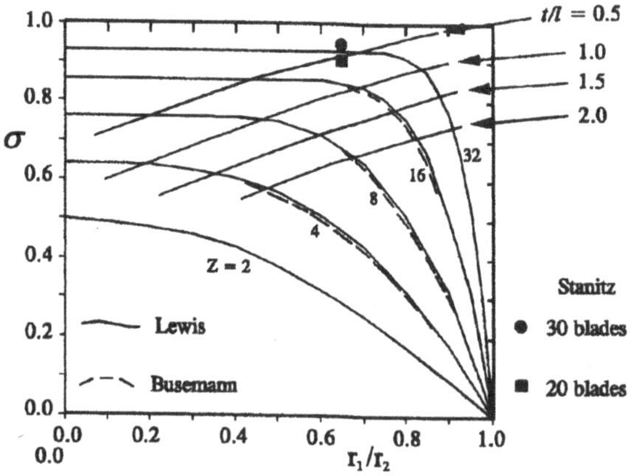

Figure 19 Predicted slip factors for thin plate radially bladed centrifugal rotors.

Exact solutions for slip factor σ by conformal transformation theory are possible for restricted geometries such as the thin plate radially bladed rotors shown in Figure 18 and were documented in detail by the classical paper of Busemann[18]. To conclude this section, predicted slip factors σ by the vortex element method are compared in Figure 19 with Busemann's results and also with the two results by Stanitz[17] calculated by finite difference modelling. The surface vorticity analysis as given above cannot accept blade profiles of zero thickness and thus a small profile thickness was superimposed. Despite this, excellent predictions were obtained for this datum test case giving confidence in the use of surface vorticity analysis for more general geometries of mixed-flow cascades. Further comparisons with other families of profiles obtained by conformal transformation theory have also been reported by Lewis[1] with equally good agreement.

4. AXISYMMETRIC FLOWS

To conclude this paper a brief summary will be given of the application of vortex element methods to the solution of axisymmetric flow problems. There have been extensive developments in this field for extension of the surface vorticity method to deal with flows past bodies of revolution and through annuli or ducts in inviscid flow, reviewed analytically in some detail in ref(1). This work has been extended to deal with the design and analysis of ducted propellers, and full meridional analysis of arbitrary turbomachines, both including the interference effects of blade vortex shedding and related gradients of stagnation pressure resulting in the presence of distributed vorticity.

The basic numerical framework will be given in outline in section 4.1. This will be followed in section 4.2 with a brief summary of work by Li Ming[19][20] to extend full vortex dynamics for simulation of real axisymmetric flows.

4.1 Surface vorticity Theory for Axisymmetric Flow

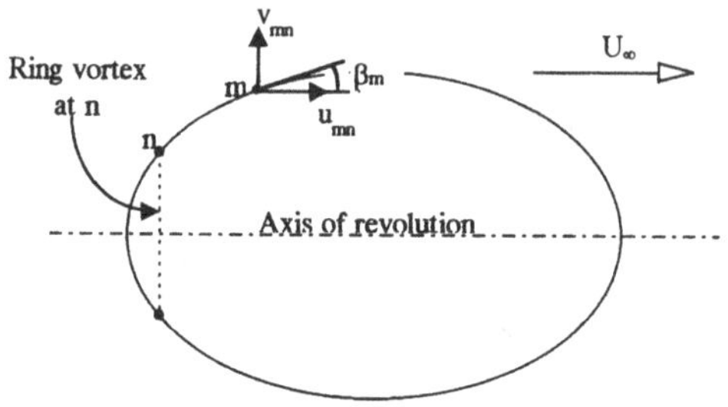

Figure 20 Flow past a body of revolution.

The linear equations(5) for plane two-dimensional flow may be applied directly to simulate axisymmetric potential flow of a uniform stream U_∞ past a body of revolution, Figure 20, namely

$$\sum_{1}^{M} K(s_m, s_n)\gamma(s_n)\Delta s_n = -U_\infty \cos\beta_m \tag{44}$$

where the coupling coefficient $K(s_m,s_n)$ represents the velocity parallel to the body surface at m due to a unit strength ring vortex at point n and is given by

$$K(s_m, s_n) = u_{mn}\cos\beta_m + v_{mn}\sin\beta_m \tag{45}$$

and where the velocity components of the unit strength ring vortex may be expressed in terms of complete elliptic integrals K(k) and E(k) of the first and second kind through

$$\begin{aligned} u_{mn} &= -\frac{1}{2\pi r_n\sqrt{\bar{x}^2+(\bar{r}+1)^2}}\left\{K(k)-\left[1+\frac{2(\bar{r}-1)}{\bar{x}^2+(\bar{r}-1)^2}\right]E(k)\right\} \\ v_{mn} &= -\frac{\bar{x}/\bar{r}}{2\pi r_n\sqrt{\bar{x}^2+(\bar{r}+1)^2}}\left\{K(k)-\left[1+\frac{2\bar{r}}{\bar{x}^2+(\bar{r}-1)^2}\right]E(k)\right\} \end{aligned} \tag{46}$$

and dimensionless coordinates $\bar{x}$ and $\bar{r}$ are defined

$$\bar{x} = (x_m - x_n)/r_n, \quad \bar{r} = r_m/r_n \tag{47}$$

the parameter k being given by

$$k = \sqrt{4\bar{r}/\{\bar{x}^2 + (\bar{r}+1)^2\}} \tag{48}$$

Procedures and computer code are given in ref(1) for evaluating K(k) and E(k) either directly or more speedily from look-up tables. The one exception to equations(46) are the self-inducing coupling coefficients $K(s_m,s_m)$ which were shown by Ryan[21][22] to reduce to

$$K(s_m,s_m) = -\frac{1}{2} - \frac{\Delta s_m}{4\pi r_m}\left\{ln\frac{8\pi r_m}{\Delta s_m} - \frac{1}{4}\right\}cos\,\beta_m - \frac{\Delta\beta_m}{4\pi} \qquad (49)$$

The second term accounts for the self-propelling velocity of a smoke ring vortex, (a feature not present in plane flows). The third term is the correction for self induced velocity due to the curvature $\Delta\beta_m$ of the body surface element Δs_m in the x,r plane at pivotal point m.

Table 2 - Prediction of flow past a sphere

Element No		Vortex elements	Exact Solution.
1,	35	0.067298	0.067297
2,	34	0.201195	0.201350
3,	33	0.333521	0.333781
4,	32	0.463153	0.463526
5,	31	0.589063	0.589538
6,	30	0.710231	0.710803
7,	29	0.825680	0.826345
8,	28	0.934477	0.935235
9,	27	1.035731	1.036594
10,	26	1.128642	1.129607
11,	25	1.212515	1.213526
12,	24	1.286588	1.287673
13,	23	1.350284	1.351453
14,	22	1.403077	1.404352
15,	21	1.444636	1.445944
16,	20	1.474587	1.475894
17,	19	1.492665	1.493962
18		1.498714	1.500000

A good test of the accuracy of this numerical vortex element analysis is its application to the flow past a sphere for which there exists an exact solution for the predicted potential flow surface velocity, namely

$$v_s / U_\infty = 1.5\,sin\,\phi \qquad \text{where } 0° \le \phi \le 180° \qquad (50)$$

The results of a computation with 35 equal length surface elements using a "look-up and interpolate" procedure for the elliptic integrals is given in Table 2 showing remarkably close agreement for all elements.

4.2 Vortex Dynamics Simulation of Flow Past a Body of Revolution.

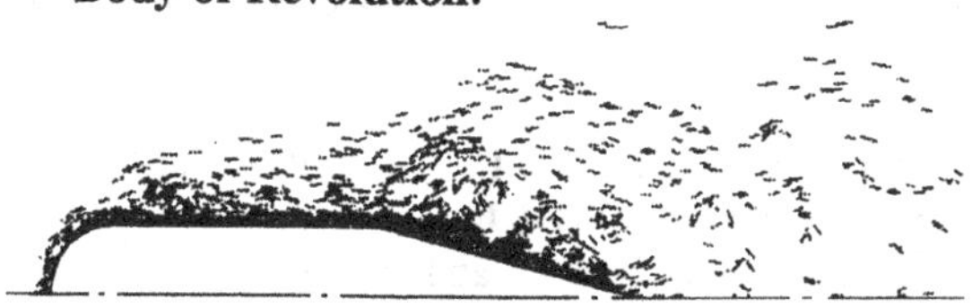

(a) Re = 100

(b) Re = 10^6

Figure 21 Vortex dynamics simulation of flow past a body of revolution.

The extension of this theory to deal with vortex dynamics simulation of axisymmetric flows has been undertaken by Li Ming and is illustrated in Figure 21 for flow past a body of revolution at low and high body Reynolds numbers $Re=U_\infty l/\nu$, where l=body length. At low Reynolds number Re=200, case(a), the predicted boundary layer is extremely thick with evidence of some separation just after the nose/body junction and then substantial separation in the wake surrounding the conical tail cone. At very much higher Reynolds number Re = 10^6, case(b), the boundary layer is much thinner and more stable as expected, with minimal and intermittent separation on the tail cone.

Now in the second of these cases extension of the foregoing axisymmetric potential flow model to vortex cloud modelling may be simply achieved following the same technique as already explained in section 1.2 for plane two-dimensional flow including the standard random walk procedure. The only significant difference is that regarding the expressions for vortex induced velocities which are now those of ring vorticity as given by equations(46). These must be used to obtain the matrix coupling coefficients as already explained but also for calculating the convection of the discrete ring vorticies shed to form the vortex cloud. Thus at high Reynolds numbers the convective process dominates over the diffusive process and the standard random walk based on the diffusion of a line vortex suffices.

At low Reynolds numbers however, Li Ming[19][20] has shown that ring vortex diffusion can no longer be modelled by the standard random walk and he has developed a revised version based on the published exact solution for a diffusing ring vortex. The particular difficulties are illustrated by Figure 22 which shows the diffusion of vorticity from an initially concentrated ring vortex.

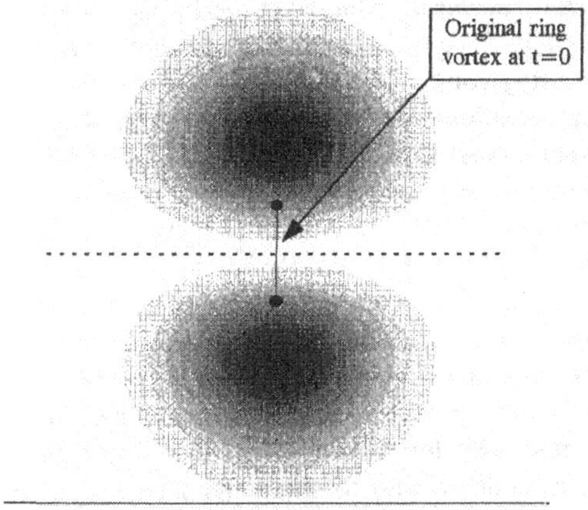

Figure 22 Growth of a ring vortex due to heavy viscous diffusion for $r_1^2/\nu t = 1.0$.

To summarise, the main features of the diffusing ring vortex are as follows:

1. The vorticity diffuses away from the centre of the core as it does in the case of the line vortex.
2. The circular core of the diffused ring vortex grows in diameter, in this case to almost twice the original size.
3. Diffusion about the core centre is no longer uniform about the core but greater away from the axis and less inwards towards the axis.
4. The ring vorticity strength on the axis remains zero.

The governing equation for vorticity diffusion of the ring vortex in the x,r plane is as follows

$$\frac{\partial\omega}{\partial t} = \nu\left\{\frac{\partial^2\omega}{\partial r^2} + \frac{1}{r}\frac{\partial\omega}{\partial r} - \frac{\omega}{r^2} + \frac{\partial^2\omega}{\partial x^2}\right\} \quad (51)$$

and its exact solution is given by

$$\omega = \frac{\Gamma}{(4\pi\nu t)^{3/2}} e^{-\frac{x^2+r^2+r_1^2}{4\nu t}} I_1\left(\frac{r r_1}{2\nu t}\right) \quad (52)$$

where r_1 is the original radius of the concentrated ring vortex Γ at time $t=0$ and I_1 is the modified Bessel function of order one. We note that the significant dimensionless parameter here is the ratio $r_1^2/\nu t$ and four categories have been identified by Li Ming[19], namely:

(a) $r_1^2/\nu t < 4$ (Heavy diffusion)
(b) $4 \le r_1^2/\nu t \le 6$
(c) $6 < r_1^2/\nu t < 300$
(d) $r_1^2/\nu t \ge 300$ (Weak diffusion)

For very large values of $r_1^2/\nu t$, case(d), flows approximate to the line vortex diffusion and the standard random walk then applies. Case(a) for very low values of $r_1^2/\nu t$ on the other hand is that of a heavily diffusing ring vortex as illustrated above for which, together with intermediate cases (b) and (c), special treatments are needed as outlined by Li Ming in full in ref.19 and in part for illustration in ref.20. These were based on analysis leading to curve fitting results such as shown in Figure22, estimating the probability that a given ring vortex element will be scattered in any given polar direction based with polar coordinates centred on its initial position at $t=0$. Although Li Ming produced a robust revised random walk method dependent on the parameter $r_1^2/\nu\Delta t$ for a time step Δt, on reflection it may have been more productive to first curve fit the growth in core radius versus parameter $r_1^2/\nu t$ and then estimate the diffusion from the new core centre r_1 instead. This would undoubtedly lead to a much simpler and more practicable random walk method and is recommended for future research. It would then be of considerable interest to study the accuracy of using a large number of small successive time steps Δt using the standard random walk (aiming at fixing the value $r_1^2/\nu\Delta t \ge 300$), instead of the single large step used in evaluating Figure22. However the quality of Li Ming's prediction method is illustrated by sample results given in Figure23 for the heavily diffusing case, $r_1^2/\nu t = 2.0$.

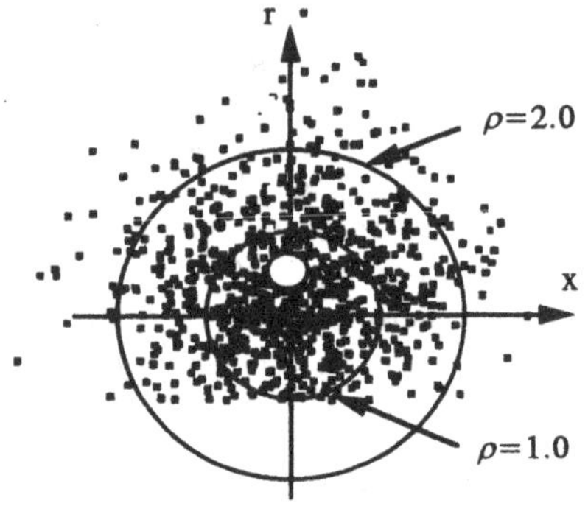

Figure 23 Random walk representation of a diffusing ring vortex for $r_1^2/\nu t=2.0$.

As shown by this simulation, the mean centre of the vortex cloud shown O has diffused outwards to a radius of approximately $r=1.5$ for this case of a heavily diffusing flow ($r_1^2/\nu t=2.0$) which is found to agree well with the exact solution including also the spread of vorticity about the core centre.

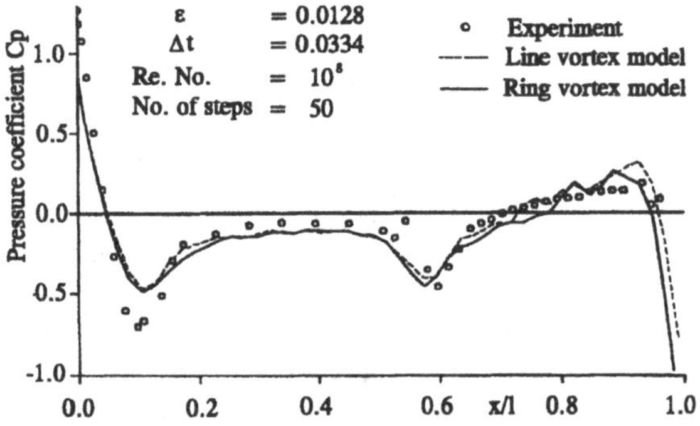

Figure 24 Surface pressure distribution for body of revolution.

To conclude this study the experimentally measured surface pressure distribution is compared in Figure 24 with predictions assuming either the line vortex random walk theory of that of Li Ming outlined above. In this case for very high Reynolds No. of 10^8, the standard random walk is perfectly adequate to handle simulation the viscous diffusion and very good predictions were obtained.

5. CONCLUSIONS AND RECOMMENDATIONS FOR FURTHER WORK

Vortex element methods and numerical methodology are now well established to deal with an enormous range of practical engineering problems. Surface vorticity potential flow modelling, originally attributable to E.Martensen(1959)[2], has been the fore-runner of full vortex dynamics and its numerical framework extends naturally and with relative ease into vortex dynamics modelling. While there is no limit to the range of potential flow problems to which surface vorticity modelling may be applied with remarkable precision, vortex dynamics deals most effectively with bluff body flows and will yield good predictions of such parameters as form drag and Strouhal number. In these respects vortex dynamics ranks high among CFD methods. Stable aerofoil flows such as turbine cascades are also handled well by vortex dynamics including simulation of blade row wake interference. Treatment of the boundary layer in less stable aerofoil flows offers room for improvement. The following areas of further work are recommended:

1. Extension of vortex cloud modelling to mixed-flow and radial turbomachine blade-to-blade flows.
2. Further development of Li Ming's random walk method for axisymmetric flows and vortex modelling for turbomachinery meridional flows and flow though ducts and annuli.
3. Further studies of hybrid methods including the use of grids to improve resolution of boundary layer flows.
4. Application of vortex dynamics to a wide range of engineering design/analysis problems linked to experimental validations and comparisons with other CDF codes.

REFERENCES

(1) Lewis, R.I. (1991), Vortex Element Methods for Fluid Dynamic Analysis of Engineering Systems. Engine Technology Series, Cambridge Univ. Press.

(2) Martensen, E.(1959), Berechnung der Druckverteilung an Gitterprofilen in ebener Potentialstromung mit einer Fredholmschen Integralgleichung. Arch. Rat. Mech., Anal., **3**, p.235-270.

(3) Wilkinson, D.H. (1967), A numerical solution of the analysis and design problems for the flow past one or more aerofoils or cascades. ARC, R & M 3545.

(4) Lewis, R.I. (1981), Surface vorticity modelling of separated flows from two-dimensional bluff bodies of arbitrary shape. J. Mech. Eng. Sci., I.Mech. E., **23**, No. 1, p.1-12.

(5) Porthouse, D.T.C. (1983), Numerical simulation of aerofoil and bluff body flows by vortex dynamics. Ph.D. thesis, Newcastle upon Tyne University, U.K.

(6) Porthouse, D.T.C. & Lewis, R.I. (1981), Simulation of viscous diffusion for extension of the surface vorticity method to boundary layers and separated flows. J. Mech. Eng. Sci., I.Mech. E., **23**, No. 3, p.157-67.

(7) Chorin, A.J. (1973), Numerical study of slightly viscous flow. J.Fluid Mech., **57**, p.785-796

(8) Lewis, R.I. & Chunjun, Ji, (1998), Extension of vortex cloud modelling to cylinder arrays and cascades in relative motion. Proc. of the Fourth European Computational Fluid Dynamics Conference, Athens, p.642-647, J. Wiley & Sons.

(9) Chunjun, Ji (1997), Vortex cloud method and its applications to turbomachine cascades. Ph.D. thesis, Newcastle upon Tyne University, U.K.

(10) Lewis, R.I. (1996), Turbomachinery Performance Analysis. W.Arnold.

(11) Gostelow, J.P. (1984), Cascade aerodynamics. Pergamon Press, Oxford.

(12) Young, L. (1958), Runners of experimental turbomachines. Engineering, London, **185**, p.376.

(13) Wilkinson, D.H., (1969), The analysis and design of blade shape for radial, mixed and axial turbomachines with incompressible flow. M.E.L. Report.No. W/M(3F), English Electric Co., Whetstone, Leicester.

(14) Fisher, E.H., (1975), Performance of mixed-flow pumps and fans. Ph.D. thesis, University of Newcastle upon Tyne, U.K..

(15) Lewis, R.I. (1999), User instructions for MIXFLO, a computer program for the design and analysis of axial, radial and mixed-flow turbomachines, cascades and blade rows.

(16) Horlock, J.H., (1958), Axial Flow Compressors. Fluid Mechanics and Thermodynamics. Butterworth Publications Ltd., London.

(17) Stanitz, J.D. (1952) Some theoretical aerodynamic investigations of impellers in radial and mixed-flow centrifugal compressors. Trans. ASME, **74**, No. 4.

(18) Busemann, A. (1928), Das Förderhöhenverhältnis radialer Kreiselpumpen mit logarithmischspiraligen Schaufeln. Z, Angew. Math. Mech., **8**(5), p.372-384.

(19) Li Ming (1995),Vortex cloud modelling for axisymmetric flows. Ph.D. Thesis, Newcastle upon Tyne University, U.K..

(20) Lewis, R.I. & Li Ming, (1998), Random walk method for diffusion of ring vorticity for extension of vortex dynamics to axisymmetric flows. Proc. of the Fourth European Computational Fluid Dynamics Conference, Athens, p.780-785, J. Wiley & Sons.

(21) Ryan, P.G. (1970) Surface Vorticity Distribution Techniques applied to ducted propellers. Ph.D. thesis, University of Newcastle upon Tyne, U.K..

(22) Lewis, R.I. & Ryan, P.G. (1972), Surface vorticity theory for axisymmetric potential flow past annular aerofoils and bodies of revolution with application to ducted propellers and cowls. J.Mech. Eng. Sci., 14, No.4.

A HYBRID VORTEX METHOD

J. Michael R. Graham and Richard H. Arkell

Department of Aeronautics, Imperial College

Prince Consort Rd, South Kensington, London SW7 2BY, UK.

e-mail: m.graham@ic.ac.uk

ABSTRACT

This paper describes a method of solving the unsteady incompressible Navier-Stokes equations, based on the Lagrangian discrete vortex technique. The technique uses a mesh to compute the velocity and diffusion fields associated with the vorticity, in contrast to the classical Discrete Vortex Method (DVM) which is mesh free. However convection of vorticity is modelled by Lagrangian motion of vortex particles as in classical DVM. The resulting hybrid method gives a smoother mesh based flux method for the diffusion while retaining the advantages of Lagrangian convection.

The development of a method for both two-dimensional and three-dimensional flows is described, in the latter case the method being based on the use of three-dimensional vortex particles known as vortons. The Eulerian part of the solution technique in both two- and three-dimensions makes use of the finite element method for the flow solution on an unstructured triangular or tetrahedral mesh. This permits complex geometries to be analysed easily.

The paper also presents some validation of the technique with examples of flow about a circular cylinder at Reynolds numbers of order 10^2 where the flow gives rise to two-dimensional laminar vortex streets which have now been computed to high accuracy. Comparisons are shown with other methods of computation. The issues of the simulation of turbulent wake flow using the method are also discussed.

In the final part of the paper a less computationally expensive method for three-dimensional cross-flows past long bodies is presented and an application of this technique to a practical case of an oil riser pipe is shown.

1. INTRODUCTION

The classical discrete vortex method (DVM) uses potential line vortices (points in a two-dimensional flow plane) to represent continuous thin layers of vorticity moving with the local velocity field. Initially the method was inviscid, so that when applied to bluff bodies the separation lines from which discrete vortices were shed had to be specified. The process of releasing vortices into the flow depended on numerical parameters but none-the-less gave a good representation of von-Karman vortex wakes and a good estimate of Strouhal number. Later Kamemoto and Bearman (1) examined the performance of this model for two-dimensional flow around a flat plate normal to an incident stream and showed that a value of a non-dimensional numerical parameter governing discrete vortex release from a separation edge could be rationally specified to provide an optimum simulation of the separated wake.

A serious problem associated with the classical DVM has been the use of the Biot-Savart integral solution of the Poisson equation relating the vorticity field to the velocity field, required for the Lagrangian convection. This procedure gives rise to an N_v^2 (N_v = number of vortices) operation count at each time step which typically limits flows to no more than the order of 10^3 vortices in the field. Two methods have been developed to counteract this. The first, known as the Vortex-in-Cell (VIC) method and derived from plasma physics, Christiansen (2), replaces the Biot-Savart integral by a finite difference solution of a Poisson equation on a mesh. This procedure considerably reduces the computational cost to $O(N_v)$ operations per timestep, but requires the vorticity field to be projected from the point vortex representation onto a mesh. It is the basis of the hybrid method to be presented in the present paper. The second technique, known as the fast multipole method, due to Greengard and Rokhlin (3) expands the integrand of the Biot-Savart integral allowing contributions to be evaluated efficiently leading to an $O(N_v.\ln(N_v))$ operation count.

In viscous flows vortex sheets diffuse to non-zero thickness and vorticity of greater than infinitesimal strength may fill significant regions of the flow field.. A number of different techniques have been incorporated into DVM to simulate diffusion. These include providing the vortices with continuously growing viscous cores, a method which does not converge to solutions of the Navier-Stokes equations, representing the vorticity by finite support with vortex-vortex interaction and exchange of vorticity (4), the random walk method (eg. 5) and the less commonly used diffusion velocity method (6). The last two represent diffusion by an addition to the Lagrangian movement of the discrete vortices which models convection. In the random walk case a random displacement is added to the convective displacement but convergence to the solution of the diffusion equation requires a very large number of vortices per unit volume of the flow. The method also has a restriction that the coefficient of diffusion should be a spatial constant limiting its use in representing eddy viscosity. This restriction has been partially overcome Smith and Stansby (5) for thin turbulent layers where the variation of eddy viscosity can be assumed to be in a single direction only for which a transformation may be used.

Using the fast multipole method of computing the velocity field from the vorticity field or VIC the flow may contain the order of 10^5 to 10^6 vortices and computations of viscous flows have been carried out with results at least as accurate as those given by conventional flux methods, eg (5), (7), (8) and (9), the latter using a mesh based diffusion. Most of these computations have been for two-dimensional flows and it is this perceived restriction which has caused the method to be less used than the primitive variable methods in industrial flow applications. However an advantage of the discrete vortex method for separated flows is the reduced numerical diffusion associated with Lagrangian convection in comparison with a flux representation of convection terms.

Three-dimensional vortex methods have been initially based on extensions of the connected vortex lattice method to the representations of free vortex sheets (10). While this method guarantees continuity of vorticity it has the serious disadvantage that distortion of vortex sheets as they evolve can lead to topological breakdown. Alternative discretisations of three-dimensional vortex fields use disconnected vortex 'sticks' or alternatively, vortex 'blobs' known also as vortons (11), the latter being used in the present work. Both methods distribute the vorticity onto a large number of discrete elements. Stick elements have finite length whereas vortons represent the vorticity field as a vector quantity at a point. Stretching of the vorticity field is handled in the vorton method by applying the terms in the vorticity transport equation directly to the strength and direction of this vector. Both methods avoid the topological problems of a free lattice but do not guarantee continuity of vorticity. One cause of the growth of divergence in the vorticity field is the discretised effect of the stretching terms in the vorticity transport equation. These adverse effects can be reduced by manipulating the terms into a symmetric form.

2. HYBRID VORTEX METHOD FOR TWO DIMENSIONAL FLOW.

In two-dimensional flow the vorticity field ω is a scalar and is governed by the vorticity transport equation:

$$\partial\omega/\partial t + u_i\, \partial\omega/\partial x_i = \nu\, \partial^2\omega/\partial x_i^2 \qquad (1)$$

The vorticity is connected to the velocity field u_i by:

$$\partial^2\psi/\partial x_i^2 = -\omega \qquad (2)$$

where ψ is the stream function, or equivalently directly by:

$$\partial^2 u_i/\partial x_j^2 = -\varepsilon_{ij3}\, \partial\omega/\partial x_j \qquad (3)$$

The continuity equation:

$$\partial u_i/\partial x_i = 0 \qquad (4)$$

is enforced by using either. The i and j directions are in the plane of the flow, the direction 3 is normal to it and the repeated suffix summation convention is assumed.

The basis of the method is operator splitting of the vorticity transport equation (1) into convection and diffusion substeps. This procedure is formally first order in time, Marchuk (12), and it is therefore consistent to solve each substep of equation (1) using first order time integration. In the present code the diffusion substep is solved over time step (n) using a fully implicit scheme in order to give maximum stability:

$$\omega^* = \omega^{(n)} + \nu\, \partial^2\omega^*/\partial x_i^2\, \Delta t \qquad (5)$$

and the subsequent convection substep

$$\partial\omega^*/\partial t + u_i\, \partial\omega^*/\partial x_i = 0 \qquad (6)$$

using a forward Euler scheme. Higher order time integration is possible but only used occasionally.

The velocity field for equation (6) is computed from either (2) or (3) using $\omega^{(n)}$ for the right hand side. At the end of the time step Δt the new vorticity field is given by:

$$\omega^{(n+1)} = \omega^* \qquad (7)$$

The convection sub-step, equation (6), for the vorticity field ω^* is satisfied by moving vortex particles with the local velocity field. Using forward Euler integration:

$$\underline{x}_k^{(n+1)} = \underline{x}_k^{(n)} + \underline{u}_k^{(n)}\, \Delta t \qquad (8)$$

for the position of the kth vortex.

The vortices carry either integrals of the vorticity field ω^* over a fixed area of fluid particles (circulation) for two-dimensional flow or analogous volume integrals ('vortonicity') for three-dimensional flow. The size of these areas or volumes is such as to cover the entire region of vortical flow without overlapping. Since the flow is incompressible and areas or volumes composed of any fixed piece of fluid are therefore conserved, this procedure satisfies equation (6) in the limit of zero particle size and time step.

The main difference in the method to be described here from other vortex methods is that the viscous diffusion substep (5) is evaluated by flux methods on a mesh, the same mesh used to compute the velocity field. The diffusion equation is elliptic and is therefore well represented by standard centred schemes of discretisation which do not introduce significant numerical viscosity. This is in contrast to the convection equations which are hyperbolic and require a significant degree of upwinding or diffusion to provide stability. In the VIC method the vorticity field is projected onto the mesh in order to compute the associated velocity field at each time step and hence is already available on the mesh for the diffusion calculation. This is performed by a standard finite difference, volume or element method and the resulting 'vorticity difference field' due to the diffusion evaluated. The difference field is projected back onto the vortex particles so that the results of diffusion appear as a change in the positions and strengths of the discrete vortices. The reverse projection is carried out using the same weighting functions as for the forward projection and is constructed to conserve integrals of the vorticity field, circulation or vortonicity, and in some versions first moments of vorticity also. The logic behind computing the difference field is to limit numerical diffusion to be small compared with real viscous diffusion. Reprojecting the difference field achieves this even if the Reynolds number is high and hence the viscous diffusion small since the numerical diffusion is in proportion to the real diffusion.

The method has similarity to that due to Chang and Chern (13) which however does not retain the identity of discrete vortices but replaces them at each time-step with a new set emitted from the mesh. The use of the mesh imposes an effective core on the vortices, removing the velocity field singularity. Vortices appear as 'blobs' of finite support with effective size depending on the shape of the elements in the mesh and the weighting functions used to project the vorticity from the particles onto the mesh. The 'mean core radius' is of order the length scale of the mesh elements.

In its original form, Graham (9), the present method used a regular mesh with 2nd order centred, finite difference spatial discretisation to solve for both the velocity field (using eqn.2) and the diffusion (eqn.5). In the most recent version eqn.3 is used for the velocity field and both velocity and diffusion are computed on an unstructured triangular mesh using a piecewise linear finite element method. This mesh and the analogous three-dimensional mesh of tetrahedral elements is constructed in the present work by the FELISA advancing front method (14) which allows much greater geometric flexibility and superior solution convergence.

The inner boundary condition used for equation (3) (or (2)) is that of zero normal flow relative to the surface:

$$\underline{u}.\underline{n} = 0 \qquad (9)$$

for a stationary solid body, where $\underline{n}$ is the outward normal. The outer boundary condition is applied either as the velocity induced by a simplified vorticity field or more simply as the unperturbed free stream.

Boundary conditions for equation (5) are derived since the no-slip condition does not give a natural boundary condition on vorticity. It has been found that the most convenient condition for the present code is:

$$\delta\omega/\delta n = u_s/(\nu\,\Delta t) \qquad (10)$$

where u_s is the slip velocity which has arisen at the body surface due to convection of the vorticity field and any other changes over the time step Δt.

The outer boundary condition on vorticity is most conveniently taken as $\omega = 0$ (exponentially small) except where vorticity is convected through a (typically downstream) boundary. Here a condition of zero vorticity gradient in the flow direction has been found to be satisfactory since in most cases this boundary is far downstream.

To solve the equations the vorticity, ω, and the velocity (u,v) fields are represented by a sum of linear shape functions N_i over the elements e_i of the mesh:

$$\omega^{(n)} = \Sigma\, \omega_i^{(n)} N_i \qquad (11)$$

$$u^{(n)} = \Sigma\, u_i^{(n)} N_i \qquad (12)$$

$$v^{(n)} = \Sigma\, v_i^{(n)} N_i \qquad (13)$$

In order to obtain the vorticity values on the nodal points of the mesh from the set of discrete vortices their strengths are projected onto the mesh using the same set of shape functions N_i as weights. Thus:

$$\omega_i = \Sigma\, N_i(x_k)\, \Gamma_k/A_i \qquad (14)$$

where Γ_k is the circulation carried by the kth discrete vortex and A_i is the area associated with the ith node.

Substituting (11), (12) and (13) into equations (3) and (5), multiplying through by the same set of functions N_j as weights, integrating by parts over the domain (Galerkin procedure) and inserting the boundary conditions results in the matrix equations:

$$[K]\{u^{(n)}\} = \{R_1(v^{(n)},\omega^{(n)})\} \quad (15)$$

$$[K]\{v^{(n)}\} = \{R_2(u^{(n)},\omega^{(n)})\} \quad (16)$$

$$[K_1]\{\omega^*\} = \{R(u^{(n)},v^{(n)},\omega^{(n)})\} \quad (17)$$

The involvement of the velocity components u and v in the right hand side vectors is due to their presence in both boundary conditions. The solution is performed using a preconditioned conjugate gradient solver which is very efficient. Because boundary conditions (15) and (16) are coupled a sub-iteration is required in the conjugate gradient procedure. Alternatively it is possible to solve equation (2) for the streamfunction ψ. This is slightly faster but results in a lower order of spatial interpolation of the velocity field and some loss of accuracy unless ψ is approximated by higher (quadratic) order shape functions. After obtaining the change in the vorticity field $\delta\omega$ due to one time integration step of the diffusion equations the change is projected back as a circulation change $\delta\Gamma$ onto the discrete vortices using the same weighting procedure:

$$\delta\Gamma_k = \Sigma\, \delta\omega_j A_j N_j/S_j \quad (18)$$

where $S_j = \Sigma\, N_j(x_k)$ (19)

is the sum of the weighting functions projected (equation 14) onto node j. At the boundaries of vortical regions it may be necessary to project the change of vorticity into cells which contain no vortex particles. This requires new particles to be created and it is this which creates all the vortex particles ultimately present in the flow.

Having obtained the velocity field at any given time, surface pressures, shear stresses and forces can be obtained. The surface shear stress is obtained directly from the surface vorticity field:

$$\tau_w = \mu\omega_w \quad (20)$$

where the suffix w indicates the wall. The pressure p can be obtained on the body surface in a number of ways. The present code solves a Poisson equation for pressure:

$$\partial^2 p/\partial x_j^2 = -\rho\,(\partial u_i/\partial x_j)(\partial u_j/\partial x_i) \quad (21)$$

together with the boundary condition along the body surface s:

$$\partial p/\partial n = -\mu\, \partial\omega/\partial s \quad (22)$$

using the same finite element procedure as for the other Poisson equations. Lift and drag forces on the body, and any other moments if required, are then easily obtained.

3. HYBRID VORTEX METHOD FOR THREE DIMENSIONAL FLOW.

The main extensions to the method for three-dimensional flow are that the vorticity transport equation now contains the additional 'stretching' term $\omega_i du_j/dx_i$ on the right hand side and that the vorticity in equation (1) is now a three component vector. The code treats the stretching term as an (explicit) extension to the diffusion term. Thus the 'diffusion' substep solves:

$$\omega_j^* = \omega_j^{(n)} + \nu\,\partial^2 \omega_j^*/\partial x_i^2\, \Delta t + \omega_i^{(n)}\, \partial u_j^{(n)}/\partial x_i\, \Delta t \quad (23)$$

the result of stretching being projected like diffusion back onto the vortex particles.

The result of vorticity and velocity both being three-dimensional vectors is that their boundary conditions couple the Poisson equations to a greater degree and hence a more extensive sub-iteration is required. More details of the method (FEMVOR) are given in Arkell (15). Continuity of vorticity and hence velocity is satisfied but not enforced by the method. In the flow simulations the divergence of the vorticity field was monitored as the flow evolved and it was found that in general there was a gradual, but small, increase with time over which the local vorticity had existed. A procedure which was used to reduce this error was to recompute the mesh vorticity field directly from the velocity field at each time step thus ensuring that it was divergence free.

Computation of three-dimensional unsteady viscous flows is computationally very expensive and a quasi-three-dimensional method has also been developed for practical applications, Giannakidis and Graham (16), Willden and Graham (17). This method is only suitable for long bodies of slowly varying section in cross-flow. The flow is represented by the two-dimensional vortex method on a number of cross-sections. These are linked via a free vortex lattice. The method offers an order of magnitude or more in speed-up compared with a full three-dimensional computation, but for a reduced accuracy in the three-dimensionality as in classical lifting-line methods. It is described in more detail in (16) and (17) and a result for cross-flow past a long, flexing circular cylinder given later in the paper.

4. TURBULENCE TRANSPORT EQUATIONS.

The hybrid method has been extended to higher Reynolds number by incorporating a turbulence model. The basic equations of the flow are taken to be the (Unsteady) Reynolds Averaged Navier-Stokes (URANS) equations. An eddy viscosity model based on the k-ε model has been used with some adjustment of the constants to take account of effects of low Reynolds numbers, separations and unsteadiness on a large scale.

Since the transport equations for the turbulence kinetic energy, k, and the dissipation, ε, are convection-diffusion equations analogous to the vorticity transport equation, the equations are solved similarly by the operator splitting procedure. The diffusion part of each equation is solved on the mesh by the same finite element method. The convection part of each equation is solved by the Lagrangian moving particle technique. Fluid particles are assumed to carry, in addition to circulation or vortonicity, similar area or volume integral quantities K and E of the turbulence kinetic energy and dissipation respectively. More particles may be required in total since some may carry K and E and not Γ, but in general there is a good correspondence between the three fields.

The same approach can be used to convect other flow quantities such as temperature or concentration if required, with, in each case, the particles carrying a conserved quantity.

5. COMPUTED RESULTS.

5.1 Two-Dimensional Flows.

The method described above has been tested on a number of basic flows including the diffusion of a single line vortex, the rolling up of a free vortex sheet and the evolution of vorticity rolling up into a diffusing spiral sheet shed from a semi-infinite sharp edge.

The two-dimensional viscous flow which has been most extensively investigated using the method is that of cross-flow past a circular cylinder at low Reynolds numbers. Figure 1 shows a vortex wake formed behind a fixed circular cylinder in impulsively started flow at a Reynolds number of 100, computed by the method on an unstructured mesh shown. Flow asymmetry was found to develop quickly naturally from the small differences associated with the use of an unstructured mesh and lead to the typical force time histories shown in figure 2. The three quantities which have been evaluated extensively for this flow are the (asymptotic) values of Strouhal number, mean drag coefficient and root mean square (RMS) lift coefficient. Bearman (18) has recently published values computed by a number of different investigators and codes for a Reynolds number of 100 assuming two-dimensional flow. Table 1 shows some results taken from this reference and a result from the present code. A significant finding from this is that it is fairly easy to predict accurately the Strouhal number, mean drag coefficient is harder, but of the quantities examined so far, fluctuating lift coefficient is the most sensitive to the accuracy of the flow simulation. A remarkably fine mesh is required to compute this quantity to two figure accuracy, even at this low Reynolds number.

The present method may also be used to compute cases in which a body oscillates in a stream. The circular cylinder, either forced to oscillate transversely or in-line, or elastically mounted and oscillating in response to the forces induced by vortex shedding has been widely studied both experimentally and computationally. In the case of a single cylinder the computation may be carried out using a mesh system moving with the cylinder with appropriate adjustment to the velocity boundary conditions. This flow is kinematically exactly the same as the case when the cylinder is fixed and the flow oscillates. The pressure fields differ only by the acceleration (Froude-Krylov) component which is easily computed.

Meneghini and Bearman (19) have studied a range of forced oscillations of a circular cylinder using a finite difference version of the method on a regular mesh. This has shown all the features of lock-in of vortex shedding and different wake flow regimes according to the amplitude of oscillation. Arkell (15) has used the unstructured mesh version to study the effects of in-line oscillation of a flow past a circular cylinder in uniform translation. This flow is a representation of a horizontal section of the flow generated by towing a vertical circular cylinder through free surface waves. The incident flow field is spatially as well as temporarily varying which generates further effects on the vortex wake as shown in figure 3.

Discrete vortex methods have also been widely used to study two-dimensional flows around arrays of bodies, using mesh-free DVM. Recent extensions of the present two-dimensional hybrid method to multiple bodies in large scale relative motion to one another have been computed by Giannakidis (20) , figure 4. The case shown here is part of a study to represent a cross-section of flow due to the passage of a wind-turbine blade past the tower. Large scale relative motion of this sort necessitates the additional complication of a moving/distorting mesh. In the present case an effective method uses overlapping (sub)-meshes attached to each body from which a global mesh is constructed, a different process from the frequently used interpolation between overlapping meshes.

At higher Reynolds numbers two dimensional simulations develop instabilities as in the flat plate boundary layers shown in figure 5, but at somewhat lower Reynolds numbers than predicted by classical stability theory.

5.2 Turbulent Flows

At Reynolds numbers above about 140 the wake of a circular cylinder in steady incident cross-flow becomes three-dimensional and eventually turbulent. Strictly such flows cannot be computed by two-dimensional methods but because of the persistence of the von-Karman vortex street a considerable number of two-dimensional computations have been undertaken to simulate the higher Reynolds number range often with an incorporated turbulence model representing the small scales of the flow. Computations of this type have been carried out with the present hybrid code for steady cross-flow past a circular cylinder at higher Reynolds numbers using the method for the k-ε model outlined above. Figures 6 and 7 show respectively the vorticity distribution and contours of the turbulence kinetic energy and dissipation for cross-flow past a circular cylinder at a Reynolds number of 10^4. This result has been obtained after adjusting a source term in the dissipation equation in order to reduce somewhat the intensity of turbulence energy in the wake.

5.3 Three-Dimensional Flows

The three-dimensional version of the method, FEMVOR, has been tested for a number of basic flows including diffusion of a line vortex and propagation of a vortex ring. In the latter case the evolution of a ring of suitably oriented vortons, placed in an unstructured tetrahedral mesh, was studied.

The velocity of propagation v of a vortex ring of radius R to lowest order in the core radius r_c is:

$$v = \Gamma/(4\pi R)[\ln(4R/r_c) - ¼] \quad (24)$$

As discussed earlier, the vortex ring projected onto the mesh has an effective core radius r_c similar to the mesh size. Hence the apparent velocity of propagation of the ring should depend on this. The computed ring positions for three different meshes and inviscid flow are compared in figure 8 with propagation given by eqn.24 suggesting that r_c take values of 1.190, 0.677 and 0.488, close to the mean mesh size in each case.

The FEMVOR code has been used to compute two practical problems in which a three-dimensional vortex wake structure is certain to appear early in the flow development.

The first was the case of a tapered cylinder in a cross-flow. Figure 9 shows an example of a computed wake for a case with taper ratio 1:12, aspect ratio of test element equal to 8 and mean Reynolds number of 200. Laminar flow was assumed throughout. The results show clearly the 'Y' shaped vortex structures which are known to be generated when a vortex wake has different frequencies of shedding at either end. The end-wall boundary conditions assumed a reflecting slip surface. This flow field was visualised experimentally at approximately the same Reynolds number using dye in water. The cylinder was similarly tapered over the test length, but joined sections of constant diameter at either end. A visualisation is shown in figure 10 where the 'Y' shaped structure is clearly visible. The difference Strouhal number is in good agreement with the mean line through the published experimental measurements of Piccarillo (21) for tapered cylinder flows.

In a second flow study a vertical cylinder moving steadily through head waves was simulated. The boundary conditions used here for the free surface assumed a linearised slip surface. Figure 11 shows the computed vortex wake for the case of waves of wavelength equal to 30 cylinder diameters and a Reynolds number of 600. This shows clearly the different vortex structures at different depths due to the variation of wave particle velocities at different depths and may be compared with the two-dimensional result shown earlier.

5.4 Quasi-Three-Dimensional Flow Calculations

Three-dimensional computations such as the ones above require meshes of at least the order of 10^5 elements and are as a result computationally expensive to carry out on a work-station size of computer. Many practical problems are at much higher Reynolds number and may require much longer flow evolution to establish such things as stability or vortex lock-in. A quasi-three-dimensional approach has been developed for these problems which is well suited to the vortex method. It is based on the use of two-dimensional sectional computations applied along the length of any long body in a cross-flow following the approach used for example by Hansen et al. (8). It has been extended (Giannakidis and Graham (16), Willden and Graham (17)) by linking the sections kinematically via a vortex lattice representation of the three-dimensional wake. Sectional computations solve the two-dimensional Navier-Stokes equations on unstructured meshes using the two-dimensional hybrid code and a reasonably fine resolution. The vorticity in these sectional flows is then projected onto a three-dimensional vortex lattice, completed by inserting components of vorticity parallel to the sectional planes to provide continuity of circulation. The resulting vortex lattice is treated inviscidly but allowed to convect and evolve as a free sheet. This in turn influences the boundary conditions on the external boundaries of the sectional meshes, thus feeding the major three-dimensional effects back. The method is very efficient but restricted to flows about long structures. Offshore oil riser pipes as shown in figure 12 and flexing bridge decks have been simulated.

6. CONCLUSIONS

The paper has described a hybrid vortex method which solves the incompressible Navier-Stokes equations by transporting the vorticity field, and other fields such as eddy viscosity or temperature if required, on a set of discrete particles, the vortices. The method involves projection of these fields in both directions between the moving particles and the mesh through which they move. In the present method this is accomplished by using the same weighting functions as are used for the finite element solution of the velocity and diffusion fields on the mesh. The mesh used was an unstructured mesh of triangular or tetrahedral elements. All parts of the calculation except convection are carried out on the mesh. The method can be used for both two- and three-dimensional flows and can accommodate relative motion of more than one body.

Source	C_{Dmean}	C_{Lrms}
Beaudan & Moin	1.35	0.24
High order finite difference		
Guschin	1.38	n/a
Finite difference		
Kravchenko	n/a	0.23
Galerkin B-spline		
Sherwin	1.36	0.24
Spectral element		
Present method	1.37	0.23

Table 1. Circular cylinder, Re.no. = 100, (from 18).

REFERENCES

1) Kamemoto K. and Bearman P.W. (1980) An inviscid model of interactive vortex shedding behind a pair of flat plates arranged side by side to approaching flow. Trans. Japan S.M.E., 46, p1299
2) Christiansen J.P. (1973) Numerical simulation of hydrodynamics by the method of point vortices. J. Comp. Phys., 13, p363
3) Greengard C. and Rokhlin V. (1987) A fast algorithm for particle simulation. J. Comp. Phys., 73, p325
4) Raviart P.A. (1986) Particle numerical models in fluid dynamics. Num. Meth. Fluid Dyn. II, (IMIA Conf. Series. No. 7), p231
5) Smith P.A. and Stansby P.K. (1988) Impulsively started flow around a circular cylinder by the vortex method. J. Fluid Mech., 194, p45
6) Ogami Y. and Akamatsu T. (1991) Viscous flow simulation using the discrete vortex model – the diffusion velocity method. Computers and Fluids, 19, p443
7) Leonard A. (1980) Vortex methods for flow simulation. J. Comp. Phys., 37, p289
8) Hansen H.T., Skomedal N.G. and Vada T. (1988) A method for computation of integrated vortex induced fluid loading and response interaction of marine risers in waves and current. Proc. Int. Conf. Behaviour of Offshore Structures, 2, 841 (Tapir).
9) Graham J.M.R. (1988) Computation of viscous separated flow using a particle method. Num. Meth. In Fluid Mech. p310, OUP.
10) Meiburg E. and Lasheras J.C. (1988) Experimental and numerical investigation of three-dimensional transition in plane wakes. J. Fluid Mech., 190, p1
11) Novikov E.A. (1975) Dynamics and statistics of a system of vortices. Sov. Phys. JETP, 41, p937
12) Marchuk G. (1971) In Numerical solution of PDE's II, Hubbard B. (ed), Acad. Press
13) Chang C.C. and Chern R.L. (1991) Flow around an impulsively started circular cylinder. J. Fluid Mech. 233, p243
14) Peiro J., Peraire J. and Morgan K. (1994) The FELISA system reference manual.
15) Arkell R.H. (1995) Wake dynamics of cylinders encountering free surface gravity waves. PhD thesis, Univ. London, UK
16) Giannakidis G. and Graham J.M.R. (1997) Prediction of loading on a HAWT rotor including effects of stall. Proc. Euro. Wind Energy Conf. p434
17) Willden R.H.J. and Graham J.M.R. (1999) Numerical simulation of the flow about deep water riser pipes. Proc. Comp. Meth. For Fluid Structure Interaction. p315, (Tapir)
18) Bearman P.W. (1998) Developments in the understanding of bluff body flows. JMSE Int. J. Series B, 41.

19) Meneghini J.R. and Bearman P.W. (1995) Numerical simulation of high amplitude oscillatory flow about a circular cylinder. J. Fluids Struct., 9, p435

20) Giannakidis G. (1999) Unpublished communication.

21) Piccirillo P.S. and Van Atta C.W. (1993) An experimental study of vortex shedding behind a linearly tapered cylinder at low Reynolds number. J. Fluid Mech., 246, p163

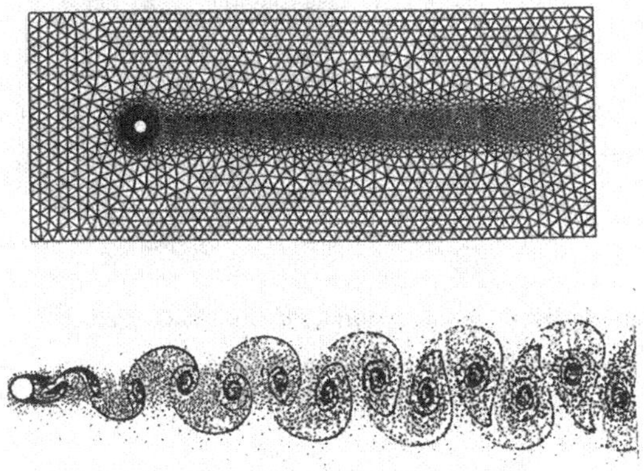

Figure 1. The wake of a circular cylinder at Re. No. = 100 and mesh used for computation

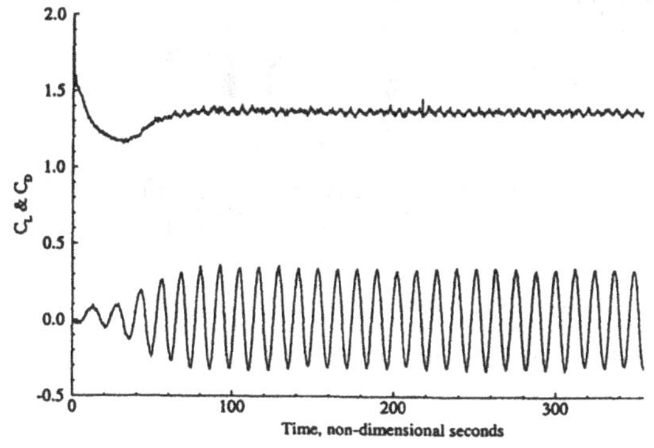

Figure 2. Force time history for flow past a circular cylinder at Re. No. = 100

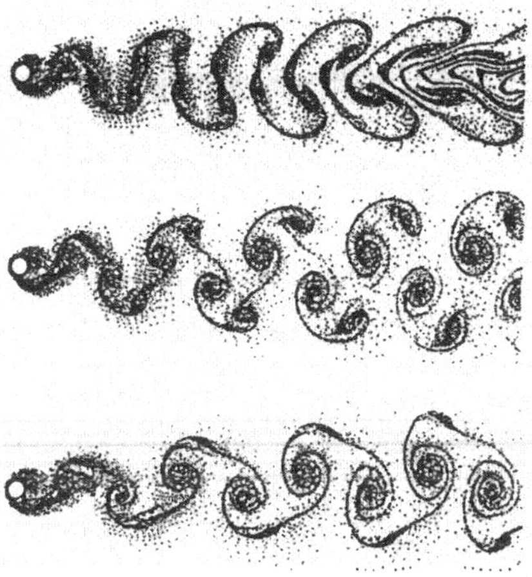

Figure 3. Two-dimensional simulations of the wake of a vertical circular cylinder in a mean current and waves. Re. No. = 150, wavelength/diameter = 30 (top), 60 (middle, bottom - larger amplitude)

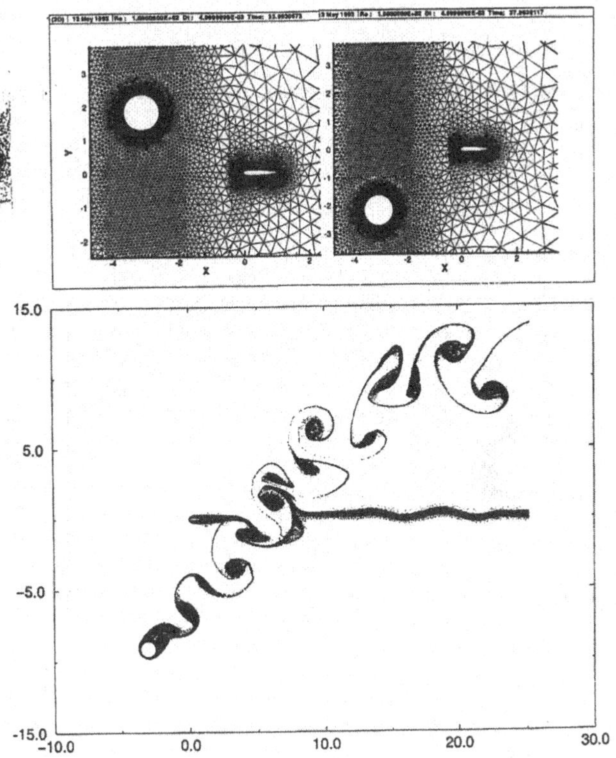

Figure 4. Interaction resulting from an aerofoil passing through the wake of a circular cylinder, computation mesh .

Figure 5. Two dimensional simulation of the vorticity in the boundary layers of a flat plate at Re. No $= 10^5$

Figure 6. Vortices in the turbulent wake of a circular cylinder at Re. No. $= 10^4$.

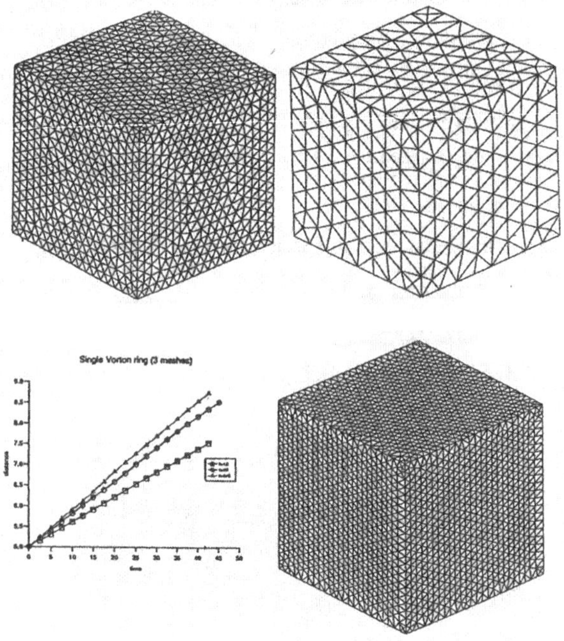

Figure 8. Displacement histories of a self-propagating vortex ring and computation meshes.

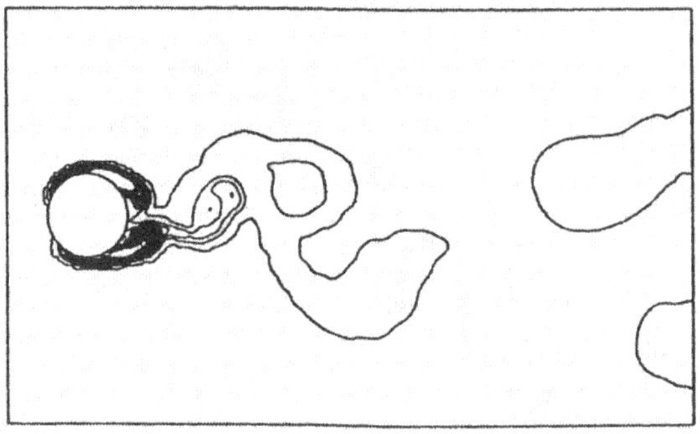

Figure 7. Turbulence intensity contours in the wake of a circular cylinder at Re. No. $= 10^4$.

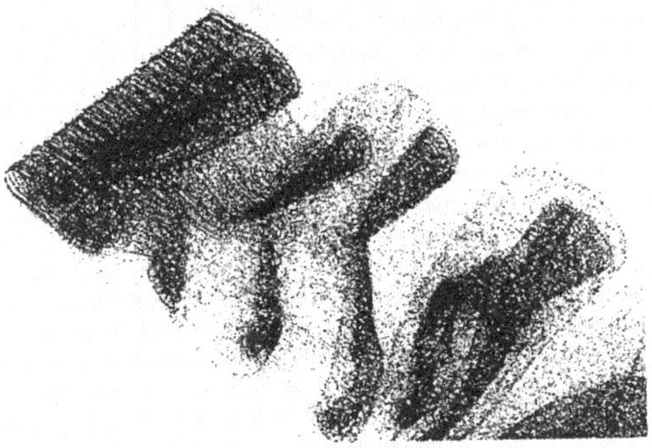

Figure 9. Numerical simulation of the vortex wake behind a tapered cylinder, mean Re. No. = 200

Figure 10. Flow visualisation in water of the vortex wake behind a tapered cylinder, mean Re. No. = 200

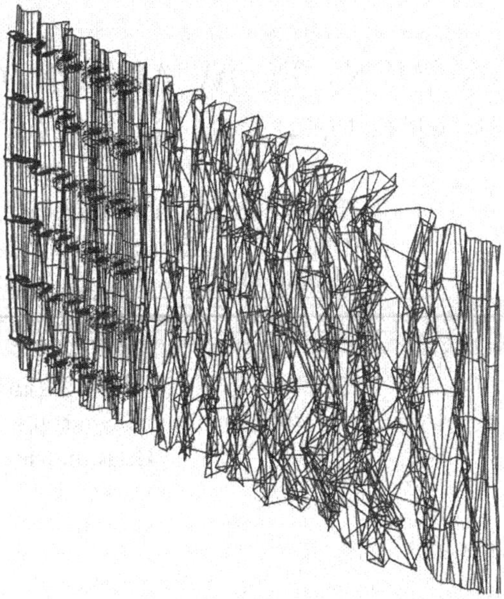

Figure 12. Quasi-three-dimensional simulation of a long flexing circular cylinder in a cross-flow, Re. No. = 100

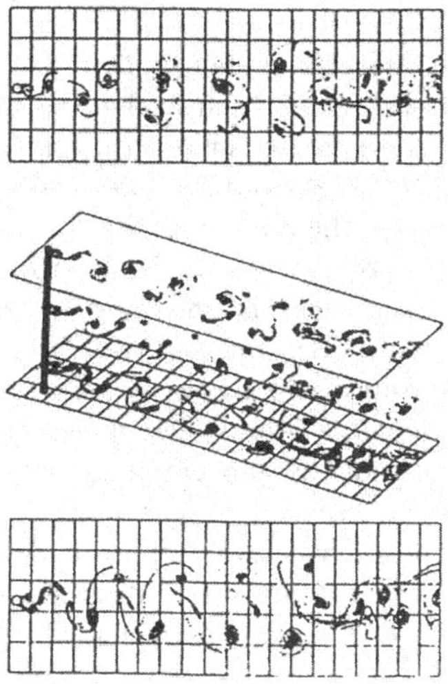

Figure 11. Three-dimensional simulation of the wake of a vertical circular cylinder in a mean current and waves. Re. No. = 200

TRANSIENT FLOW AROUND A CIRCULAR CYLINDER NEAR THE MOVING AND RIGID GROUND BY A VORTEX METHOD

Teruhiko Kida, Daisuke Okamoto*, Mitufumi Wada**
Department of Energy Systems Engineering, Osaka Prefecture University,
Sakai, Osaka 599-8531, Japan / Email: kida@energy.osakafu-u.ac.jp
Present:*EICOM, **Daikin Industries, LTD.

Tekanori Take
Department of Mechanical Systems Engineering, Shiga Prefecture University,
Hikone, Siga 522-8533, Japan

Abstract

The transient flow around a circular cylinder placed above a plane wall is simulated by a vortex method, which is the combination of vortex blob method and vortex sheet method. In this paper two cases are treated to know the effect of the boundary layer on the wall: (1) The cylinder suddenly moves over the fixed wall and (2) it starts suddenly over the moving wall with the same velocity. The present numerical result shows: (1) Shedding vortices are a little different between two cases. (2) The lift and drag distributions against time are almost the same for $h \geq 0.5$, where h is the gap ratio between the cylinder and the wall normalized with the diameter of the circular cylinder, but it is different for $h \leq 0.25$.

INTRODUCTION

The present paper treats transient flow around a circular cylinder placed above a plane boundary: The circular cylinder starts suddenly over the plane wall. In order to know the effect of boundary layer on the wall, two cases of the plane wall are treated in this paper: one is the fixed wall and another is the moving wall with the same velocity as the circular cylinder.

This problem is of fundamental interest in some applications, such as the design of ground effect wings, the interference effect between road vehicles or large structures, and the design of pipelines. We must design for the large structures so as to avoid potentially disastrous wind-induced large amplitude oscillations due to vortex-shedding characteristics. In the ground effect wing or the road vehicle, knowledge of the effect of the ground plane is also crucial in the design of the transient flight during the take-off and landing of the wing or the stability of the road vehicle. In spite of the importance of the ground effect flows, relatively little is known about the effect of boundary conditions of the ground to the vortex-shedding characteristics of bluff bodies. In the design of the large structure, the wall plane and the bluff body are fixed, but in the ground effect wing or the ground effect of road vehicles, the bodies move over a fixed wall. For the fixed coordinate systems with the body, the latter case is that the wall slides with the same velocity as of the body.

In potential flow theory flow around a cylinder above a plane wall is equivalent to flow around two cylinders in a side-by-side arrangement. In real flow around bluff bodies above the wall, boundary layer will grow on the wall and the flows about the two cylinders may not always be mirror images of each other, in particular the shedding vortices induce complicated flows. Bearman & Wadcock [1] showed that there are four distinct flow regimes around two circular cylinders in a side-by-side arrangement dependent on the ratio of the gap between the two cylinders to the cylinder diameters: (1) for the case where h/d is greater than 2, where h and d are the half of the gap between two cylinders and diameter of the cylinder respectively, the two cylinders shed vortices independently, (2) for $0.5 < h/d < 2$, the shedding of vortices from one cylinder was the mirror image of the shedding from the other, (3) for $0.05 < h/d < 0.5$ an unstable flow regime exists, and (5) for $h/d < 0.05$, a repulsive force acts between the two cylinders.

For the rigid wall, Bearman & Zdravkovicj [2] showed that there are two regimes: (1) for $h/d > 0.3$, regular vortex shedding is persisted and (2) for $h/d < 0.3$, strong regular vortex shedding was suppressed. Kamemoto et al. [3] studied flows around a normal flat plate and a triangular cylinder and showed

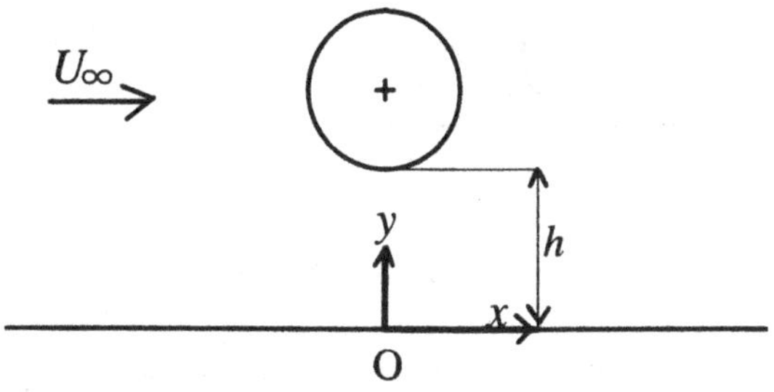

Figure 1: Physical plane.

that vortex shedding in these two bodies is suppressed when the gap ratio is less than about 0.6 and 0.37, respectively.

The effect of sliding or moving wall to aerodynamic forces has been studied in the field of road vehicles [4]: For a cube, Bearman [5] reviewed about pressure distribution on the ground and the aerodynamic forces. Since there was no detailed investigation of influence of the sliding wall to shedding process, Arnal et al. [6] studied time-dependent, two-dimensional flow around square cylinder in contact with a fixed and sliding wall by using a finite volume difference scheme. They showed that while the sliding wall causes the flow past the body to become unsteady at a lower Reynolds number than in the fixed wall case, it also acts to stabilize the vortex shedding frequency. However, there is little works of shedding vortices of transient flows around smooth bluff bodies such like a circular cylinder above the sliding or fixed wall. This problem is fundamental for knowing vortex shedding mechanism of bluff bodies due to the ground effect.

1 Physical state and vortex method

The present paper treats a transient flow around a circular cylinder above plane walls, which is fixed, or sliding. Figure 1 shows the physical state fixed with the circular cylinder. The present paper treats two cases: (1) Uniform flow U_∞ is impulsively imposed to the cylinder over the fixed wall and (2) uniform flow is impulsively imposed to the cylinder and at the same time as the wall begins to slide with the same speed.

These situations are corresponding to the cases: (1) the circular cylinder and the wall impulsively starts with constant speed and (2) the circular cylinder impulsively starts with constant speed above the fixed wall. Therefore, the case (1) and (2) are corresponding to an experiment by using fixed and moving ground equipment, respectively.

$$Case(1) \quad : \quad \begin{cases} \vec{u} = (0,0), & \text{on } y = 0 \text{ and on } S, \\ \vec{u} = (U_\infty, 0), & \text{as } |\vec{x}| \to \infty, \end{cases}$$

$$Case(2) \quad : \quad \begin{cases} \vec{u} = (U_\infty,0), & \text{on } y = 0 \text{ and on } S, \\ \vec{u} = (U_\infty, 0), & \text{as } |\vec{x}| \to \infty, \end{cases}$$

where S is the surface of the circular cylinder. The coordinate system (x, y) is taken as shown in Fig.1: the origin is taken as the ground position just below the center of the circular cylinder and the gap between the ground and the surface of the circular cylinder is defined by h, as shown in Fig.1. Here, lengths are normarized with the diameter of the circular cylinder d, and speeds are done with the uniform velocity U_∞. Time is also normarized with d/U_∞.

In the present paper, viscous splitting method is used: The Euler flow is first solved and second viscous effect is taken into account. To solve the Euler flow, a vortex method and a panel method are used and the viscous effect is simulated by random walk method. In the present paper, vortex blob and vortex sheet methods are used as the vortex method. The original method is given by Cheer [7] and Kida & Kurita [9] is revised the boundary condition imposed to Euler flow. The panel method which proposed by Kida et al. [8] is used to solve the potential flow.

1.1 Vortex blob method

The original idea of vortex blob is given by Chorin [10]. Several generic theorems for the validity of the vortex blob method to solutions of the incompressible Euler equations have been established (see Ying & Zhang [11]). In the present paper, Chorin's original cut-off function is used. The velocity $\vec{u}$ induced by a vortex blob with circulation Γ is given by

$$\vec{u}(\vec{x}) = (-y, x) \begin{cases} \frac{\Gamma}{2\pi|\vec{x}|\delta}, & |\vec{x}| < \delta, \\ \frac{\Gamma}{2\pi|\vec{x}|^2}, & |\vec{x}| \geq \delta. \end{cases} \tag{1}$$

Here δ is the cut-off radius.

1.2 Vortex sheet method

The vortex sheet method is based on the boundary layer theory:

$$\frac{\partial \omega}{\partial t} + \vec{u} \cdot \nabla \omega = \nu \frac{\partial^2 \omega}{\partial y^2}, \tag{2}$$

$$\omega = -\frac{\partial u}{\partial y}, \tag{3}$$

$$\frac{\partial u}{\partial x} + \frac{\partial v}{\partial y} = 0, \tag{4}$$

$$\vec{u} = 0 \quad \text{on } y = 0, \tag{5}$$

$$\lim_{y \to \infty} u = U_o, \tag{6}$$

where (x, y) is the local coordinates on the surface of the cylinder or on the wall, $\vec{u} = (u, v)$, and U_o is the velocity of the outer flow of the boundary layer.

The vortex sheet method proposed by Chorin [12] is expressed as

$$\tilde{\omega} = \sum_j \omega_j b_{l_j}(x - x_j)\delta(y_j - y), \qquad (7)$$

where $b_l = b(x/l)$,

$$b(x) = \begin{cases} 1 - |x|, & |x| \leq 1, \\ 0, & \text{otherwise.} \end{cases}$$

The velocity components u and v are obtained numerically by using Eqs.(3) and (4).

1.3 Panel method

The panel method proposed by Kida et al. [8] is used in this paper. The delta-function and the uniform function on each panel are used for the distribution of source and vorticity, respectively. Here, the strength of the vorticity is the same value on all panels. Their strength is solved from the flow tangency condition $u_n = v$, where u_n is the normal component of velocity vector on the boundary (see Kida & Kurita [13]):

$$v = -\int_0^{h_i} \frac{\partial u_s}{\partial x} dy, \qquad (8)$$

where h_i is the height of the boundary layer and u_s is the induced tangential velocity due to the uniform flow, the vortex blobs and sheets.

Further, Kelvin theorem is used for two vorticity fields generated by the surface of the circular cylinder and the surface of the wall:

$$\sum_j \Gamma_{cj} + \gamma_c = 0, \quad \sum_j \Gamma_{wj} + \gamma_w = 0, \qquad (9)$$

where Γ_j and γ are the circulation of j-th vortex blob and uniform vortices distribution on the boundary surface, and suffix c and g denote ones created from the circular cylinder and the wall, respectively.

From the capacity of computer, the length of the plane wall is takes as a finite length in the present calculation: $-4 \leq x \leq 12$. The typical arrangement of panels is shown in Fig.2. Here n_1 and n_2 are the number of panels on the surface of the circular cylinder and on the plane wall, which are taken as $n_1 = 40$ and $n_2 = 80$ in the present calculation.

1.4 Pressure distribution

The pressure distribution is solved by surface singularity method. For the high Reynolds number flows,

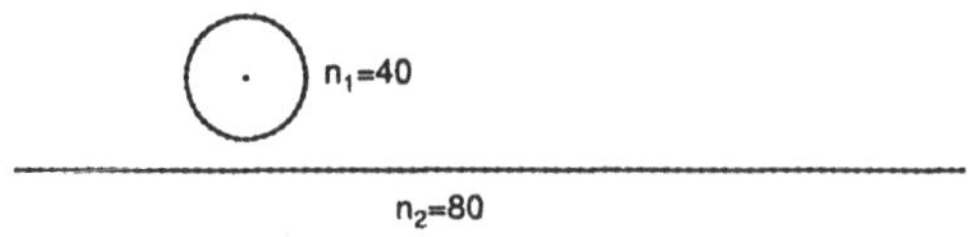

Figure 2: Panel cnstruction.

the pressure field outside the boundary layer is approximately governed by the Euler equations. From the two-dimensional Euler equations, we have

$$\nabla^2 H = \nabla \cdot (\vec{u} \times \vec{\omega}), \qquad (10)$$

where $H = \frac{p}{\rho} + \frac{1}{2}|\vec{u}|^2$. From this equation and Green theorem, we have

$$\begin{aligned} H &= J - \frac{1}{2\pi}\int_C H \frac{\partial}{\partial n} \log|\vec{x} - \vec{x}'|ds \\ &\quad + \frac{\nu}{2\pi}\int_C \nabla^2 u_n \log|\vec{x} - \vec{x}'|ds, \end{aligned} \qquad (11)$$

where C is some closed curve, n is the normal direction of C and $\vec{x}'$ is on C. J is defined by

$$J = \frac{1}{2\pi}\int_D \left\{\frac{\partial}{\partial x}(v\omega) - \frac{\partial}{\partial y}(u\omega)\right\} dx'dy'. \qquad (12)$$

where D is the region closed by C. The value of J is obtained by Kida et al. [13]:

$$\begin{aligned} J &= -\sum_j \frac{\Gamma_j}{2\pi|\vec{x} - \vec{x}_j|^2}\left(u_{1j}(y - y_j) - u_{2j}(x - x_j)\right) \\ &\quad \times \left(1 - \exp(-|\vec{x} - \vec{x}_j|^2/\delta^2)\right) \\ &\quad - \sum_j \frac{\Gamma_j^2}{4\pi^2\delta^2}\Big[E_1(|\vec{x} - \vec{x}_j|^2/\delta^2) \\ &\quad - E_1(2|\vec{x} - \vec{x}_j|^2/\delta^2)\Big], \end{aligned} \qquad (13)$$

where (u_{1j}, u_{2j}) is the velocity of j-th blob and $E_1(x) = \int_x^\infty \frac{e^{-x}}{x} dx$.

In the present paper, the value of H on the surface of the circular cylinder and the wall is solved from Eq.(11) by using the similar method to the panel method. The lift and drag coefficients, C_D and C_L, are defined by $D/(\frac{1}{2}\rho U_\infty d)$ and $L/(\frac{1}{2}\rho U_\infty d)$, respectively, where D and L are drag and lift forces respectively. In the present calculation, the viscous force is neglected because it is very small for high Reynolds number flows.

2 Numerical results

In this section, the numerical results will be shown in the case of $R_e = 3000$, where R_e is the Reynolds number defined by $U_\infty d/\nu$. The thickness of the boundary layer, h_i, is set as $h_i = 2/R_e^{1/2}$ and the allowance maximum strength of vortex sheet ξ_{max} normalized by the uniform flow is taken as 0.2.

Three cases are calculated:

- Case (1): a circular cylinder above the fixed wall.
- Case (2): a circular cylinder above sliding wall.
- Case (3): side-by-side arrangement of two circular cylinders.

2.1 Vortex blob distribution and velocity field

Figures 3, 4 and 5 show the time development of vortex blob distribution in the cases (1), (2) and (3) for gap height $h = 1/8$.

In the case (1) and (2), the vortices shed from the lower surface of the cylinder form a cluster of vortices at the beginning of start, but vortices from the upper surface of the cylinder suppress the separation from the surface. This process will be discussed in more detail. The difference between the cases (1) and (2) is the vortices created from the surface of the wall. In the case (3), the interaction of shedding vortices from upper and lower cylinders induces a sinusoidal gap flow, such as a biased gap flow, as already shown in earlier works (see Bearman & Wadcock [1]).

Figure 6 shows the development of clusters of shedding vortices at time $t = 3$, 5, and 10 in the case of $h = 1/4$. The circulation of the clusters A_1 and B_1 created from upper and lower surface is positive and negative, respectively. At $t = 3$, the induced velocity due to B_1 acts to depress the separation of A_1 from the cylinder and the circulation of A_1 becomes large, so that the velocity in the gap is depressed and the lift force becomes large, as shown in Fig.8. On the other hand, the created vortices on the wall gradually concentrate downstream from B_1. At $t = 5$, a cluster of these vortices C_1 is formed and its circulation is negative, so that B_1 and C_1 are lifted up due to the induced velocity of these clusters. The new cluster A_2 is formed on the upper surface of the cylinder and A_1 and B_2 are separated from the cylinder. With further time development $t = 10$, amalgamation process of $A_1 \sim A_3$, denoted as AA_1, is almost completed and B_2 is lifted up by its large induced velocity, so that upward jet like flow from the gap is formed as shown in this figure. Thus, the gap velocity becomes large and the lift force becomes minimum.

The velocity vector at $t = 3$, 5 and 10 is shown in Fig.7. At $t = 10$, the jet-like flow is clearly shown and the amalgamation/pairing of vortices is found in the wake. Further, we see that two vortex streets are not distinctive due to the existence of the wall.

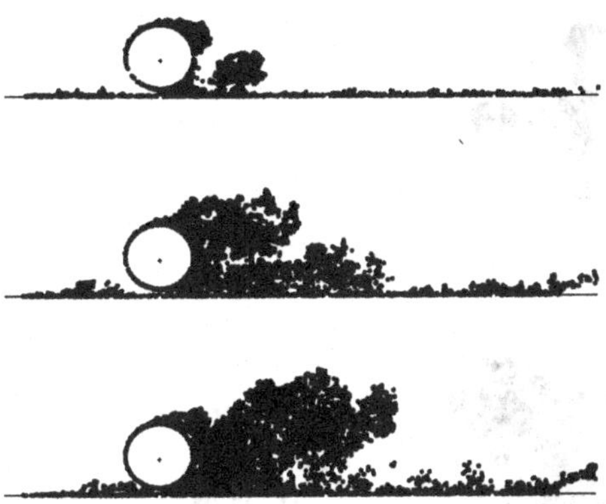

Figure 3: Vortex blob distribution in the case (1) for $h = 1/8$ and $t = 3$(top), 10(middle), 15(bottom).

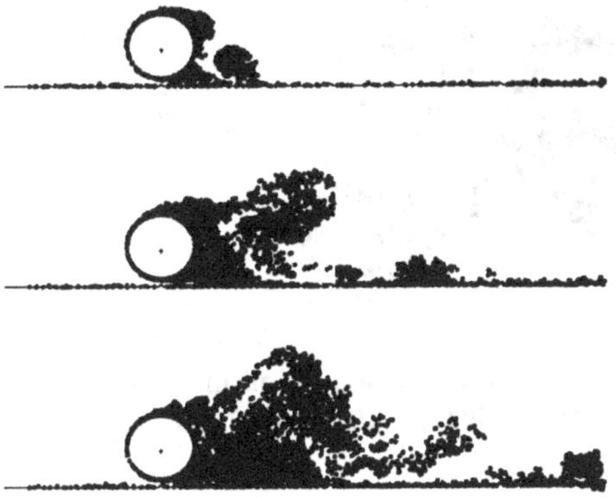

Figure 4: Vortex blob distribution in the case (2) for $h = 1/8$ and $t = 3$(top), 10(middle), 15(bottom).

2.2 Lift and drag coefficients

Figures 8 and 9 show the time development of aerodynamic forces in three cases for $h = 1/8$ and $h = 1/2$. The solid line, dotted line, and dashed line denote the case (1), (2) and (3), respectively. We see that the lift force increases at the beginning of time and arrives at the maximum, after then it changes sinusoidal but time when the lift coefficient becomes maximum or minimum is different in these cases for $h = 1/8$. The drag coefficient is also different in these cases: The time history in the case (1) is almost antphase of in the case (2). This causes the shedding process of vortices. For the rigid wall, the velocity of cluster of vortices shed from the lower surface of the cylinder is smaller than that for the sliding wall due to the boundary layer on the wall, therefore, its separation

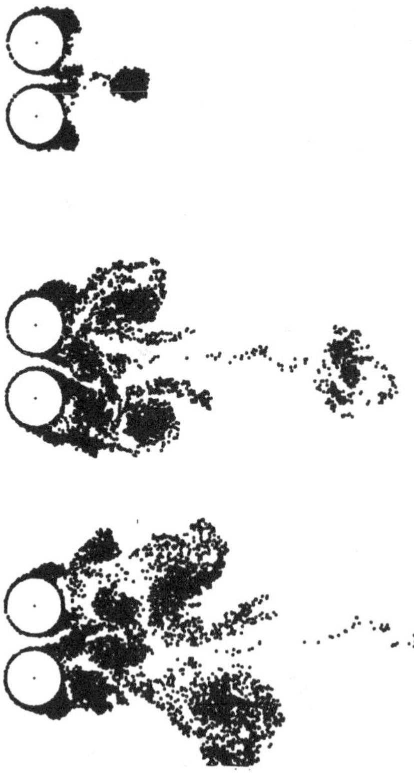

Figure 5: Vortex blob distribution in the case (3) for $h = 1/8$ and $t = 3$(top), 10(middle), 15(bottom).

time from the cylinder becomes late. However, for $h = 1/2$, we see from Fig.9 that aerodynamic forces are almost the same in three cases. Thus, the wall effect is very small for $h \geq 1/2$.

Figures 10 and 11 show the effect of gap against aerodynamic forces. The solid line, dotted line, dashed line and dash-dotted line denote $h = 1/8$, $1/4$, $1/2$ and 1, respectively. We see from these figures that in spite of gap height the lift force increases at the beginning of time and arrives at maximum, after then sinusoidal fluctuation begins. We see that the sinusoidal fluctuation is depressed by taking small gap and the time when the lift force arrive at the maximum is related with the separation of the cluster B_1 from the cylinder (see Fig.6): The gap velocity becomes small with increase of the gap height, so that the time when the cluster B_1 separates from the cylinder becomes late with increase of gap height. Comparing with Figs.10 and 11, we see that there is not so remarkable difference in the global feature between the case (1) and the case (2). The time when the lift coefficient is minimum for $h = 1/8$ and $1/2$ is different between the cases (1) and (2). The drag coefficient is different in a sense of time when C_D becomes maximum and minimum.

2.3 Pressure distribution

Figure 12 shows the pressure distribution on the surface of the cylinder. C_p is the pressure coefficient and n denotes the panel number labeled as $\theta = n\pi/20$, where θ is the angular coordinate from the uniform flow direction. $n = 20$ is the leading point and $n = 0$ and 40 are the rear point of the circular cylinder. The small dotted line, small dashed line, small solid line, dotted line, dashed line, and solid line denote time $t = 3$, 5, 8, 10, 13, 15, respectively. We see that the base pressure is not uniform due to the unsteadiness of the wake. In the leading point, the pressure coefficient is almost unity.

We see that the stagnation point at the front side is a little low side from the center of the cylinder. Further, we see from these figures that the pressure distribution of the lower surface is different in these cases: In the case (1), the pressure of the lower surface at $t = 3$ is less than that in the case (2) due to the sliding wall and in the case (3) the fluctuation of the pressure on the lower surface is the smallest among them.

Figure 13 shows the contribution of dynamic pressure to the pressure distribution. The distribution of H implies the effect of unsteadiness. We see from this figure that the dynamic pressure is important role in the pressure distribution, the effect of the unsteadiness is not so large and it affects to the back pressure coefficient.

3 Conclusions

The present paper simulates the transient flow around a circular cylinder near the wall by a vortex method. Two boundary conditions on the wall are mainly treated; (1) fixed wall and (2) sliding wall. By comparing with the case of a side-by-side arrangement of two cylinders, the wall effect of these two cases are discussed: The vorticity distribution of two cases is different from that of the side-by-side arrangement, but the different between the fixed wall and the sliding wall is not so remarkable. In particular, upward jet-like flow from the gap is clearly generated in the region between the lower surface of the cylinder and the wall for the case (1) and $h = 1/4$. This flow causes the cluster generated from the boundary layer of the wall surface. Aerodynamic forces are also calculated: The global feature is almost the same in these two cases, but the time history is different due to the wall effect.

References

[1] P.W. Bearman & A.J. Wadcock, The interaction between a pair of circular cylinders normal to a stream, J. Fluid Mech., 61, (1973), 499-511.

[2] P.W. Bearman & M.M. Zdravkovich, Flow around a circular cylinder near a plane boundary, J. Fluid Mech., 89, (1978), 33-47.

[3] K. Kamemoto, Y. Oda, & M. Aizawa, Characteristics of the flow around a bluff body near a plane surface, Bulltin of JSME, 27-230, (1984), 1637-1643.

[4] M. Sardar, "Reynolds effect" and "Moving ground effect" tested on a quarter scale wind tunnel over a high speed moving belt, J. Wind Engr. Industrial Aerodynamics, 22, (1986), 245-270.

[5] P.W. Bearman, Review-Bluff body flows applicable to vehicle aerodynamics, Trans. ASME, J. Fluid Engr., 102, (1980), 265-274.

[6] M.P. Arnal, D.J. Goering & J.A.C. Humphrey, Vortex shedding from a bluff body adjacent to a plane sliding wall, Trans. ASME, J. Fluid Engr., 113, (1991), 384-398.

[7] A.Y. Cheer, Unsteady separated wake behind an impulsively started cylinder in slightly viscous fluid, J. Fluid Mech., 201, (1989), 485-505.

[8] T. Kida & M. Kurita, High Reynolds number flow past an impulsively started circular cylinder (Time marching of random walk vortex method), Com. Fluid Dynamics J., 4-4, (1996), 489-508.

[9] T. Kida, T. Nagata, & T. Nakajima, Accuracy of the panel method with distributed sources applied to two-dimensional bluff bodies, Com. Fluid Dynamics J., 2-1, (1993), 73-90.

[10] A.J. Chorin, Numerical study of slightly viscous flow, J. Fluid Mech., 157, (1973), 785-796.

[11] L-a. Ying & P. Zhang, Vortex methods, Science Press, Beijing, 1997.

[12] A.J. Chorin, Vortex sheet approximation of boundary layers, J. Comp. Physics, (1978), 428-442.

[13] T. Kida, H. Sakate & T. Nakajima, Pressure distribution obtained using two-dimensional vortex method, Trans. JSME, Sr.B, 63-6060, (1997), 378-386 (in Japanese).

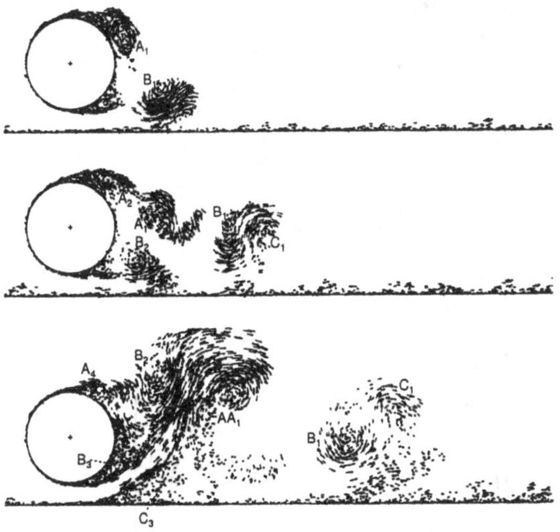

Figure 6: Shedding vortices in the case (1) for $h = 1/4$ and $t = 3$(top), 5(middle), 10(bottom).

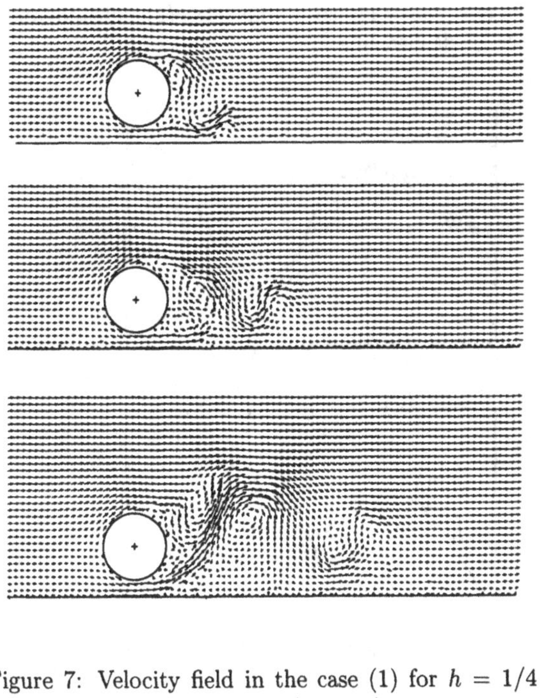

Figure 7: Velocity field in the case (1) for $h = 1/4$ and $t = 3$(top), 5(middle), 10(bottom).

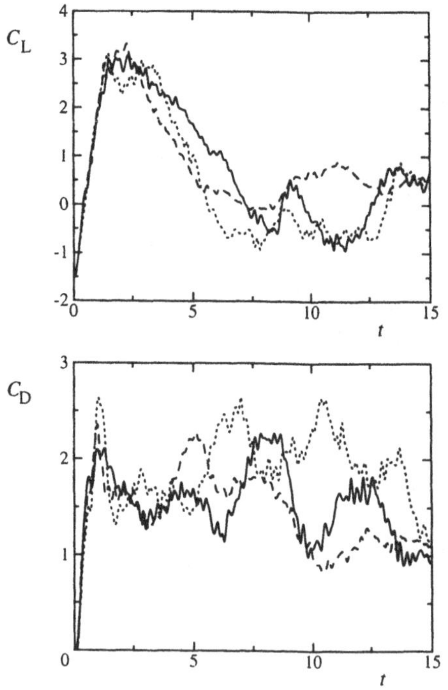

Figure 8: Comparison of lift and drag coefficients for $h = 1/8$. Case(1): solid line, Case(2): dotted line, Case(3): dashed line.

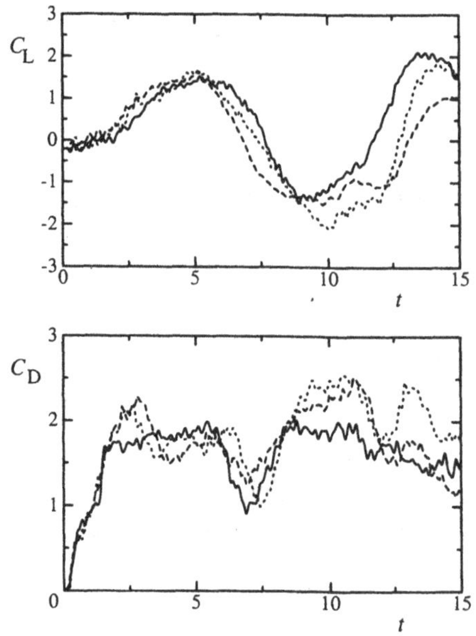

Figure 9: Comparison of lift and drag coefficients for $h = 1/2$. Case(1): solid line, Case(2): dotted line, Case(3): dashed line.

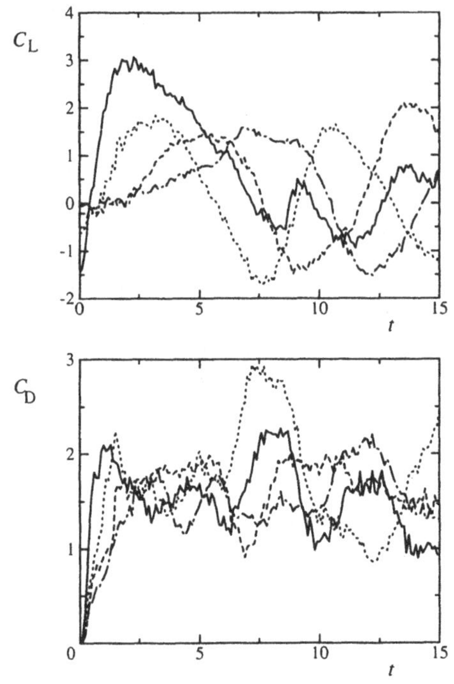

Figure 10: Lift and drag coefficients in the case (2). $h = 1/8$: solid line, $h = 1/4$: dotted line, $h = 1/2$: dashed line, $h = 1$: dash-dotted line.

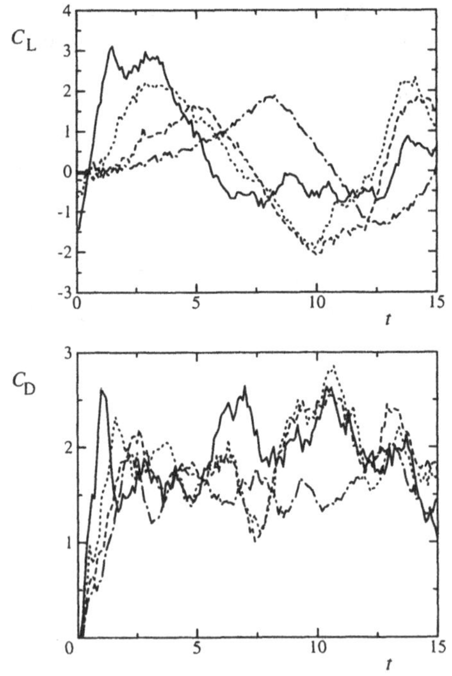

Figure 11: Lift and drag coefficients in the case (3). $h = 1/8$: solid line, $h = 1/4$: dotted line, $h = 1/2$: dashed line, $h = 1$: dash-dotted line.

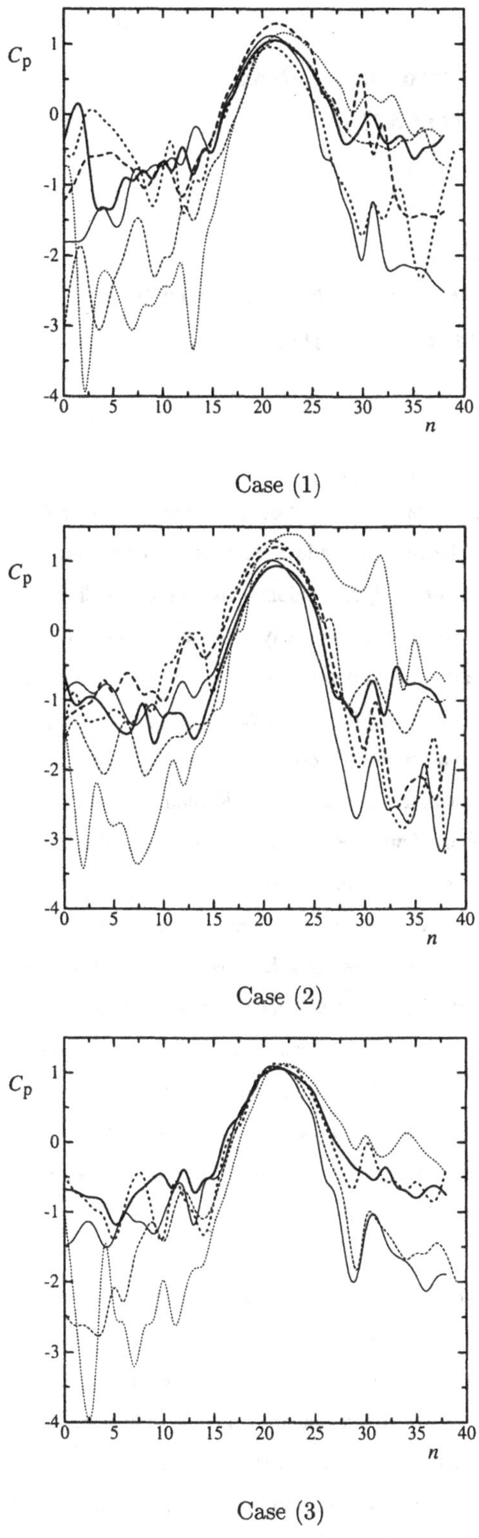

Figure 12: Pressure distribution on the surface of the cylinder for $h = 1/8$. $t = 3$: small dotted line, $t = 5$, small dashed line, $t = 8$: small solid line, $t = 10$: dotted line, $t = 13$: dashed line, $t = 15$: solid line.

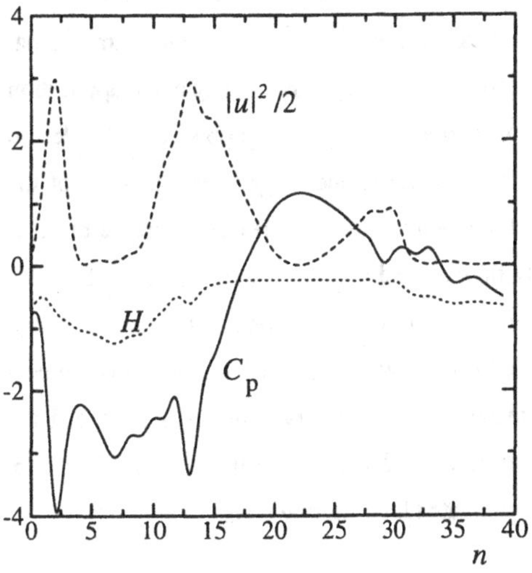

Figure 13: Contribution of dynamic pressure to pressure distribution in the case (1) for $h = 1/8$.

Numerical Simulation of Unsteady Flow around a Sphere by a Vortex Method for Re Number from 300 to 1000

Akira OJIMA and Kyoji KAMEMOTO

Department of Mechanical Engineering and Materials Science, Yokohama National University

79-5 Tokiwadai Hodogaya-ku Yokohama, 240-8501 Japan / Email: d98ja001@ynu.ac.jp

ABSTRACT

Recently, the present authors have developed an advanced vortex method for two and three-dimensional analysis of a viscous unsteady flow, aiming at the application of it toward various engineering problems. In order to investigate validity and applicability to the analysis of the flow around three-dimensional bodies, numerical simulation of unsteady flow around a sphere in the range of Reynolds number from 300 to 1000 is carried out with the vortex method based on a new scheme of introduction of nascent vortex elements. The boundary layer separation and the development of vortical wake from the body surface were reasonably simulated for a starting flow.

1. INTRODUCTION

Although the recent progress of computational fluid dynamics is quite rapid, the numerical analysis of high Reynolds number flow is not very easy in terms of engineering applications. The vortex methods have been developed and applied to analysis of complicated, unsteady and vortical flows related with problems in a wide range of industries, because they consist of simple algorithm based on physics of flow. Leonard[1] summarized the basic algorithm and examples of its applications. Sarpkaya[2] presented a comprehensive review of various vortex methods based on Lagrangian or mixed Lagrangian-Eulerian schemes and the Biot-Savart law or the vortex in cell methods. Recently, Kamemoto[3] summarized the mathematical basis of the Biot-Savart law methods.

In considering the expression of viscous diffusion of vorticity in the Biot-Savart law methods, the core spreading method by Leonard[1], the random walk method by Chorin[4,5] and the integrated vorticity equation method by Wincklemans and Leonard[6] were proposed. Although Greengard[7] criticized the core spreading method, Nakajima and Kida[8] mathematically confirmed the validity of the core spreading method. Nakanishi and Kamemoto[9] developed a Biot-Savart law method combined with the three-dimensional core spreading method.

Many investigations, related to two-dimensional analysis with vortex methods, have been done. However, only a few studies related to three-dimensional analysis with vortex methods have been reported. For examples, Ojima & Kamemoto[10] analyzed three-dimensional unsteady flow around a sphere and a prolate spheroid with a three-dimensional vortex method based on a new scheme of introduction of nascent vortex elements. Kiya et al.[11] applied a three-dimensional vortex blob method to the simulation of an impulsively started round jet. Dimas et al.[12] calculated the flow past a prolate spheroid by a three-dimensional vortex method. Gharakhani et al.[13] applied a three-dimensional vortex-boundary method to the simulation of the compression stroke in combustion engines.

The developments of three-dimensional analysis are necessary to understand the complicated vortical structures of the wake behind three-dimensional bodies. A lot of experimental works and several numerical simulation works have been done on the flow around a sphere in order to clarify three-dimensional vortical flow. Experimental investigations of the flow around a sphere have been carried out on the formation mechanism and the characteristics of the flow related to the wake structure by Margarvey & Bishop[14], Achenbach[15], Taneda[16] and Sakamoto & Haniu[17]. In particular, it has reported by Margarvey & Bishop[14] and Sakamoto & Haniu[17] that axisymmetric vortical wake structure like a vortex ring begins to oscillate and the wake forms hairpin-shaped vortex loops when a Reynolds number is reached about 300. Recently, Tomboulidos[18] has presented the numerical results of the flow around a sphere for $25<Re<10^3$ with direct simulation and with large-eddy simulation at $Re=2\times10^4$. Shirayama[19] has carried out the numerical simulation of unsteady flow past a sphere for the case of the flow accelerating from rest to Re=500. These results have showed the same structures as visualized in previous experiments.

In this study, the vortex method based on a new scheme of the introduction of nascent vortex elements is proposed and it is applied to the unsteady flow around a sphere, in order to investigate validity and applicability to the analysis of the flow around three-dimensional bodies.

2. ADVANCED VORTEX METHOD

2.1 Mathematical basis

The governing equations of viscous and incompressible flows are the vorticity transport equation and pressure Poisson equation which can be derived by taking the rotation and divergence of Navier-Stokes equations, respectively

$$\frac{\partial \omega}{\partial t}+(\boldsymbol{u}\cdot grad)\,\omega=(\omega\cdot grad)\,\boldsymbol{u}+\nu\nabla^2\omega \qquad (1)$$

$$\nabla^2 p=-\rho\, div(\boldsymbol{u}\cdot grad\ \boldsymbol{u}) \qquad (2)$$

Where $\boldsymbol{u}$ is a velocity vector and the vorticity $\boldsymbol{\omega}$ is defined as

$$\omega = rot\ \boldsymbol{u} \qquad (3)$$

As explained by Wu and Thompson[20], the Biot-Savart law can be derived from the definition of vorticity as follows.

$$\boldsymbol{u}=\int_V \omega_0\times\nabla_0 G dv-\int_S[(\boldsymbol{n_0}\cdot\boldsymbol{u_0})\cdot\nabla_0 G+(\boldsymbol{n_0}\times\boldsymbol{u_0})\times\nabla_0 G]ds \qquad (4)$$

Here, subscript "$_0$" denotes variable, differentiation and integration at a location $\boldsymbol{r}_0$, and $\boldsymbol{n}_0$ denotes the normal unit vector at a point on a boundary surface S. G is the fundamental solution of the scalar Laplace equation with the delta function $\delta(r-r_0)$ in the right hand side, which can be written for a three-dimensional field as follows.

$$G=\frac{1}{4\pi R} \qquad (5)$$

Here, $\boldsymbol{R}=\boldsymbol{r}-\boldsymbol{r}_0$, $R=|\boldsymbol{R}|=|\boldsymbol{r}-\boldsymbol{r}_0|$. In Eq. (4), the inner product $\boldsymbol{n}_0\cdot\boldsymbol{u}_0$ and the outer product $\boldsymbol{n}_0\times\boldsymbol{u}_0$ stand for normal and tangential velocity components on the boundary surface, and they correspond to source and vortex distributions on the surface, respectively. Therefore, it is mathematically understood that velocity fields of viscous and incompressible flows are obtained from the field integration concerning vorticity distributions in the flow field and the surface integration concerning source and vortex distributions around the boundary surface.

The pressure in the field is obtained from the integration equation formulated by Uhlman[21], instead of the finite difference calculation of the Eq.(2) as follows.

$$\beta H+\int_S H\frac{\partial G}{\partial n}ds=-[\int_V \nabla G(\boldsymbol{u}\times\omega)dv + \int_S G\cdot\boldsymbol{n}\cdot\frac{\partial \boldsymbol{u}}{\partial t}ds+\nu\int_S \boldsymbol{n}\cdot(\nabla G\times\omega)ds] \qquad (6)$$

Here, $\beta=1$ in the flow field and $\beta=1/2$ on the boundary S. G is the fundamental solution given by Eq.(5). H is the Bernoulli function defined as follows.

$$H=\frac{p}{\rho}+\frac{u^2}{2} \qquad (7)$$

2.2 Vorticity layer spreading from the solid surface

The vorticity near the solid surface must be represented by proper distributions of vorticity layers and discrete vortex elements so as to satisfy the non-slip condition on the solid surface. In the present method, a thin vorticity layer with thickness of h is considered along the solid surface and the surface of outer boundary of the thin vorticity layer is expressed by a number of vortex and/or source panels as shown in Fig.1. If a linear distribution of velocity in the thin vorticity layer is assumed, the normal velocity V_n on a panel can be expressed using the relation of continuity of flow and non-slip condition on the solid surface for the element of the vorticity layer

$$V_n = \frac{1}{\Delta S_p}\sum_{i=1}^{4}\int_{\Delta S_i} u_{si}\, ds$$
$$u_{si} = \boldsymbol{u}_i \cdot \boldsymbol{n}_{si} \qquad (8)$$
$$\Delta S_i = h \cdot \Delta l_i$$

Where, ΔS_p, $\boldsymbol{u}_i$ and $\boldsymbol{n}_{si}$ denote the panel area, the velocity vector and the normal vector on the side sectional planes of the element of the vorticity layer, respectively. Using the normal velocity for each panel expressed by Eq.(8), the strength of the vortex and/or source panel for the following step can be calculated numerically from Eq.(4). On the other hand, the vorticity of the thin layer diffuses through the panel into the outer flow field with the diffusion velocity. In order to consider this vorticity diffusion, the diffusion velocity is employed in the same manner as the Vorticity Layer Spreading Method proposed by Kamemoto[22], which is expressed as given following equation.

$$V_d = \frac{c^2 \nu}{2h}, \quad (c=1.136) \qquad (9)$$

Here, ν is kinematic viscosity of the fluid. If V_n+V_d becomes positive, a nascent vortex element is introduced into the flow field, where the thickness and vorticity of the element are given from the relation of the strength of vorticity conservation as follows.

$$\omega_{vor} = \frac{\int_V \omega\, dv}{V + V_{vor}}$$
$$h_{vor} = (v_n + v_d)\cdot dt \qquad (10)$$
$$V_{vor} = \Delta S_p \cdot h_{vor}$$

Here, ω is the vorticity originally involved in the element of the vorticity layer, V and V_{vor} are the volume of the vorticity element and the nascent vortex element. Every vortex element is introduced at the distance of $0.5h_{vor}$ from the panel as a vortex plate. Every vortex plate element, which moves beyond a boundary at the distance of four times h from the solid surface, is replaced with a vortex blob of the core-spreading model proposed by Nakanishi and Kamemoto[9]. In this scheme, by the assumption of a linear distribution of velocity in the thin vorticity layer, shearing stress on the wall surface is evaluated from the following equation.

$$\tau_w = \mu\frac{\partial u}{\partial y} = -\mu\omega \qquad (11)$$

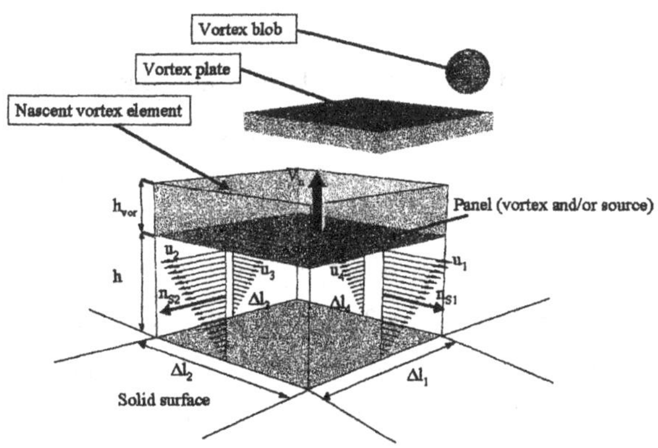

Fig.1 Introduction of a nascent vortex element.

2.3 Vortex blob model

In this study, vortex blob model is employed. Vorticity distribution around the vortex blob is represented by

$$\omega(r) = \omega\ p(r/\varepsilon)\,\varepsilon^{-3} dv$$
$$p(\xi) = 15/8\pi\,(\xi^2+1)^{-7/2} \qquad (12)$$

Here, the location is $\boldsymbol{r}$, vorticity is ω, volume is dv, core radius is ε and $p(\xi)$ is smoothing function proposed by Winckelmans & Leonard[6]. The motion of the vortex blob at a location $\boldsymbol{r}$ is represented by Lagrangian form of a simple differential equation.

$$\frac{d\boldsymbol{r}}{dt} = \boldsymbol{u} \qquad (13)$$

Then, the trajectory of the fluid particle over a time step dt is approximately computed from the Adams-Bashforth method.

On the other hand, the evolution of vorticity is calculated with three-dimensional core spreading method modified by Nakanishi & Kamemoto[9]. In this study, the stretch term and diffusion term of Eq.(1) are separately considered. The change of core radius due to the stretch is calculated according to the following equations.

$$\frac{d\omega}{dt} = (\omega \cdot grad)\, \boldsymbol{u} \tag{14}$$

$$\frac{dl}{dt} = \frac{l_t}{|\omega_t|} \cdot \left|\frac{d\omega}{dt}\right| \tag{15}$$

$$\left(\frac{d\varepsilon}{dt}\right)_{stretch} = -\frac{\varepsilon_t}{2 \cdot l_t} \frac{dl}{dt} \tag{16}$$

Here, ε and l are core radius and length of the vortex blob model as shown in Fig.2. The viscous term of Eq.(1) is expressed by a core spreading method. The core spreading method is based on the Navier-Stokes equation for viscous diffusion of an isolated two-dimensional vortex filament in a rest fluid, and the rate of core spread is represented as follows.

$$\left(\frac{d\varepsilon}{dt}\right)_{diffusion} = \frac{c^2 \nu}{2\varepsilon_t}, \quad (c=2.242) \tag{17}$$

Taking two factors into account, variables of a new element are obtained from the following equations.

$$\varepsilon_{t+\Delta t} = \varepsilon_t + \left[\left(\frac{d\varepsilon}{dt}\right)_{stretch} + \left(\frac{d\varepsilon}{dt}\right)_{diffusion}\right] \cdot \Delta t \tag{18}$$

$$l_{t+\Delta t} = l_t + \frac{dl}{dt} \cdot \Delta t \tag{19}$$

$$|\omega_{t+\Delta t}| = |\omega_t| \cdot \left(\frac{\varepsilon_{t+\Delta t}}{\varepsilon_t}\right)^2 \tag{20}$$

In this study, the new element is replaced into a vortex blob that has equivalent volume.

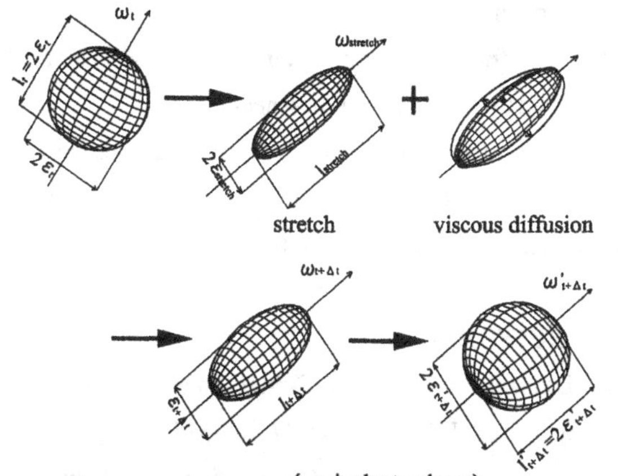

Fig.2 Mechanism of core spreading method for vortex blob.

3. CALCULATION RESULTS AND DISCUSSIONS

3.1 Calculation condition

Calculations are performed for an impulsively started sphere with a constant speed U_∞ in a rest fluid at Reynolds number $Re=U_\infty D/\nu$=300, 600 and 1000. Here, D and ν are diameter of a sphere and kinematic viscosity, respectively. The sphere is divided into 360 source and vortex panels. The strength per unit area of vortex panels can be obtained by dividing the opposite component of total circulation introduced into the flow field during a time interval by the surface area of the sphere. The other parameters were set as: vorticity layer thickness h/D is $0.5(Re)^{1/2}$ and time interval $\Delta t U_\infty/D$ is 0.05, respectively.

3.2 Development of the wake behind a sphere

The evolution of vortical wake behind a sphere after its impulsively started in a rest fluid is explained by using vorticity distributions. For each Reynolds number, figures 3 to 5 show contours of z-component of vorticity ω_z in the (x,y)-plane at time tU_∞/D=5.0 and the streamwise velocity distribution u/U_∞ along the x-axis at time tU_∞/D=2.0, 3.0, 4.0 and 5.0.

Axisymmetric vortex structure like a vortex ring is formed behind a sphere at each Reynolds number as shown in Figs.3(a) to 5(a). In the initial stage, a vortex ring characterizes the axisymmetric flow field. For the case of Re=300, a large vortex ring is formed behind a sphere. At Re=600 and 1000, a large-scale vortex structure and initial vortex rings, which have small diameters, are formed behind a sphere as shown in Figs. 4(a) and 5(a). According to Figs.3(b) to 5(b), similar velocity profiles at each Reynolds number are seen until tU_∞/D=4. As shown in Fig.4(b), the fluctuations of the streamwise velocity are seen at x/D=1.5 and 2.5 at tU_∞/D =5.0 for Re=600. In this study, the vortex method based on a new scheme of the introduction of nascent vortex elements is proposed and it is applied to the unsteady flow around a sphere in order to investigate validity and applicability to the analysis of the flow around three-dimensional bodies. Velocity fluctuations are also seen at Re=1000 as shown in Fig.5(b). These fluctuations are caused by initial vortex rings that are formed at initial stage.

The axisymmetric vortex structure has been developed as time goes on, and the wake gradually has been deformed non-axisymmetric structures. Figures 6 to 8 show instantaneous flow patterns represented by discrete vortices and streamwise vorticity ω_x in the wake near the sphere at time tU_∞/D =18.75 for each Reynolds number. According to experiments, the flow past a sphere is known to result in a periodic (and eventually chaotic) wake as the Reynolds number increases. For the case of Re=300, vortex loops are formed behind a sphere as shown in Fig.6(a). The longitudinal vortex structures in the wake, which have parallel to each other, are clearly observed in Fig.6(b). At Re=600, as shown in Fig.7(a), the basic wake structure consists of interconnected vortex loops like the one at Re=300. In this case, small-scale structures begin to appear in the large-scale wake structures as shown in Fig.7(b). For larger Reynolds number Re=1000, vortex loops do not appear and the more complicated wake structures, which has been consisted of small-scale vortical structures, are observed in Fig.8(a) and (b).

3.3 Total fluid force and frictional force coefficients

Figures 6(c) to 8(c) show time histories of fluid force coefficients acting on the sphere at each Reynolds number. It is identified that each component of the total fluid force coefficients (C_x, C_y, C_z) fluctuates with the change of the wake structure for all Reynolds number cases. It is confirmed that the wake structure is planar symmetry for x-z plane, because the fluctuation of y-component of total fluid force coefficient is dominant. In previous papers, the characteristics of the Strouhal number with vortex shedding from a sphere have been well investigated by Sakamoto & Haniu[17]. However, it should be noted that the comparisons of Strouhal number between numerical and experimental results have not been made since the calculation time is not sufficiently long in the present study. At Re=300, from non-dimensional time 10.0 to 20.0, the average of total drag coefficient is 0.71 and the value of frictional drag coefficient is 0.10. For the case of Re=600 and 1000, we calculated the average of total drag and frictional drag coefficients as well. In Fig.9, we plotted mean values of total drag coefficient from our results and experimental results reproduced from Schlichting[23]. There is similar tendency for total drag coefficient between our numerical results and experimental results. In the same way, we plotted mean values of frictional drag coefficient from our results and general solution quoted from Churchill[24]. Our numerical results show reasonable agreement with general solution. In this study, velocity distribution around the solid surface has been approximately calculated by linear distribution. Therefore, it is expected that the values of frictional drag coefficient are less than the real values.

4. CONCLUSIONS

A three-dimensional vortex method based on a new scheme of the introduction of nascent vortex element was proposed and applied to the simulation of unsteady flows around a sphere at Re=300, 600 and 1000, and the following conclusions were obtained.

1. In the initial stage, initial vortex rings, which have small diameters, were formed behind a sphere at Re=600 and 1000.
2. At Re=300, hairpin-shaped vortex loops were formed behind a sphere and longitudinal vortical structures were clearly observed. At Re=600 and 1000, the complicated wake structures, which have been consisted of small-scale vortical structures, were confirmed.
3. The drag coefficient from our result had similar tendency with experimental results.
4. The frictional force coefficients could be easily calculated by using the present scheme.

It is confirmed that the method used in the present study is very useful and convenient for investigation of three-dimensional unsteady separated flow around a sphere.

5. REFERENCES

(1) Leonard, A., (1980) Vortex methods for flow simulations. J. Comp. Phys. 37, 289-335
(2) Sarpkaya, T., (1989) Computational methods with vortices-the 1988 freeman scholar lecture. J.Fluids Eng. 111, pp.5-52
(3) Kamemoto, K., (1995) On attractive features of the

vortex methods. Computational Fluid Dynamics Review 1995.ed. M.Hafez and K.Oshima, JHON WILLY & SONS, 334-353

(4) Chorin, A.J., (1973) Numerical study of slightly viscous flow. J. Fluid Mech. 57, 785-796

(5) Chorin, A.J., (1978) Vortex sheet approximation of boundary layers. J. Comp. Phys. 74, 283-317

(6) Winkelmans, G. and Leonard, A., (1988) Improved vortex methods for three-dimensional flows. Proc. Workshop on Mathematical Aspects of Vortex Dynamics. 25-35, Leeburg. Virginia

(7) Greengard, C., (1985) The core spreading vortex method approximates the wrong equation. J. Comp. Phys. Vol. 61, pp.345-348

(8) Nakajima, T. and Kida, T., (1990) A remark of discrete vortex method (Derivation from Navier-Stokes equation). Trans. JSME. B. 56-531, 3284-3291

(9) Nakanishi, Y. and Kamemoto, K., (1992) Numerical simulation of flow around a sphere with vortex blobs. J. Wind Engng. and Ind. Aero., 46 & 47, pp.363-369

(10) Ojima, A., and Kamemoto, K., (1999) Numerical simulation of unsteady flows around three dimensional bluff bodies by an advanced vortex method. Proc. 1999 ASME FEDSM. FEDSM99-6822.

(11) Kiya, M., Nagatomi, M. and Mochizuki, O., (1997) Simulation an impulsively started round jets by a 3d vortex method. Proc. JSME Centennial Grand Congress, Int. Conf. on Fluid Engineering, Tokyo, Vol.1, pp.135-140

(12) Dimas, A.A., Collins, J.P. and Bernard, P. S., (1998) A first, parallel vortex method for turbulent flow simulation. Proc. 1998 ASME FEDSM. FEDSM98-5000.

(13) Gharakhani, A. and Ghoniem, A.F., (1998) 3D vortex simulation of flow in engine during compression. Proc. 1998 ASME FEDSM. FEDSM98-5001.

(14) Margarvey, R.H. & Bishop, R.L., (1961) Transition ranges for three-dimensional wakes. Can. J. Phys. Vol.39, pp.1418-1422

(15) Achenbach, E., (1974) Vortex shedding from spheres. J. Fluid Mech. Vol.62, Part2, pp.209-221

(16) Taneda, S., (1978) Visual observations of the flow past a sphere at Reynolds numbers between 10^4 to 10^6. J. Fluid Mech. Vol.82, Part1, pp.187-192

(17) Sakamoto, A. and Haniu, H., (1990) A study on vortex shedding from spheres in a uniform flow. Trans. ASME, J.Fluid Eng., 112

(18) Tomboukides, A.G., Orszag, S.A. & Karniadakis, G.E., (1993) Direct and large-eddy simulation of axisymmetric wakes. AIAA paper 93-0546.

(19) Shirayama, S., (1992) Flow past a sphere: topological transitions of the vorticity field. AIAA J. Vol30, pp.349-358

(20) Wu, J.C. and Thompson, J.F., (1973) Numerical solutions of time-dependent incompressible Navier-Stokes equations using an integro-differential formulation. Computers & Fluids. 1, 197-215

(21) Uhlman, J.S., (1992) An integral equation formulation of the equation of motion of an incompressible fluid. Naval Undersea Warfare Center T.R., 10, 086

(22) Kamemoto, K., (1994) Development of the vortex methods for grid-free Lagrangian direct Numerical Simulation; Proc. 3rd JSME/KSME Fluids Engineering Conference, Sendai, pp.542-547

(23) Schlichting, H., (1979) Boundary-layer theory, 7th ed., McGraw-Hill, pp.17

(24) Churchill, S.W., (1988) Viscous flows, the particle use of theory, Butterworth Publishers, pp.359-409

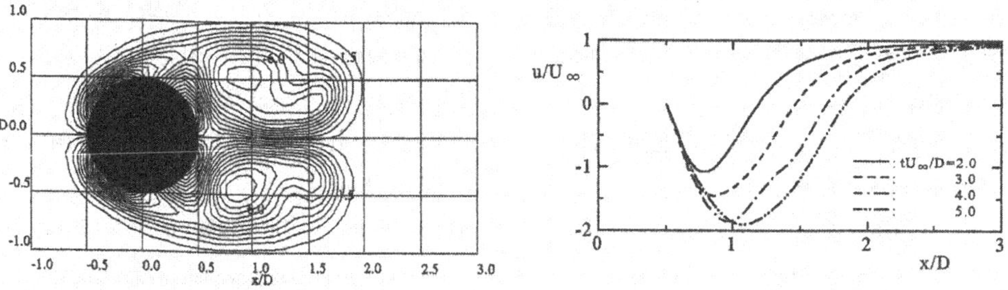

(a) Contours of z-component of vorticity ω_z at tU_∞/D =5.0 (b) Streamwise velocity profiles along the x-axis

Fig.3 Characteristics of axisymmetric vortical wake structure at Re=300

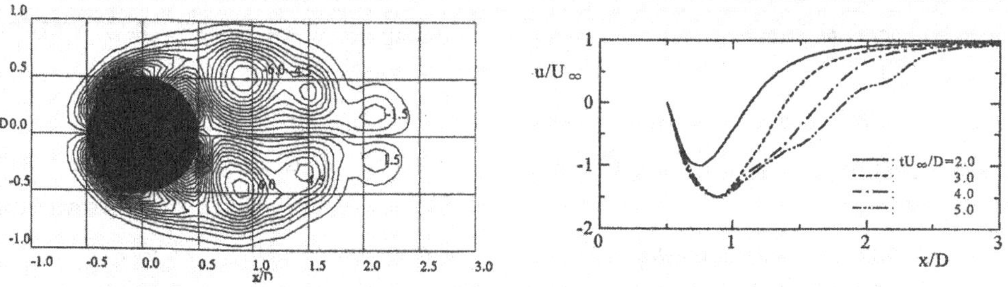

(a) Contours of z-component of vorticity ω_z at tU_∞/D =5.0 (b) Streamwise velocity profiles along the x-axis

Fig.4 Characteristics of axisymmetric vortical wake structure at Re=600

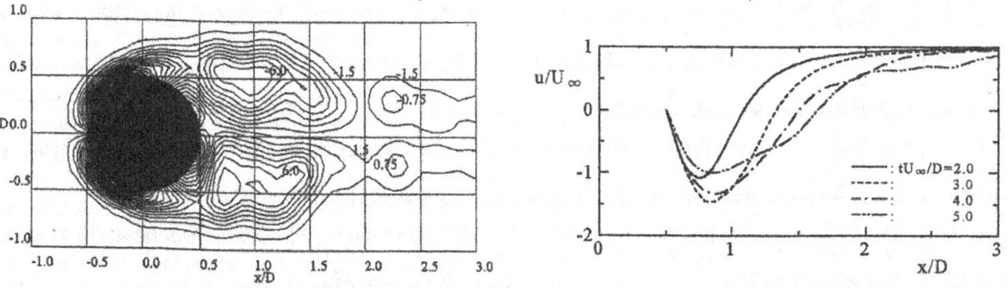

(a) Contours of z-component of vorticity ω_z at tU_∞/D =5.0 (b) Streamwise velocity profiles along the x-axis

Fig.5 Characteristics of axisymmetric vortical wake structure at Re=1000

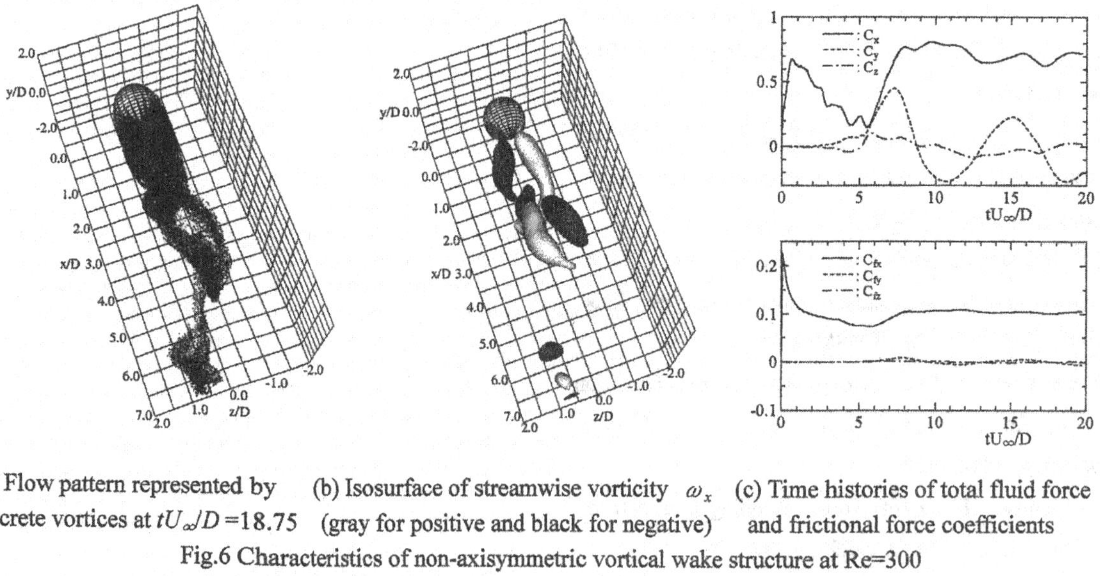

(a) Flow pattern represented by discrete vortices at tU_∞/D =18.75 (b) Isosurface of streamwise vorticity ω_x (gray for positive and black for negative) (c) Time histories of total fluid force and frictional force coefficients

Fig.6 Characteristics of non-axisymmetric vortical wake structure at Re=300

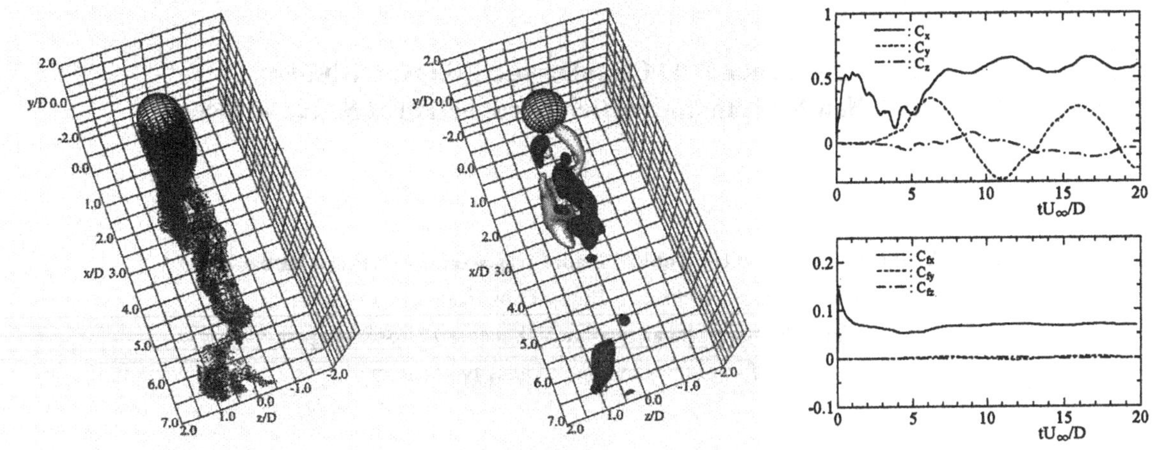

(a) Flow pattern represented by discrete vortices at tU_∞/D =18.75 (b) Isosurface of streamwise vorticity ω_x (gray for positive and black for negative) (c) Time histories of total fluid force and frictional force coefficients

Fig.7 Characteristics of non-axisymmetric vortical wake structure at Re=600

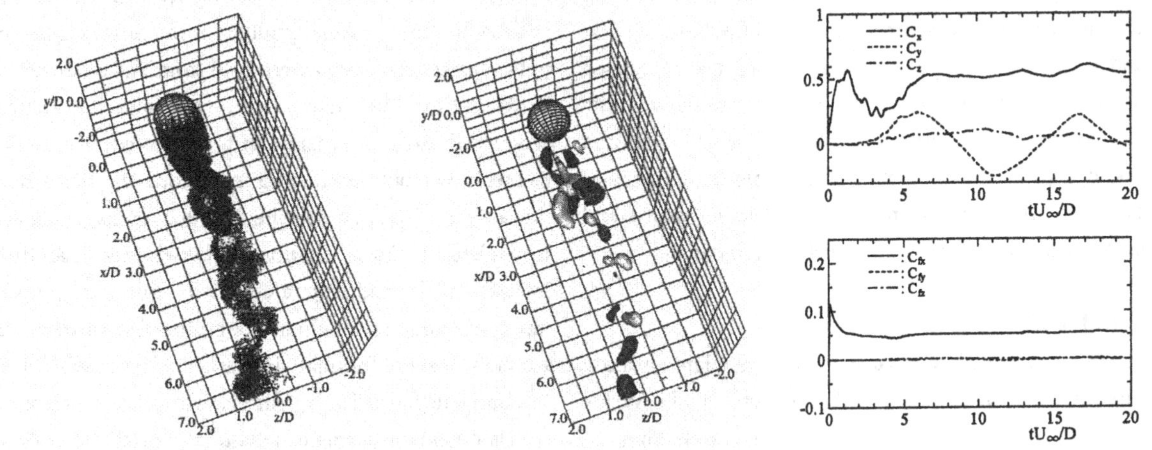

(a) Flow pattern represented by discrete vortices at tU_∞/D =18.75 (b) Isosurface of streamwise vorticity ω_x (gray for positive and black for negative) (c) Time histories of total fluid force and frictional force coefficients

Fig.8 Characteristics of non-axisymmetric vortical wake structure at Re=1000

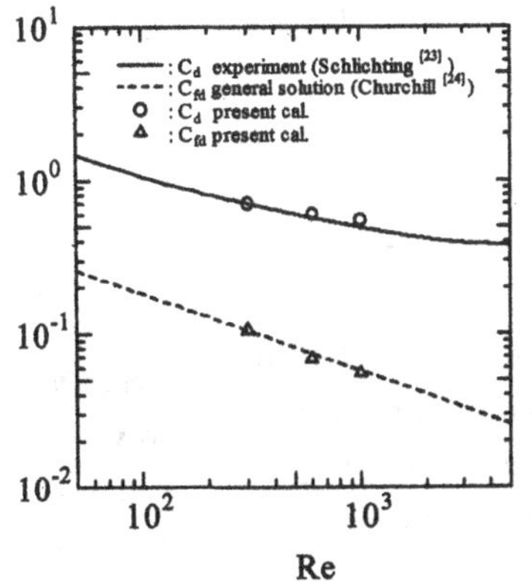

Fig.9 Relation between the total and frictional drag coefficients and the Reynolds number.

Flows around 2D Circular and Elliptic Cylinders using New Scheme for Vorticity Production at Solid Surface

Yuji Nakanishi, Koichi Tobita, Naoto Hagihara & Mamoru Kubokawa
Department of Mechanical Engineering, Kanagawa University
Rokkakubashi, Kanagawa-ku, Yokohama 221-8686, Japan
/E-mail: nakanish@cc.kanagawa-u.ac.jp

ABSTRACT

This study proposes a new scheme for the vorticity production at a solid surface to determine the circulation of a nascent vortex element. The flows around circular and elliptic cylinders impulsively started are calculated using the combination of the vortex method and the boundary element method with the newly proposed vorticity production scheme. Reasonable fluid forces of drag and lift were obtained independent of source and vorticity models of the boundary element method.

1. INTRODUCTION

The vortex method has been used to simulate unsteady viscous flows at high Reynolds numbers. In the vortex method, continuous distribution/production/shedding of vorticity in the flow field is repreresented by introducing a number of discrete vortex elements. The time evolution of the vortical field is simulated by tracking the vortex elements in the Lagrangian manner. So far, various manners have been proposed to introduce vortex elements into flow field, e.g. the various manners related to the introduction of vortex elements can be known in terms of the approximation levels for representation of the boundary layer on the solid surface classified by Kamemoto(1).

In this study, a new scheme for the vorticity production at a solid surface is proposed to enhance the applicability of the vortex method to a wide range of engineering problems. The recent studies enabled us to easily calculate the pressure distribution in the flow field(2). On the other hand, the pressure distribution along a solid surface is related to the vorticity flux at the solid surface and the pressure gradient in the direction tangent to the surface has been determined from this relation(3,4). In this study, the new scheme calculates the vorticity production from the change in pressure over the surface elements without considering a spurious slip velocity on the surface. The new scheme for the vorticity production is applied to the simulation of two-dimensional flows around circular and elliptic cylinders. Since the pressure has continuous distribution along the solid surface, the Kelvin's theorem for total circulation in the flow field can be completely satisfied by this scheme. The source model of the boundary element method is, therefore, tried to use for representing the solid surface of the elliptic cylinder.

2. VORTICITY PRODUCTION

The two-dimensional Navier-Stokes equation can be reduced to the following equation at a stationary solid surface.

$$\frac{\partial p}{\partial s} = -\nu \frac{\partial \omega}{\partial n} \tag{1}$$

where, $\boldsymbol{n}$ denotes normal unit vector at the solid surface and $\boldsymbol{s}$ unit vector tangent to the surface. Eq.(1) gives the relation between the vorticity flux and pressure gradient. This equation can also be generalized to a moving wall as given by Morton (5) and to 3D by Panton(6). The vorticity production over the a small lenght of the solid surface varys the circulation in the flow field at the following rate,

$$\frac{d\Gamma}{dt} = -\nu \int_{\delta s} \frac{\partial \omega}{\partial n} ds = \int_{\delta s} \frac{\partial p}{\partial s} ds = \delta p \tag{2}$$

The circulation $\delta\Gamma$ corresponding to the nascent vorticity during the small time step δt is, therefore, calculated by the following equation under the assumption of constant vorticity flux during a time step δt.

$$\delta\Gamma = \delta p \delta t \tag{3}$$

The total circulation is preserved its initial value of zero for the case of the flow around a cylinder impulsively started due to the continuous distribution of the pressure along the solid surface.

3. CALCULATION METHOD

3.1 Mathematical Model

The governing equation of a two-dimensional incompressive viscous flow is given as follows.

$$\frac{D\omega}{Dt} = \nu\nabla^2 \boldsymbol{u} \tag{4}$$

where, $\boldsymbol{u}$ and ω denote a velocity vector and the vorticity defined as ω=rot $\boldsymbol{u}$, respectively.

The vorticity distribution is represented by a number of discrete vortex elements. The flow field at the next time step is obtained by convecting the vortex elements using the following differential equation.

$$\frac{D\boldsymbol{r}}{Dt} = \boldsymbol{u} \tag{5}$$

where, $\boldsymbol{u}$ denotes the velocity at the vortex element under consideration, which is given by the generalized Biot-Savart law as follows.

$$\boldsymbol{u} = \frac{1}{2\pi} \int_V \frac{\omega \times \boldsymbol{R}}{R^2} dv + \boldsymbol{u}_0 \tag{6}$$

In this equation, The first term of the right hand side is the contribution of the vorticity distribution in the flow field. The second term includes the uniform stream U and the contribution of vorticity/source distribution on the boundary surface. In this study, the vorticity/source strength is determined to ensure the flow parallel to the body surface.

The i-th vortex element at a position $\boldsymbol{r}_i$ has the following vorticity distribution proposed by Leonard (7).

$$\omega(R) = \frac{\Gamma_i}{\pi\varepsilon_i^2} \exp\left(-\frac{R^2}{\varepsilon_i^2}\right) \tag{7}$$

where, Γ_i and ε_i denote the circulation of the vortex element and the cut-off radius, respectively. R denotes the distance from the i-th vortex element, defined as $R = |\boldsymbol{R}| = |\boldsymbol{r}-\boldsymbol{r}_i|$. The induced velocity $\boldsymbol{u}_v$ by the vortex element is given in vector form as follows.

$$\boldsymbol{u}_v = \frac{\Gamma_i \times \boldsymbol{R}}{2\pi R^2}\left(1 - \exp\left(-\frac{R^2}{\varepsilon_i^2}\right)\right) \tag{8}$$

The second term on the right hand side of Eq.(6) is represented using the boundary element method, in which the body surface is divided into a number of vorticity/source-distributed panels. In this study, representing the body by vorticity-distributed panels is called as the vorticity model and by source-distributed panels as the source model.

The effect of the viscous diffusion is taken into account by the core-spreading method. The instantaneous pressure distribution on the solid surface is calculated to determine the circulation of the nascent vortex element to be introduced at every time step. In this study, the pressure distribution is obtained using the integral equation derived from the pressure Poisson equation as follows.

$$cH + \int_S H \frac{\partial G}{\partial n} ds = -\int_V \nabla G \cdot (\boldsymbol{u} \times \omega)\, dv - \frac{1}{Re} \int_S (\nabla G \times \omega) \cdot \boldsymbol{n}\, ds \tag{9}$$

where, c=1/2 for the boundary surface, and G is the fundamental solution of Laplace equation given as $G = 1/(2\pi)\log R^{-1}$. H is defined as $H=p+u^2/2$.

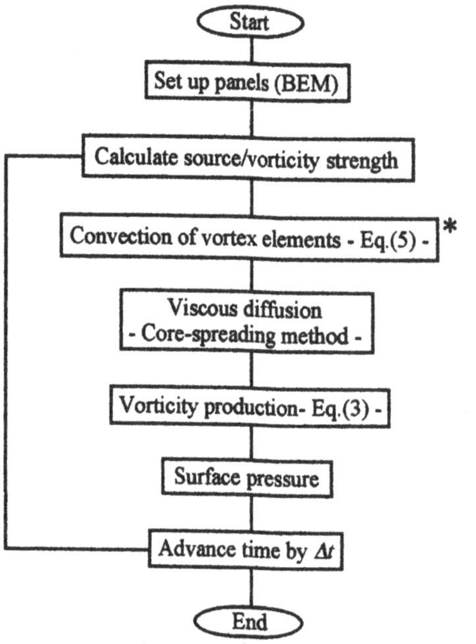

Fig. 1 Flowchart for calculation with vorticity production scheme (* includes the parallel computing).

3.2 Flowchart with Vorticity Production Scheme

The present vortex method is implementd by following the flowchart shown in Fig. 1. The flowchart consists of the following steps:

(i)Set up panels for the boundary element method
(ii)Calculate source/vorticity strength on the solid surface
(iii)Convection of vortex elements
(iv)Consider the effect of the viscous diffusion
(v)Vorticity production and shedding
(vi)Calculate surface pressure
(vii)Advance time by Δt

The time evolution of the flow fields can be simulated by implementing the steps from (ii) to (vii) repeatedly.

4. PARALELL COMPUTING

Since the number of vortex elements increases step by step of time, a long time calculation is needed to compute the induced velocities due to the vortex elements at the large time-step number. The Spalart & Leonard have reduced the computation time by calculating the longer distance interaction between groups of vortex elements instead of from element to element. In this study, parallel computing with MPI (Message Passing Interface) is implemented using six processing elements (PE) to reduce the computation time. As the calculation of the vortex movements during one (1) time step takes huge time at the large step number, the parallel processing allots a portion of the computation mainly for the movement of the vortex element to each PE at every time step as shown in Fig.1. In the vortex method, it is quite easy to achieve good load balancing at each PE without introducing any special algorithm for parallel computing, different from grid-based CFD codes. In the present calculation, the vortex elements introduced are divided into six groups, and the convection of the vortex element in each group is simultaneously computed.

5. CONDITIONS

Two flows around circular and elliptic cylinders are simulated using the newly proposed vorticity production scheme. The vortex elements are introduced from all surface panels at every time step, whose circulation is given by Eq.(3). The circular and elliptic cylinders are represented by 100 panels using the boundary element method. In the present calculation, two cases of the source and vorticity models of the boundary element method are used to represent the solid surface for the circular cylinder. In the case of the vorticity model, the condition of circulation-free is imposed on the body instead of on the flow field. For the elliptic cylinder, only the source model of the boundary element method is used to represent the solid surface. The flow around 2:1 elliptic cylinder is calculated at 30° incidence. The Reynolds numbers are defined using the uniform velocity U and diameter d for the cylinder, and double the major axis a for the elliptic cylinder. The Reynolds number is 5000 for all cases in the present calculations. The time step Δt is $0.05(d/U)$ for the circular cylinder and $0.05[(2a)/U]$ for the elliptic cylinder.

6. RESULTS

Figure 2 shows the flow pattern around the circular cylinder calculated using the vorticity model of the boundary element method at T=50 near a positive peak of lift coefficient. This is similar to the corresponding flow pattern calculated using the source model at T=49 shown in Fig. 3. The distant views are also shown in Figs. 4 and 5. The time histories of drag and lift coefficients are

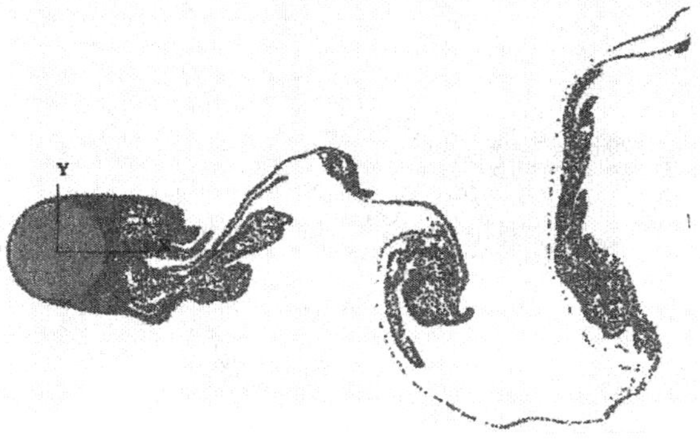

Fig. 2 Flow pattern around circular cylinder (Vorticity model, T=50)

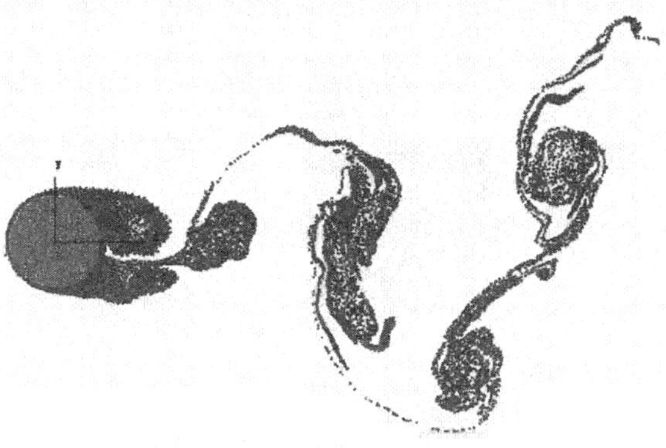

Fig. 3 Flow pattern around circular cylinder (Source model, T=49)

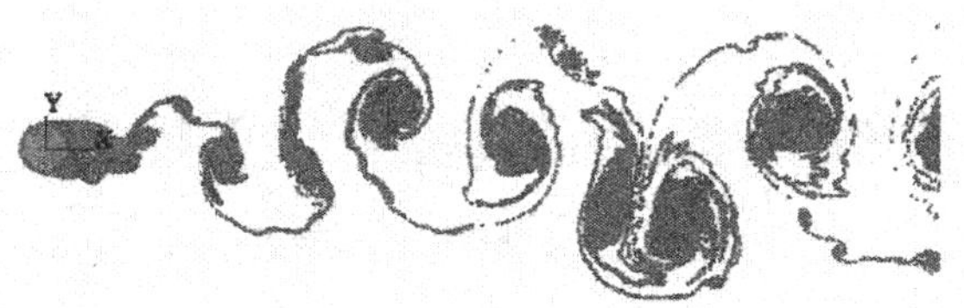

Fig. 4 Flow pattern around circular cylinder (Distant View, Vorticity model, T=50)

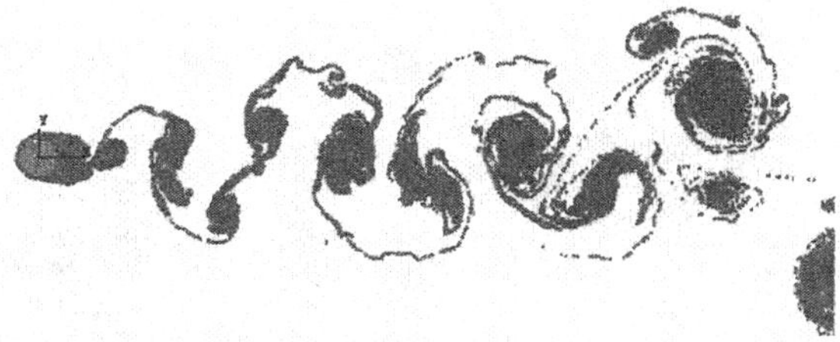

Fig. 5 Flow pattern around circular cylinder (Distant View, Source model, T=49)

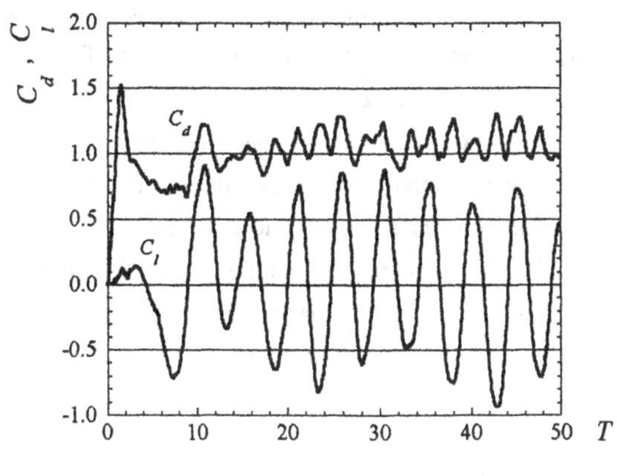

Fig. 6 Time history of C_d and C_l (Circular cylinder - Vorticity model)

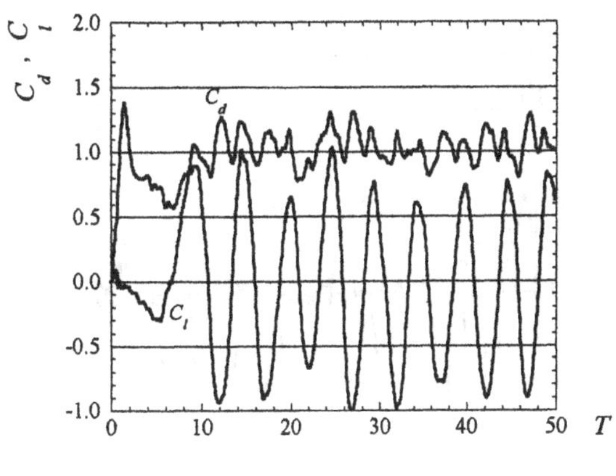

Fig. 7 Time history of C_d and C_l (Circular cylinder - Source model)

shown in Figs. 6 and 7. There is no remarkable difference in both the variations of drag and lift coefficients between Figs. 6 and 7. It seems that the difference in the surface model of vorticity or source makes little qualitative difference in the flow patterns and the drag and lift coefficents.

In the case of the elliptic cylinder, only the source model of the boundary element method was used to represent the solid surface. Figure 8 shows the instantaneous flow pattern at a small time after starting of T=1.5. A starting vortex is formed and convected downstream. After shedding a few isolated vortices as shown in Fig.9, the flow pattern becomes steady state with dead water region on the upper side of the cylinder

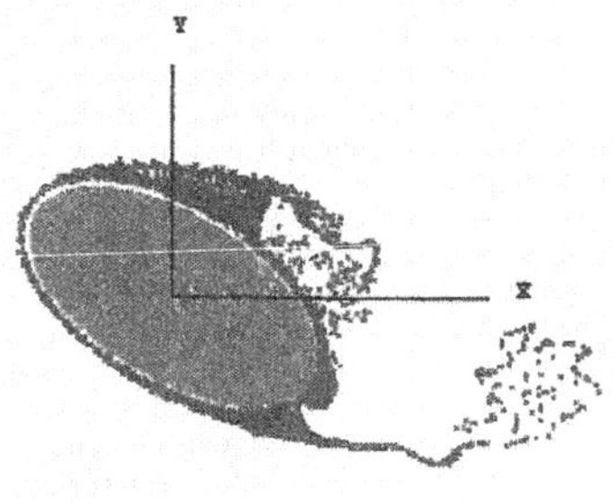

Fig. 8 Flow around elliptic cylinder
(Source model, T=1.5)

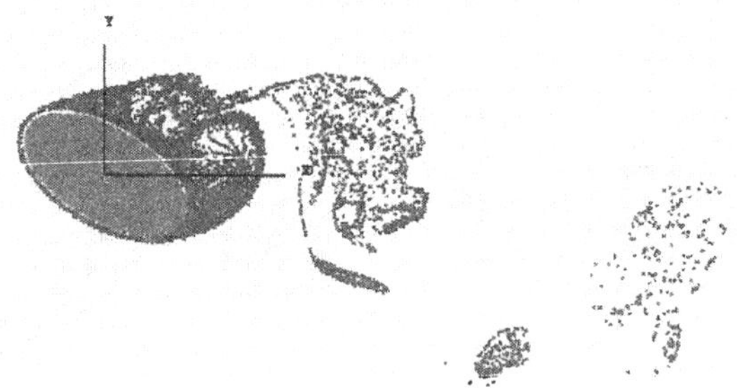

Fig. 9 Flow around elliptic cylinder
(Source model, T=4.0)

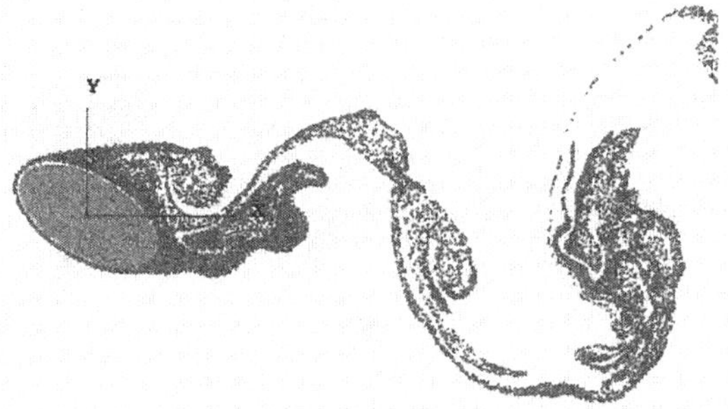

Fig. 10 Flow around elliptic cylinder
(Source model, T=50.0)

Fig. 11 Flow around elliptic cylinder
(Distant view, Source model, T=50.0)

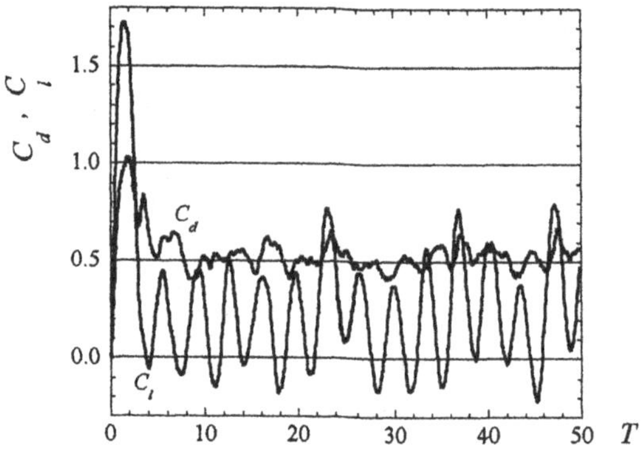

Fig. 12 Time history of C_d and C_l
(Elliptic cylinder, Source model)

as shown in Fig. 10. Figure 11 shows the distant view of the flow pattern at the same elapsed time as Fig.10. The corresponding time history of the drag and lift coefficients is shown in Fig. 12. The drag coefficient has almost constant value of 0.5. On the other hand, the lift coefficient has the maximum peak at a small time after starting. After that, the lift coefficient fluctuates approximately between zero and 0.5.

In the present calculation, parallel computing using six processing elements was implemented, the computation time has been reduced to about 20 percent compared with that of the single processing element computation. It is found that the parallel computing is very effective for the vortex method.

7. CONCLUSIONS

In this study, a new scheme for vorticity production at a solid surface was proposed to enhance the applicability of the vortex method to a wide range of engineering problems. The new scheme calculates the vorticity production from the change in pressure over the surface elements without considering a spurious slip velocity on the surface. The two-dimensional flows around circular and elliptic cylinders were calculated using the new scheme. In the case of the circular cylinder, the reasonable fluid forces were obtained, and the difference in the surface model of vorticity or source made little qualitative difference in the flow patterns and the drag and lift coefficents. The flow around the 2:1 elliptic cylinder at 30° incidence was also calculated using the source model. The variation of the lift coefficient had a positive mean value. It is found that lifting bodies can be represented by the source model with the vorticity production scheme. Parallel computing was implemented using six processing elements, and it is found that parallel processing for vortex convection is very effective to reduce the computation time.

References

(1) Kamemoto, K. (1995), Analysis of Turbulent Flows, University of Tokyo Press, p137-160.

(2) Uhlman, J. S. (1992), "An Integral Equation Formulation of the Equation of Motion of an Incompressive Fluid, Naval Undersea Warfar Center T.R. 10,086.

(3) Spalart P. R. & Leonard A., "Computation of Separated Flows by a Vortex-Tracing Algorithm", AIAA-81-1246.

(4) Speziale C. G., Sisto F. & Jonnavithula S., "Vortex Simulation of Propagating Stall in a Linear Cascade of Airfoils", Trans. ASME, Vol. 108, p304-312.

(5) Morton B. R., (1984), "The Generation and Decay of Vorticity", Geophys. Astrophys. Fluid Dynamics, Vol. 28, p277-308.

(6) Panton R. L., (1984), Incompressive Flow, Wiley, p335-338.

(7) Leonard, A. (1980), "Vortex Methods for Flow Simulation", J. Comp. Phys., Vol. 37, p289-335.

TWO-DIMENSIONAL TRANSIENT FLOWS AROUND A RECTANGULAR CYLINDER BY A VORTEX METHOD

Masahiro OTSUKA, Teruhiko KIDA, Mitsufumi Wada*
Department of Energy Systems Engineering, Osaka Prefecture University,
Sakai, Osaka 599-8531, Japan / Email: ohtsuka@energy.osakafu-u.ac.jp
Present:*Daikin Industries, LTD.

Mitsuo KURATA
Department of Mechanical Engineering, Setsunan University,
Neyagawa, Osaka 572-8508, Japan

Abstract

This paper is concerned with flow around a two-dimensional rectangular cylinder by a vortex method. The vortex method used in the present paper consists of the combination of vortex blob and sheet method with a panel method and viscous effect is simulated by random walk method. The Reynolds number is taken as 500 and 3000. The results shown in this paper are vortex distribution, velocity vector and velocity profile in the wake. To know utility of this method, we calculate flows around the rectangular cylinder with the width-to-height ratios of 3 and 6 and compare them with earlier results.

1 Introduction

A rectangular cylinder is one of basic shape of bluff bodies which are very often utilized in large structures. The flow around the body is very complicated, though it has simple shape. Many investigators have studies this problem experimentally and numerically. Davis & Moore [1] presented numerical solutions for time-dependent flow about rectangular cylinders by using third-order upwind difference for convection term. The regime of the Reynolds number is from 100 to 2800. They showed flow visualization of vortex shedding development. Okajima [2] presented the experimental results in a wind tunnel and in a water tunnel. The regime of the Reynolds number is from 70 to 2×10^4. He found that there is a certain range of the Reynolds number for rectangular cylinder with the width-to-height ratios of 2 and 3 where flow pattern abruptly changes with a sudden discontinuity in Strouhal number. Davis, Moore & Purtell [3] treated the effect of confining walls numerically and experimentally. Okajima, Ueno & Sakai [4] computed flows around the rectangular cylinder with various width-to-height ratios from 0.2 to 10, and showed that the flow pattern changes critically about the ratios 2.1 and 6. Hwang & Sue [5] treated a shear flow around the rectangular cylinder and showed that the Strouhal number, the mean drag and the amplitude of fluctuating force tend to decrease as the shear rate increase. Barton [6] also calculated flow around the rectangular cylinder for the Reynolds number 250 by using Simple and Piso-type scheme and showed the transient flow behavior from an impulsive start. But there is not so many papers that are calculated by vortex method. Sarpkaya & Ihring [7] used a discrete vortex method and investigated an impulsively started flow around the rectangular cylinder. In their vortex method, four point vortices are shed from corners to simulate the separated shear flows. Chiu and Ko [8] studied the interaction of shedding vortices in the case of a side-by-side arrangement of two rectangular cylinders by using a vortex method based on Ogami & Akamata [9] .

The main purpose of this work is to construct a vortex method, comparable with earlier numerical method for a two-dimensional, viscous, incompressible time-dependent flow about a rectangular cylinder. The combination of vortex blob and sheet method with a panel method has been shown to be the comparable method for a circular cylinder(see Kida & Kurita [10]). However this method has not been confirmed for flows around bluff bodies with a corner such as rectangular cylinders. In the present paper, hence, the vortex method used for the circular cylinder will be used. For the corner flow it will be revised and numerical results will be compared with other numerical ones.

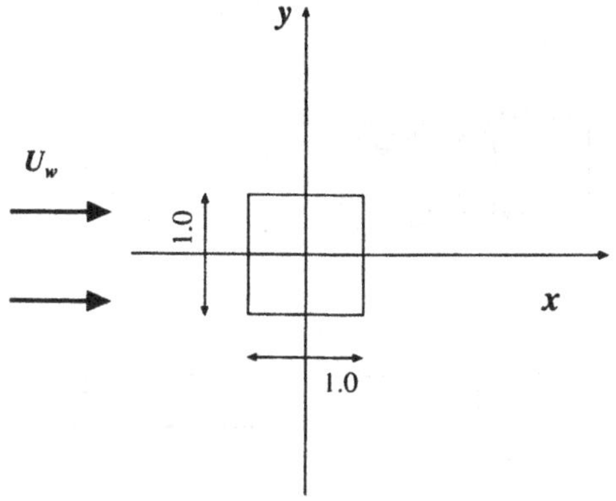

Figure 1: Geometry and coordinates

2 Numerical Method

The physical problem considered in this paper is a two-dimensional, viscous, incompressible time-dependent flow around a rectangular cylinder. Figure 1 shows the geometry of the body and coordinate system (x, y) fixed with the cylinder is also shown in this figure. The moving velocity is U_w, x-axis is the moving direction, and the origin is the geometry center, as shown Fig.1. In this paper, U_w is constant for $t \geq 0$, where t is time, and the initial flow $(t < 0)$ is at rest.

2.1 Panel Method

In the present paper, the fractional method is used, so that we have to solve the potential flow. To solve this flow, we use the panel method proposed by Kida, et al [11]. The complex velocity potential $f(= \phi + i\psi$, where ϕ and ψ are velocity potential and stream function, respectively) due to the surface singularity and the uniform flow U_w is given by

$$f(z) = U_w z - \frac{i}{2\pi}\int_C \gamma(s)\log(z - z(s))ds + \frac{1}{2\pi}\int_C \sigma(s)\log(z - z(s))ds, \quad (1)$$

where $z = x + iy$, σ is the *source* placed on each panel, γ is the *vortex* placed on each panel, C is boundary of the body and z(s) is on C. Then the complex velocity is

$$u - iv = \frac{df}{dz} = U_w + \frac{1}{2\pi}\int_C \frac{\sigma(s)}{z - z(s)}ds - \frac{i}{2\pi}\int_C \frac{\gamma(s)}{z - z(s)}ds, \quad (2)$$

In the actual calculation, $\gamma(s)$ is constant on every panels and $\sigma(s)$ is distributed as Dirac delta function on each panels. These values are obtained by using the following two conditions; Kelvin's circulation theorem (3) and the boundary condition (4) that normal velocity on the body surface is zero.

$$\sum_j \Gamma_j + \gamma = 0, \quad (3)$$

$$\vec{u}_c \cdot \vec{e}_n = 0 \quad on\ C, \quad (4)$$

where $\vec{u}_c(=\vec{u}_p + \vec{u}_v$, where $\vec{u}_p = (u, v)$ and $\vec{u}_v$ is velocity due to vortex blob and sheet) is velocity on the boundary surface, $\vec{e}_n$ is the unit vector perpendicular to the panel.

2.2 Vortex Blob & Sheet Method

We can not impose no-slip condition by the panel method. To satisfy the no-slip condition, we have to create vorticity on the surface of the body, that is, we have to create vortex sheets on the surface of the body. Strength of the created vortex sheet is given by

$$\xi = -2(u_s + \sum_j \xi_j d_j), \quad (5)$$

where u_s is tangential velocity on the surface of the body, d_j is degree of a pile of vortex sheets (see Kida & Kurita [10]). If vortex sheet is shed out of numerical boundary layer, the vortex sheet is changed to a vortex blob as shown in Fig.2. The velocity induced by a vortex blob is assumed to be (see Chorin)

$$\vec{u}_j = \begin{cases} \dfrac{\Gamma_j(\vec{e}_k \times \vec{R})}{2\pi|\vec{R}|^2} & (|\vec{R}| \geq \varepsilon) \\ \dfrac{\Gamma_j(\vec{e}_k \times \vec{R})}{2\pi\varepsilon|\vec{R}|} & (|\vec{R}| < \varepsilon) \end{cases} \quad (6)$$

where ε is cut off radius, $\vec{e}_k$ is unit vector normal to space, $\vec{R}$ is a position vector from the vortex blob and Γ_j is circulation of the vortex blob. In the present paper cut off radius is taken as $\varepsilon = l_j/\pi$, where l_j is the typical panel length.

To conserve circulation between vortex sheet and vortex blob, we use the following relation, when the vortex sheet changes to vortex blob;

$$\Gamma_j = \xi_j \times l_j. \quad (7)$$

The thickness of the boundary layer is of the order of $Re^{-1/2}$, so the thickness of numerical boundary layer is assumed to be $2/\sqrt{Re}$.

The induced velocity of vortex sheets is almost tangential of the surface of the cylinder and the flow separates from the corners. Therefore, vortex sheets which separate from the corners are assumed to shed out tangentially and to become vortex blob. The method which we propose for corner flow are illustrated in Fig.2.

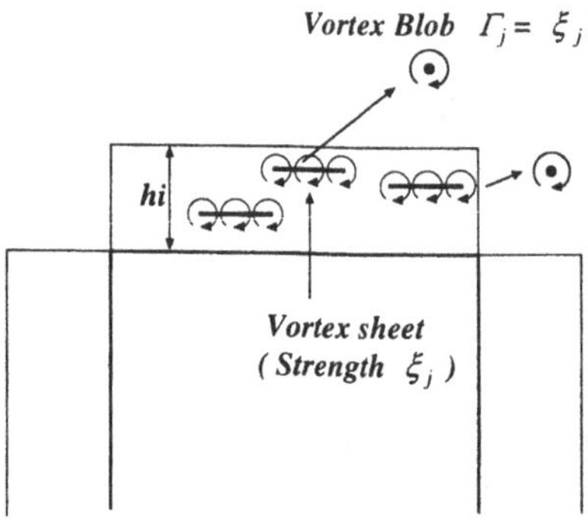

Figure 2: Numerical boundary layer

We use random walk method to express viscous of fluid. The distance of movement of the vortex, Δr, is given as

$$\Delta r = \sqrt{\frac{2\Delta t}{Re}} \times \Omega, \tag{8}$$

where Ω is regular random number possessing $N(0,1)$. Vortex blobs are randomly moved in the directions of x and y axes and vortex sheets are randomly moved in only direction of normal to panel.

3 Results

In the present calculation, time step is 0.1 and maximum allowance vorticity is 0.2. The present numerical results of the square cylinder are shown in Fig.3~9. In Fig.3~9, the number of panels is 80 and the Reynolds number is 3000 and 500. Figure 3 shows a pair of recirculation zone in the wake for $t = 5$, where t is dimensionless time: (a) is the vortex blob distribution and (b) is the velocity vector. The Reynolds number is 3000. From Fig.3(b), we see that the flow is symmetry with respect to x-axis. Barton [6] also shows the transient flow from the start for $Re = 250$. His numerical results shows the existence of a pair of recirculation zone within $t = 20 \sim 30$. Davis & Moore [1] shows the transient flow in the case of width-to-height ratio 1.7 and $Re = 1000$. At $t = 12$, the flow is almost symmetry and a pair of recirculation zone exists at this time. In Fig.3(a), vortex blobs are shown. We see that vortex blobs near the front corners of the cylinder are shed along the front surface, however its effect to the velocity field is not so remarkable as shown in Fig.3(b). Figure 4 shows the time history of vortex blob distribution at $Re = 3000$ and we see that von Karman's vortex street is growing in wake as time advances. Further, pairing of vortices is seen. Comparing with the results for $Re = 500$ (see Fig.6), we see that the cluster of vortices is more stretched at $Re = 3000$ than at $Re = 500$. In the case of $Re = 500$ (see Fig.6), the viscous effect is larger than the case of

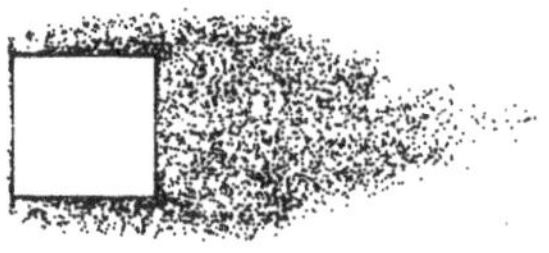

(a)

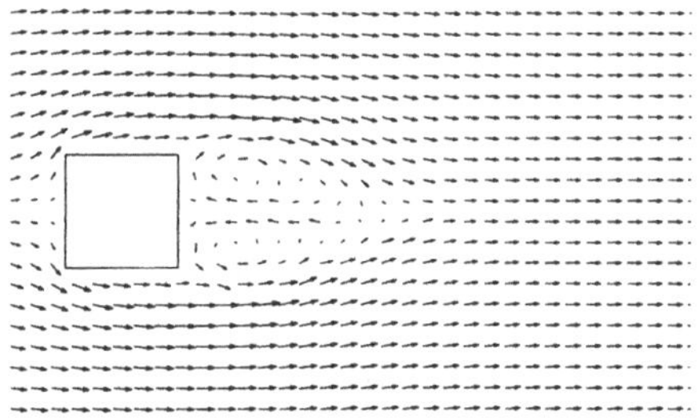

(b)

Figure 3: Flow pattern of the square cylinder at $t = 5$ and $Re = 3000$: (a) vortex blob distribution (b) velocity vector.

$Re = 3000$, so the coherent structure of vortices at far downstream is less clear than in case of $Re = 3000$. This feature is also shown by Davis & Moore [1].

Figure 5 is the profile of center line velocity near the wake. U_x is the x-component of velocity on the x-axis. These results is not averaged with respect to time, so there is some fluctuation due to the use of random walk method. From this figure, we can estimate the length of the recirculation zone, L, which is equal to the length of the zone that U_x is negative. We easily see that the length, L, is almost linear with time. However, the minimum center line velocity is almost the same after $t = 2$. This fact implies: the vortices shed from the cylinder concentrates in the near wake and becomes a pair of cluster, and the circulation of the cluster becomes large with time until $t = 2$, after $t = 2$, the cluster grows up to the x-direction and it becomes unsteady at a certain time. Figure 7 shows the time development of U_x in case of $Re = 500$. In this Reynolds number, the length, L, of the near wake becomes also large with time but it is not proportional to time, as shown in Fig.7. The minimum center line of velocity is also nearly the same after $t = 3$: the viscous effect implies that the near wake extends to y-direction.

Figure 8 shows the profile of U_x at a given x in case of $Re = 3000$. The upper and lower figures show it at $x = 1$ and $x = 2$, respectively. The velocity profile is clearly shown that the velocity is almost symmetry with respect to x-axis till $t = 7$, but the break down of the symmetry is seen at $t = 15$. For $Re = 500$, its

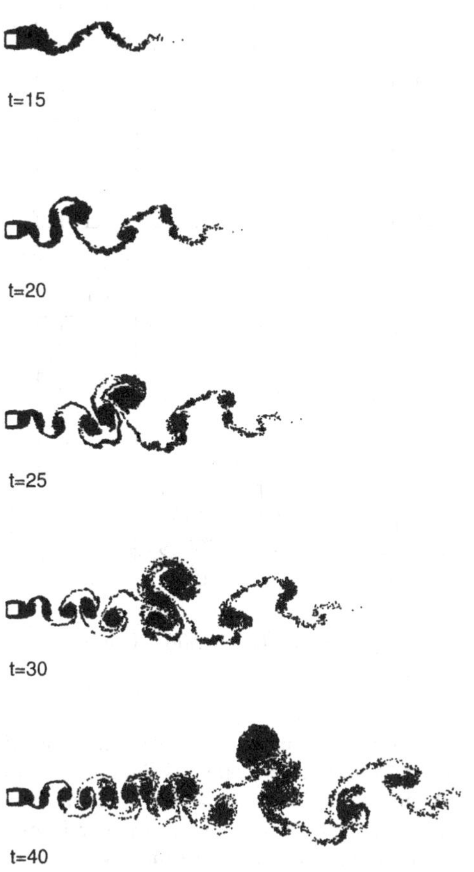

Figure 4: Time history of vortex blob distribution in case of $Re = 3000$.

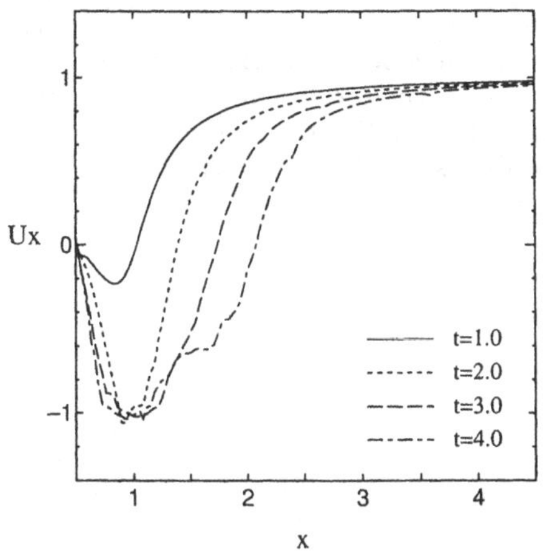

Figure 5: Profile of x-component of velocity on the center line of wake in $Re = 3000$.

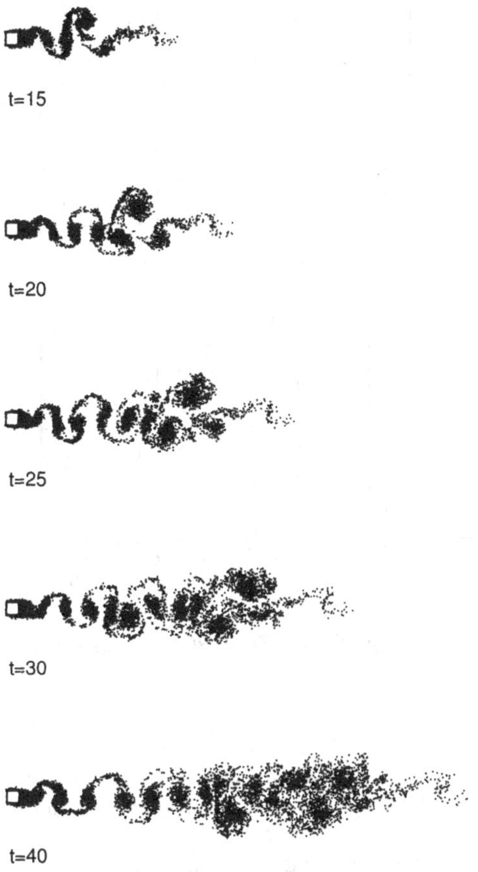

Figure 6: Time history of vortex blob distribution in case of $Re = 500$.

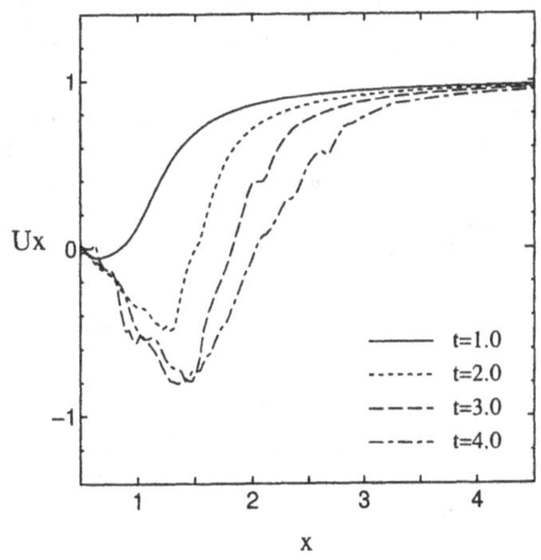

Figure 7: Profile of x-component of velocity on the center line of wake in $Re = 500$.

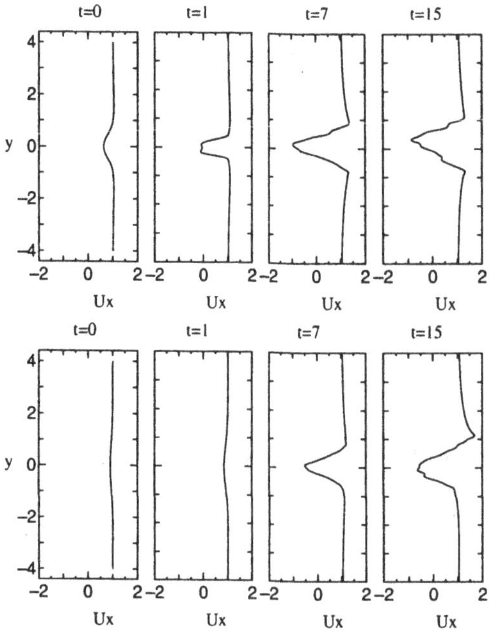

Figure 8: U_x profile at Re = 3000. Upper: $x = 1$, lower: $x = 2$.

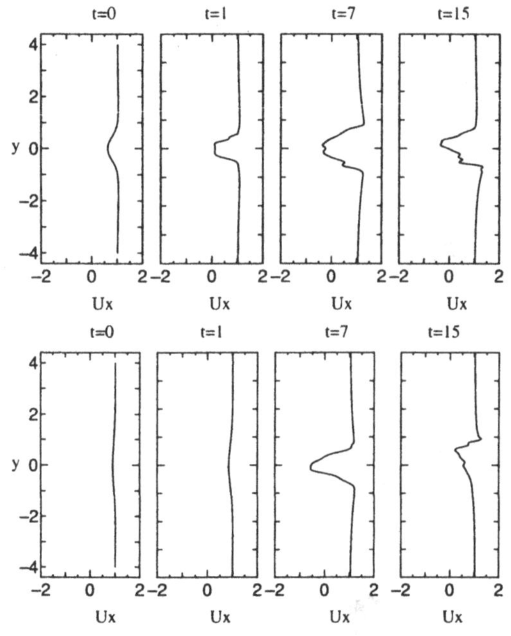

Figure 9: U_x profile at Re = 500. Upper: $x = 1$, lower: $x = 2$.

profile is shown in Fig.9. The fluctuation of velocity is more clear due to random walk method than at Re = 3000. In this Reynolds number, the recirculation zone begins to break down at $t = 15$.

To examine the validity of the present numerical approach, the cases of the width-to-height ratios 3 and 6 are calculated for $Re = 3000$. Figures 10 and 12 show the time development of vortex blob distribution with the ratios 3 and 6, respectively. The shear flow shed from the front edges reattaches to the side surface of the cylinder. This feature is shown by Ozono et al. [12].

Figures 11 and 13 shows the center-line velocity in the near wake with the ratios 3 and 6, respectively. The length of the near wake L is different between the ratios 3 and 6. Okajima [4] shows the numerical results of aerodynamic force: the drag force decreases at beginning of time and after arriving at the minimum it increases and begins to oscillate. The time when it begins to oscillate decreases with the increases the ratio. This time when L dose not increase with time is considered to be corresponding to the beginning of force oscillation. Therefore, the above mentioned result is corresponding to this feature.

4 Conclusion

The two-dimensional transient flow around a rectangular cylinder is calculated by a vortex method. The vortex method used in the present paper consists of the combination of vortex blob and sheet method with a panel method and viscous effect is simulated by random walk method. From numerical results, we see that the vortex method used in the paper is useful to know the global feature: the time history of vortex blob distribution is almost the same as in earlier works. The recirculation zone is formed in the wake at beginning of start, the recirculation flow because unsteady shed out the wake with increase of time and von Karman's vortex street is growing as the advance of time. The profile of velocity is also predicted and global feature is obtained. But from these results we can not arrive the conclusion that the vortex method can simulate detail of the flows. To conclude it, we have to discuss further about the drag and lift coefficient, streamline and pressure field.

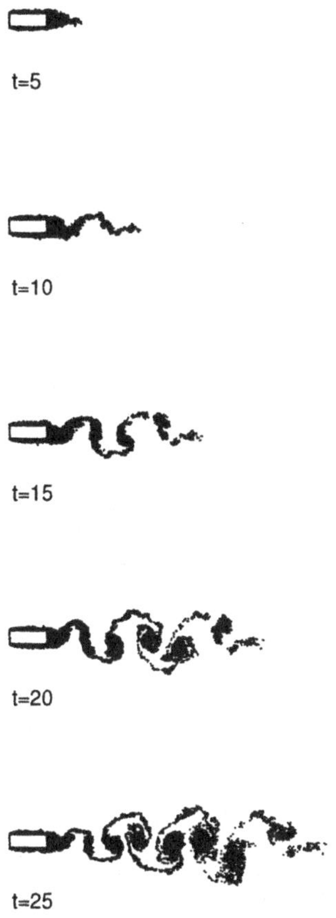

Figure 10: Time history of vortex blob distribution with the ratio 3 in case of $Re = 3000$.

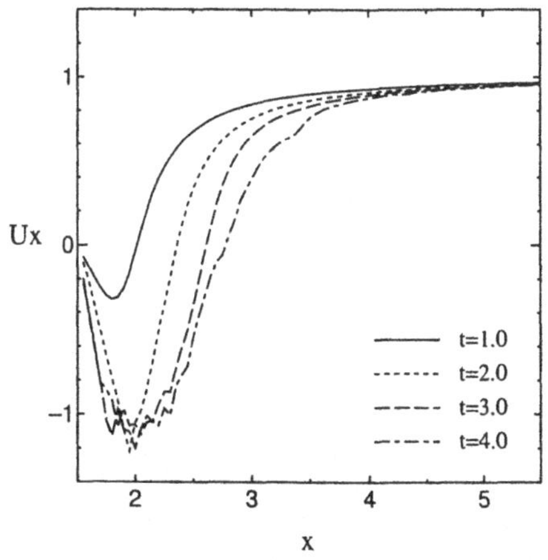

Figure 11: Profile of x-component of velocity on the center line of wake with the ratio 3 in $Re = 3000$.

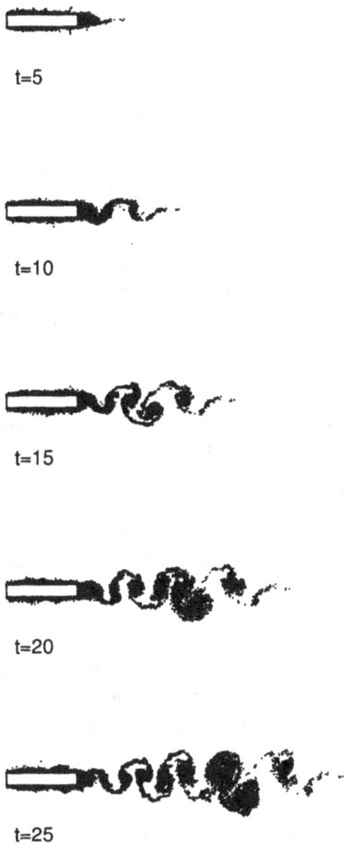

Figure 12: Time history of vortex blob distribution with the ratio 6 in case of $Re = 3000$.

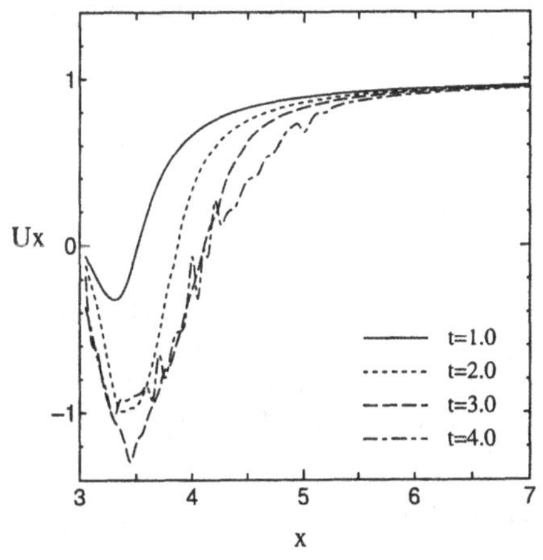

Figure 13: Profile of x-component of velocity on the center line of wake with the ratio 6 in $Re = 3000$.

References

[1] R.W. Davis and E.F. Moore, "A numerical study of vortex shedding from rectangles", J. Fluid Mech. (1982), vol.116, pp.475-506

[2] A. Okajima, "Strouhal numbers of rectangular cylinders", J. Fluid Mech. (1982), vol.123, pp.379-398

[3] R.W. Davis, E.F. Moore and L.P. Purtell, "A numerical-experimental study of confined flow around rectangular cylinder", Phys. Fluids, 27(1), (1984), pp.46-59

[4] A. Okajima, H. Ueno and H. Sakai, "Numerical simulation of laminar and turbulent flows around rectangular cylinders", Int. J. Num. Meth. in Fluids, vol.15, (1992), pp.999-1012

[5] R.R.Hwang and Y.C.Sue, "Numerical simulation of shear effect on vortex shedding behind a square cylinder", Int. J. Num. Meth. in Fluids, vol.25, (1997), pp.1409-1420

[6] I. E. Berton, "Comparison of simple and piso type algorithms for transient flow", Int. J. Num. Meth. in Fluids, vol.26, (1998), pp.459-483

[7] T.Sarpkaya and C.J.Ihrig, "Impulsively started steady flow about rectangular prisms: Experiments and discrete vortex analysis", Tran. ASME.J.Fluids Engineering, vol.108, (1986), pp.47-54

[8] A.Y.W.Chiu and N.W.M,Ko, "Numerical simulation of proximity interference of two square cylinders", ASME/JSME Fluids Engineering Conference, (1999)

[9] Y.Ogami T.Akamatsu, "Viscous flow simulation using the discreate vortex model -the diffusion velocity method", Com.Fluid, vol.19, (1991), pp.433-441

[10] T.Kida and M.Kurita, "High Reynolds number flow past an impulsively started circular cylinder (time marching random walk vortex method)", Com.Fluid Dynamics Journal, vol.4,no.4, (1996), pp.489-508

[11] T.Kida, T.Nagata and T.Nakajima, "Accuracy of the panel method with distributed sources applied to two-dimentional bluff bodies", Com.Fluid Dynamics Journal, vol.2,no.1, (1993), pp.73-90

[12] S.Ozono,Y.Ohya and Y.Nakamura, "Stepwise increase in the Strouhal number for flows around flat plates", Int. J. Num. Meth. in Fluids, vol.15, (1992), pp.1025-1036

[13] Jang.Wissink, "Dns of 2D turbulent flow around a square cylinder", Int. J. Num. Meth. in Fluids, vol.25, (1997), pp.51-62

SIMULATION OF THREE DIMENSIONAL SEPARATED FLOW AROUND AN ELLIPSOID BY DISCRETE VORTEX METHOD

M. Fevzi Ünal
Department of Aeronautics, Istanbul Technical University
80626 Maslak, İstanbul, Turkey / Email: munal@itu.edu.tr

Kyoji Kamemoto
Department of Mechanical Engineering and Materials Science, Yokohama National University
79-5 Tokiwadai Hodogaya-ku Yokohama, 240-8501, Japan / E-mail: kame@post.me.ynu.ac.jp

ABSTRACT

Using a three-dimensional discrete vortex method, separated flow around an ellipsoid having axes ratio of three is simulated. The method is Lagrangian and thus does not require a grid around the ellipsoid. Separated flow from the ellipsoid, surface of which is represented with vortex panels is simulated by tracking vortex panels and vortons shed from the surface. Evolution in time of structure of flow and pressure distribution around the ellipsoid are calculated.

1. INTRODUCTION

Understanding and realistic numerical simulation of separated flow around three dimensional bodies is central to a variety of important problems, including flow induced vibration, noise generation, heat transfer and mixing processes, with applications in energy conservation, propulsion and manufacturing industries.

In contrast to the view expressed in many computational studies of bluff-body wakes, even the fundamental flow case of a nominally two-dimensional body, e.g. a circular cylinder in a uniform stream exhibits several three-dimensional features (1) and at low Reynolds numbers, maintaining parallel vortex shedding to the cylinder axis requires very carefully controlled end conditions by means of suitably angled end plates (2) or with other types of end constraint (3). At high Reynolds numbers, two-dimensional shedding is possible only with a strong external influence such as body or flow oscillation in synchronisation with shedding (4). Strictly, two-dimensional geometries and two-dimensional flows can only exist in theoretical models, since in reality, all bodies have finite lateral extents and, above some critical Reynolds number, all flows generate instabilities with some spanwise wavelength.

Consequently, in recent years increasing experimental and numerical research effort is being directed towards three-dimensional features of nominally two dimensional, mildly or strongly three-dimensional bluff bodies. Since key three-dimensional features of the wake of a nominally two-dimensional body, such as vortex dislocations, appear apparently randomly in time and space, Tombazis and Bearman (5) attempted to control their locations by applying a mild geometric disturbance in the form of a wavy trailing edge. They made interesting observations about a number of distinct vortex shedding patterns. Several other authors (e.g. 6) have reported results of investigations of bluff body flows influenced by geometric disturbances that might also be described as mild.

Increasing appreciation of the importance of three dimensional effects aroused interest also in numerical studies. Several researchers have developed three-dimensional simulation programs based on the discrete vortex method by employing e.g. a fully Lagrangian approach (7) or an Eulerian-Lagrangian hybrid method (8). In addition, a number of other authors have developed finite difference (e.g. 9), finite element and spectral element codes to simulate three-dimensional bluff body flows.

In the present study, a fully Lagrangian discrete vortex method is employed to simulate separated flow around an ellipsoid with axes ratio of three. Surface representation of the bodies is performed by means of vortex panels allowing satisfaction of the tangential velocity condition at the surface of the ellipsoid. As previously applied by Nakanishi and Kamemoto (7) the vorticity field around bodies is represented by a number of discrete vortices with a spherically symmetric distribution of vorticity, i.e. using vortons. In order to take into account viscous diffusion, the core spreading method based on the exact solution of Navier-Stokes equations for viscous diffusion of an isolated vortex is introduced into the model.

2. NUMERICAL METHOD

In the discrete vortex method, the basic equations are the vorticity transport,

$$\frac{D\vec{\omega}}{Dt} = \frac{\partial\vec{\omega}}{\partial t} + (\vec{U}.\nabla)\vec{\omega} = (\vec{\omega}.\nabla)\vec{U} + \nu\nabla^2\vec{\omega} \qquad (1)$$

and the vorticity definition,

$$\vec{\omega} = \nabla \wedge \vec{U} \qquad (2)$$

equations. On the other hand, the pressure Poisson equation is stated as:

$$\nabla^2 p = -\rho\nabla.(\vec{U}.\nabla\vec{U}) \qquad (3)$$

From the vorticity definition equation, Biot-Savart law can be determined (10) as,

$$\vec{U}(r) = \frac{1}{4\pi}\int_V \frac{\vec{\omega}_O \wedge \vec{R}}{|\vec{R}|^3} dV_O - \frac{1}{4\pi}\int_S \left(\frac{\sigma_O \vec{R}}{|\vec{R}|^3} + \frac{\vec{\gamma}_O \wedge \vec{R}}{|\vec{R}|^3}\right) dS_O \qquad (4)$$

Therefore, once the vorticity distribution $\vec{\omega}_o$ in a flow field (V) is approximated by a number of discrete vortices, source, σ_O and/or vortex, $\vec{\gamma}_o$ distribution on a body surface can be calculated so as to satisfy the normal or tangential velocity conditions at a finite number of control points on it, by solving a set of equations, in the form of,

$$[K]\{\beta\} = \{b\} \qquad (5)$$

where, [K] is coefficient matrix depending on surface geometry, {b} is induced velocities at the control points and, {β} is unknown source and/or vortex strengths around the surface. The surface of the ellipsoid with axes ratio of 3 is divided into 12 and 24 along the major axis and the azimuthal directions respectively and thus is represented by 288 panels (Figure 1). Tangential velocity condition is satisfied at the panel corners and by means of equation 5 and vortex strengths along panel edges are determined there. This potential flow calculation is followed by introduction of those panels into the flow field to simulate separated flow around the ellipsoid. The surface vortex panels are introduced from a distance,

$$\delta_{dif} = 1.136(\nu\Delta t)^{0.5} \qquad (6)$$

which corresponds to the displacement thickness vorticity layer in the Rayleigh problem. Consequently, the vortex panels are transported to their new positions with convective velocities induced by virtue of the Biot-Savart law and at the same time diffusive velocities in normal direction to the surface. Changing right hand side of equation 5 by newly introduced vortex panels is recalculated and an updated vortex strength distribution around the body surface is determined. After the previously introduced ones are displaced to their new locations, those vortex panels with updated strengths are introduced into the flow field. Then the calculation procedure returns to the first step in which an updated vortex strength distribution around the surface of the body is determined again.

In order to cut back on the computational time, vortex panels convected outside of an envelope during evolution of the separated flow are converted into vortons of equivalent strengths. The envelope is in the form of an ellipsoid with axes ratio of 3 but its small axis length is twice that of ellipsoidal body.

Trajectories of vortex panels and vortons are calculated by means induced velocities due to the Biot-Savart law and, from,

$$\frac{d\vec{r}_0}{dt} = \vec{U}_0 \qquad (7)$$

equation with Euler method, i.e.

$$\vec{r}_0(t+\Delta t) = \vec{r}_0(t) + \vec{U}_0 \Delta t$$

On the other hand, the distribution function and the resultant induced velocity of ith vorton having a spherically symmetric vorticity distribution are given (11) as,

$$\vec{\omega}_{oi} = \frac{\vec{\alpha}_i}{\varepsilon_i^3} f(\chi) \quad , \quad \vec{U}_i = \frac{\vec{\alpha}_i \wedge \vec{R}}{4\pi|\vec{R}|^3} g(\chi) \qquad (8)$$

where, ε_i and α_i represent radius and strength of the vorton, and

$$f(\chi) = \frac{15}{8\pi} \frac{1}{(\chi^2+1)^{7/2}} \quad ; \quad g(\chi) = \frac{5\chi^3 + 2\chi^5}{2(\chi^2+1)^{5/2}} \quad ;$$

$$\chi = \frac{|\vec{R}|}{\varepsilon_i} \; ; \; \vec{\alpha}_i = \frac{4}{3}\pi \varepsilon_i^3 \vec{\omega}_i$$

In order to calculate variation of vorticity in time due to three dimensional convection and diffusion, the vorticity transport equation 1 is solved by the spherical vortex blob method originally proposed by Nakanishi and Kamemoto (7). According to this model, a spherical vortex blob is tentatively replaced by an equivalent cylindrical vortex stick in the consideration of the longitudinal elongation and the stick elongated and diffused during a small time interval is replaced again by a new equivalent spherical blob with a different core radius.

After evolution in time of the separated flow around the ellipsoid is determined, pressure Poisson equation is solved by a boundary integration method (10) at selected times.

3. RESULTS

Three dimensional separated viscous flow around an ellipsoid with axes ratio of 3 is calculated using vortex panel method in combination with a Lagrangian discrete vortex method. The ellipsoid, surface of which is represented by 288 panels, is placed into an impulsively started uniform stream, and evolution of separated flow around it is examined for angles of attack of 30° and 90°. Non dimensional time step of calculation based on the free stream velocity U_∞ and major axis length of the ellipsoid L is $\Delta T = U_\infty t / L = 0.1$. Reynolds number of the simulation is $Re = U_\infty L / \nu = 1000$.

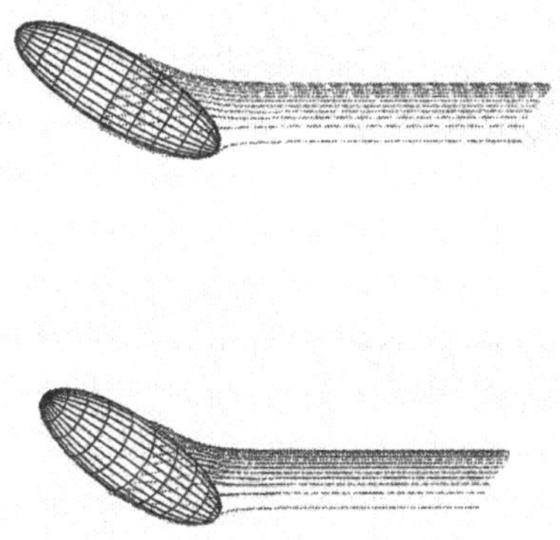

Figure 1: Potential flow streamlines around ellipsoid at angle of attack of 30°

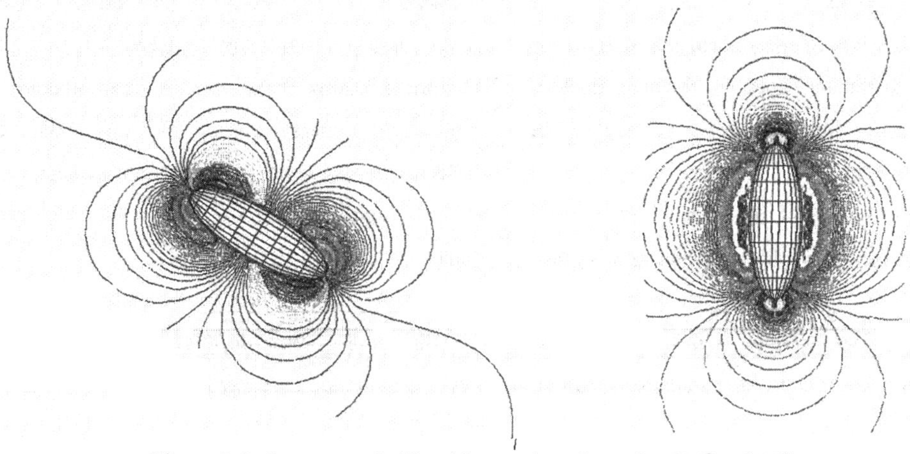

Figure 2: Isobars around ellipsoid at angles of attack of 30° and 90°

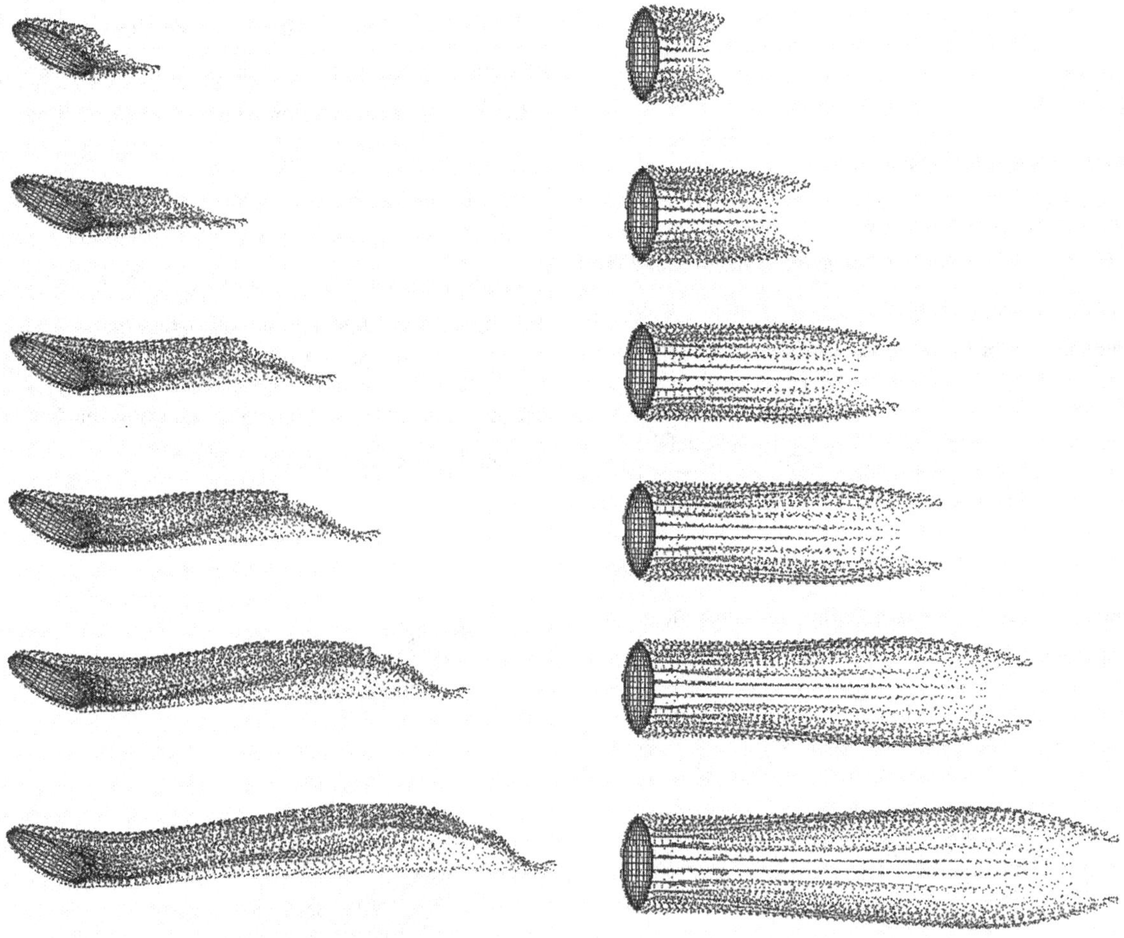

Figure 3: Evolution in time (T) of wake behind ellipsoid at 30° and 90°. From top to bottom T= 1, 2, 3, 3.5, 4.5 and 5.5. Re=1000

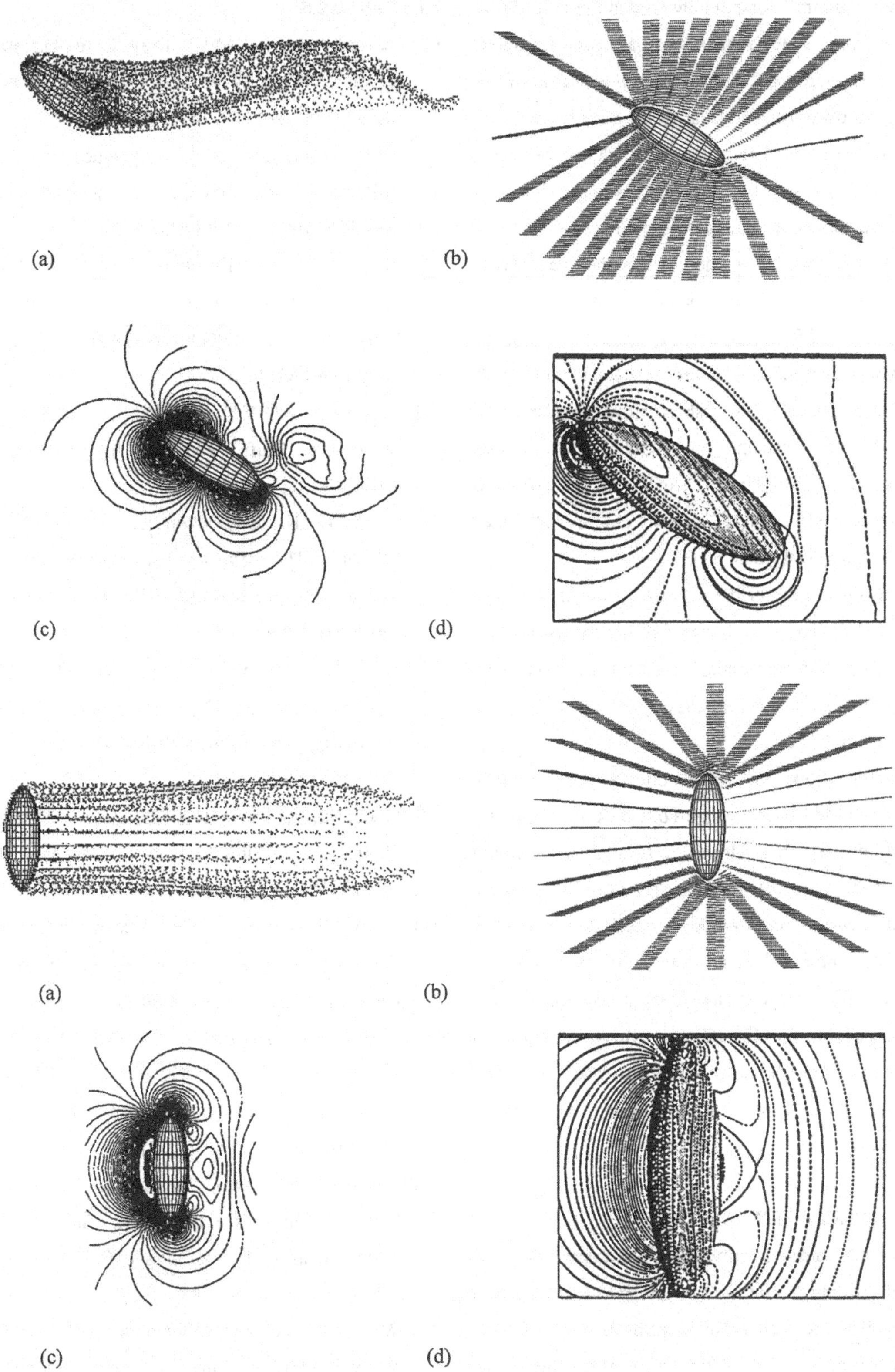

Figure 4: At 30° (up) and 90° (down) angles of attack, at T=4, wake behind ellipsoid (a); velocity distribution (b); isobars Re=1000 (c), Re=300 (d) (Dallmann et al., 1995)

To visualise the potential flow around the ellipsoid at angle of attack of 30° a number of particles with initial locations on a circle 0.05L away from the surface of the body in the plane perpendicular to the major axis at its center are traced in virtual time with induced velocities due to Biot-Savart law. Figure 1 shows side and perspective views of a number of streamlines.

Isobars obtained by solution of the pressure Poisson equation for potential flows around ellipsoid at angles of attack of 30° and 90° are shown in Figure 2. Calculation of pressure distributions requires a computation mesh around the ellipsoid. The mesh employed in this study has 60 constant spacings of 0.025L in local normal direction; it is divided into 12 in the direction of the major axis and into 24 around circles in planes perpendicular to that axis.

Evolution in time of the separated flow at Re=1000 for angles of attack of 30° and 90° is demonstrated in Figure 3. At 30° wake structure indicates an asymmetric development as expected, whereas at 90° no asymmetry is developed until T=5.5.

Figure 4 shows the wake structures, the velocity distributions and isobars at angles of attack of 30° and 90° for T=4 and Re=1000. Also demonstrated in this figure are the isobars calculated by Dallmann et al (9) using a finite difference code for angles of attack of 30° and 90° at Re=300. These isobars correspond to steady flows at 30° and 90° with one and two separation bubble formation behind the ellipsoid respectively. Instantaneous isobars at Re=1000 reflect changes due to vortex shedding with respect to those steady state pressure distributions.

ACKNOWLEDGMENTS

The first author acknowledges the supports of the Istanbul Technical University Research Fund through grant no 978 and the NATO Scientific Affairs Division through grant no.CRG972960. Support of the Japan Society for the Promotion of Science (JSPS L95511) which initiated this study is also acknowledged.

REFERENCES

(1) Gerrard, J.H.(1966), "The Three-Dimensional Structure of the Wake of a Circular Cylinder", J. Fluid Mech., Vol.25,p.143-164.

(2) Williamson,C.H.K.(1989), "Oblique and Parallel Modes of Vortex Shedding of a Circular Cylinder", J. Fluid Mech., Vol. 206, p.579-627.

(3) Eisenlohr,H. and Eckelmann, H.(1989), "Vortex Splitting and its Consequences in the Vortex Street Wake of Cylinders at Low Reynolds Numbers", Phys. Fluid A1: p.189-192.

(4) Bearman,P.W.(1984), "Vortex Shedding from Oscillating Bluff Bodies", Ann. Rev. Fluid Mech.,Vol.16, p.195-222.

(5) Tombazis, N. and Bearman, P.W. (1997), "The Effects of Three-dimensional imposed disturbances on Bluff Body Near Wake Flows",J. Fluid Mech., Vol. 330,p. 85-112.

(6) Nuzzi, F., Magnes,C. and Rockwell, D.(1992), "Three-Dimensional Vortex Formation from an Oscillating Non-Uniform Cylinder",J. Fluid Mech., Vol.238, p. 31-54.

(7) Nakanishi,Y.,Kamemoto,K.(1992), "Numerical Simulation of Flow around Sphere with Vortex Blobs",J. Wind Eng.&Ind. Aero.,Vol.46,p.363-369.

(8) Arkell,R.H.(1993), "Wake Dynamics of Cylinders Encountering Free Surface Gravity Waves",PhD thesis University of London,UK.

(9) Dallmann, U., Herberg, Th., Gebing, H., Su, W.H., Zhang, H.Q.(1995), "Flow Field Diagnostics: Topological Flow Changes and Spatio-Temporal Flow Structure",AIAA Paper 95-0791.

(10) Kamemoto, K. (1995), "On Attractive Features of the Vortex Methods",Computational Fluid Dynamics Review (Editors Hafez, M. and Oshima, K.), John Wiley & Sons.

(11) Winckelmans,G. and Leonard, A.(1988), "Improved Vortex Methods for Three-Dimensional Flows",Proc. Workshop on Mathematical Aspects of Vortex Dynamics, Leeburg, Virginia, p. 25-35.

Numerical Simulation of Vortex Shedding from an Oscillating Circular Cylinder using a Discrete Vortex Method

J. R. Meneghini*, F. Saltara*, and C.R. Siqueira**
*University of São Paulo, EPUSP, Department of Mechanical Engineering, CEP 05508-900, São Paulo, SP Brazil
EMail: jmeneg@ usp. br
**University of São Paulo, EPUSP, Department of Naval Engineering, CEP 05508-900, São Paulo, SP Brazil

Abstract

The interaction between cylinder oscillation and the shedding of vortices is investigated numerically in this paper. The near wake structure is presented for different values of reduced velocity of a cylinder free to oscillate transversely. One of the objectives of this paper is to compare the numerical results with experimental data obtained by Parra [11] in the water tank facility of IPT/University of São Paulo. The attraction of applying numerical methods to this problem is that the way the flow is modified can be studied in closer detail. In the computer it is possible to investigate many different flow conditions more easily.
The method used for the simulation is based on the Vortex-in-Cell formulation incorporating viscous diffusion. The Navier-Stokes equations are solved using the operator-splitting technique, where convection and diffusion of vorticity are treated separately. The convection part is modelled assuming that the vorticity field is carried on a large number of discrete vortices. Force coefficients are calculated by considering the normal gradient of vorticity at the wall to evaluate the pressure contribution and the vorticity at the wall to obtain the skin friction.

1 Introduction

Many investigations of the effect of transverse oscillations on vortex shedding can be found in the literature. It is observed that sinusoidal transverse oscillations are characterised by the capture of the vortex shedding frequency by the oscillation frequency over a range of cylinder oscillation amplitudes. This phenomenon is called *lock-in*. Meneghini and Bearman [8] investigated square, saw-tooth and parabolic wave forms of cylinder transverse oscillations, and found that only for a parabolic wave did *lock-in* occur in a similar way to that observed with sinusoidal oscillation.

With a cylinder free to oscillate, the *lock-in* phenomenon is characterised by the capture of the vortex shedding frequency by the natural frequency of the cylinder, over a range of reduced velocities. Results by Brika and Laneville [3] and Parra [11] show that in the region of *lock-in* large amplitudes of oscillation are observed for high mass parameter values, with the mass parameter defined by equation (11). According to Blevins [2] and others, large amplitude vibration increases the correlation of vortex shedding along the cylinder axis. With this consideration, two-dimensional numerical simulations should be reliable in terms of analysing flow details and wake structures in the *lock-in* regime.

In this work the vortex shedding from a cylinder free to oscillate transversely is investigated numerically. The results are compared with the experimental data obtained by Parra [11] and Khalak and Williamson [5]. The method used for the simulations is based on the Vortex-in-Cell formulation incorporating viscous diffusion. The Navier-Stokes equations are solved using the operator-splitting technique, where convection and diffusion of vorticity are treated separately. The convection part is modelled assuming that the vorticity is carried on a large number of discrete vortices. Force coefficients are calculated by considering the normal gradient of vorticity at the wall to evaluate the pressure contribution and the value of the vorticity at the wall to evaluate the skin friction. For each time step, once the force coefficients were calculated, the second order ordinary differential equation for the transverse motion of the cylinder is solved through a Runge-Kutta method. The resulting cylinder velocity is used to obtain the relative free-stream velocity for the next time step.

2 Numerical Method

The Vortex-in-Cell formulation incorporating viscous diffusion has been applied by Meneghini and Bearman to investigate the effect of large amplitude oscillations on vortex shedding from an oscillating circular cylinder [7] and to investigate the effect of displacement wave form, also on vortex shedding from a circular cylinder [8]. Arkell et al. [1] used the method to study the effects of free surface waves on the far wake behind a circular cylinder. This approach has been developed by Graham [4] and details about the method can be found in Meneghini and Bearman [6]. A thorough review of vortex methods has been published by Sarpkaya [12].

In order to study the flow about a circular cylinder a conformal transformation $(x,y) \rightarrow (\xi, \eta)$ is used. The cylinder wall is specified by a line $\eta = 0$ in the transformed plane. The two-dimensional Navier-Stokes equations in vorticity (ω) stream function (ψ) formulation in the transformed plane can be written as:

$$J\frac{\partial\omega}{\partial t} - \frac{\partial\psi}{\partial\eta}\frac{\partial\omega}{\partial\xi} + \frac{\partial\psi}{\partial\xi}\frac{\partial\omega}{\partial\eta} = \nu\left(\frac{\partial^2\omega}{\partial\xi^2} + \frac{\partial^2\omega}{\partial\eta^2}\right) \tag{1}$$

$$\frac{\partial^2\psi}{\partial\xi^2} + \frac{\partial^2\psi}{\partial\eta^2} = -J\omega \tag{2}$$

where ν is the kinematic viscosity and J is the Jacobian of the transformation. Equation (2) represents Poisson's equation for the stream function in the transform plane. Equation (1) is solved using the operator-splitting technique, where convection and diffusion of vorticity are treated separately:

$$\left[J\frac{\partial\omega}{\partial t}\right]_{convection} = -\frac{\partial\psi}{\partial\eta}\frac{\partial\omega}{\partial\xi} + \frac{\partial\psi}{\partial\xi}\frac{\partial\omega}{\partial\eta} \tag{3}$$

$$\left[J\frac{\partial\omega}{\partial t}\right]_{diffusion} = \nu\left(\frac{\partial^2\omega}{\partial\xi^2} + \frac{\partial^2\omega}{\partial\eta^2}\right) \tag{4}$$

The convection part is modelled assuming that the vorticity field ω is carried on a large number of discrete vortices. The vorticity is represented by a distribution of discrete vortices in the form:

$$\omega(\xi,\eta,t) = \sum_{k=1}^{Nv}\Gamma_k\delta(\xi - \xi_k(t))\delta(\eta - \eta_k(t)) \tag{5}$$

where Γ_k is the circulation of the kth point vortex, and δ is the Dirac delta function. Poisson's equation (2) is solved at each time step on a grid which is uniform in the ξ direction so that a Fast Fourier Transform algorithm may be used. A stretched mesh is employed in the η direction in order to resolve accurately the cylinder boundary layer. For the purpose of solving Poisson's equation, circulation of the kth discrete vortex in a mesh cell is projected to the four surrounding mesh points according to a bilinear area weighting scheme. Equation (2) results in a tridiagonal set of equations for the transform of ψ on the η = constant grid lines, after taking a fast discrete Fourier Transform in the ξ direction and using a central difference scheme. The solution of this tridiagonal set of equations gives ψ at every mesh point (i,j) . The velocity components at these points are then calculated by a finite difference scheme applied to the relation between velocity and stream function.

Boundary conditions on ψ are $\psi = 0$ at the body surface and the value of ψ is evaluated by Biot-Savart integration at the outer boundary of the computational domain. The contribution of the free stream is considered separately. The values of vorticity at the mesh points are considered for the Biot-Savart integration rather than the circulation of each discrete vortex. This is done in order to have a more efficient procedure in terms of computational time.

The diffusion part of equation (1), which is given in (4), is solved by a finite difference scheme in a semi-implicit form carried out on the same fixed expanding mesh as used for convection. The wall vorticity is calculated in order to satisfy the no-slip boundary condition. The solution of (4) gives the change in vorticity due to diffusion at every mesh point. The change in vorticity is projected back on to a point vortex in a similar manner to that used by the area weighting scheme. The convection part of the Navier-Stokes equations is satisfied by convecting the point vortices in a Lagrangian way. The velocity components of the kth discrete vortex are found by interpolation of the velocities at the four mesh points surrounding this vortex.

3 Force Evaluation

Force coefficients are calculated by suitably integrating the pressure and skin friction contributions. After considering the contributions from skin friction and pressure, the force components are resolved in the two directions (x,y) in the physical plane , yielding F_x and F_y. These forces are then non-dimensionalised as follows:

$$Cl = \frac{2F_y}{\rho U^2 D} + \frac{\pi D}{2U^2}\frac{d^2 y}{dt^2} \tag{6}$$

$$Cd = \frac{2F_x}{\rho U^2 D} \tag{7}$$

where ρ is the fluid density, U is the free stream velocity D is the circular cylinder diameter, and y is the position of the cylinder in the transverse direction . The second term on the right side of equation (6) is the correction due to the acceleration of the cylinder in the transverse direction. As our mesh is fixed to the body, this correction must be carried out to take into account the inertia effect.

4 Equations for Vortex-induced Vibration

The equation of motion for a cylinder free to oscillate in the transverse direction is:

$$m\ddot{y}+2\beta m\omega_n \dot{y}+m\omega_n^2 y=Cl(t)\rho\frac{U^2}{2}D \tag{8}$$

where m is the mass of the cylinder per unit length, ω_n is the natural frequency of the cylinder, and β is the fraction of critical viscous damping. According to Parkinson [10], if the non-dimensional transverse displacement of the cylinder given is given by $Y=y/D$ and the non-dimensional time is given by $\tau=Ut/D$, then equation (8) can be rewritten as:

$$\ddot{Y}+2\beta\dot{Y}+Y=C_l n\frac{1}{4\pi^2}V_r^2 \tag{9}$$

where:

$$\ddot{Y}=\frac{d^2Y}{d\tau^2} \quad \text{and} \quad \dot{Y}=\frac{dY}{d\tau}\;, \tag{10}$$

n, the mass parameter, is given by:

$$n=\frac{\rho D^2}{2m} \tag{11}$$

and V_r, the reduced velocity, is given by:

$$V_r=\frac{UT_n}{D} \tag{12}$$

where T_n is the natural period of the cylinder, $T_n=2\pi/\omega_n$.

Once a lift coefficient for a time step is calculated using (6), equation (9) can be solved through a fourth order Runge-Kutta method to give the velocity of the cylinder in the transverse direction. With the cylinder considered fixed on the grid, this velocity is applied to the free stream for the next time step.

5 Discussion of Results and Conclusions

In all numerical results shown in this paper, Re = 200, with the Reynolds number defined in terms of cylinder diameter (D) and free stream velocity (U), $Re = UD/\nu$. A mesh with 170 points in the radial direction and 128 points in the angular direction has been used in all simulations. There are about 30 points in the boundary layer with this mesh. A non-dimensional time step, Ut/D, equal to 0·005 has been used. Lift and drag coefficients, for the case of a fixed circular cylinder at a Reynolds number equal to 200, are shown in figure 1. The wake structure, represented by the point vortices, is shown in figure 2.

In the simulations for Re = 200 vortex shedding occurs with a Strouhal number of about 0·2. This value is very close to those observed in experiments for Re between 200 and about 2×10^5. For Re = 200, the wake is still laminar and hence no turbulence model is needed. The non-dimensional parameters n, β and V_r in the simulations have been kept equal to those measured in the experiments of Parra [11]. It should be noted that in the experiments the frequency used in V_r is the frequency measured in still water and not the true natural frequency. Hence, in the presentation of the computational results, the frequency employed in V_r is also the one that relates to small amplitude oscillations in still water.

The experimental results of Parra [11] were obtained in a water tank facility, with a circular cylinder free to oscillate transversely with β = 0·01710 and n = 0·34681 . The experiments were conducted with a circular cylinder of D = 0·11m and ω_n = 2·238 rad/s. The flow velocity U was varied in order to change the reduced velocity. The Reynolds Number Re of the experiments varied in the range 14410 to 50380. In the numerical simulations, the velocity U and diameter D of the cylinder have been fixed for all calculations, and the natural period T_n has been varied to change the reduced velocity. The non-dimensional amplitude of the cylinder transverse oscillation A/D has been plotted as a function of the reduced velocity UT_n/D. The experimental and numerical results can be seen in Figure 9.

Simulations have been carried out for values of reduced velocity, V_r, from 2·0 to 14·0. In figures 3, 4, and 5 force time histories and cylinder displacements are show for V_r equal to 5·0, 7·5 and 12·5, respectively. The highest amplitude of oscillation occurred for a reduced velocity equal to 5·0, and this is also the reduced velocity for the highest value of the mean drag coefficient. As the reduced velocity is increased above 5, the amplitude and mean drag coefficient decrease. The phase angle by which the lift coefficient leads the cylinder displacement changes dramatically as the reduced velocity varies from 5·0 to13·0. This result has been observed in experiments (as can be seen in the review by Parkinson [11]), and also in simulations where the cylinder is forced to oscillate (Meneghini and Bearman [7]). Plots of the wake structure for these cases are shown in figures 6, 7, and 8. The plots are for the moment when the cylinder is in its upper most position. The wake structure for V_r equal to 5·0 has a distinctive pattern with the vortices in the wake exhibiting a large lateral spacing.

The maximum amplitude, non-dimensionalised with the cylinder diameter ($D =$ 2·0 in our simulations), versus the reduced velocity is plotted in the graph shown in figure 9. The experimental results by Parra [11] and Khalak and Williamson [5] are also shown and compared with the present simulations. As can be noticed, the maximum amplitude from the simulations is considerably lower than those found in the experiments. The reason for this disagreement is not yet known. The explanation could be related to the difference in Reynolds number in the experiments and in our simulations. However, Brika and Laneville [3] and Khalak and Williamson [5] have shown from experiments that there may be two possible values for the maximum amplitude associated with either an upper or a lower branch to the amplitude versus reduced velocity curve. Also there is known to be a hystersis associated with moving between the two branches. The values of mass and damping in Khalak and Williamson's experiments are reasonably close to the ones used in the simulations and it is interesting to note that the maximum amplitude for their lower branch is about 0·6, which is similar to the maximum value computed here. Brika and Laneville [3] contend that the mode of shedding is different in the two branches and, following the nomenclature of Williamson and Roshko [13], find the 2S mode in the lower branch and the 2P mode in the upper branch. In the 2S mode two vortices are generated per oscillation cycle and in the 2P mode two vortex pairs are formed per cycle. It is clear from figures 6, 7 and 8 that the computed flow is in the 2S mode, which is compatible with the levels of amplitude predicted. More work is required to determine if the vortex shedding can be encouraged to change into the 2P mode and whether this results in larger amplitudes.

The results shown in this paper are part of a research project that is still being carried out. The next steps will include investigating whether it is possible to cause the vortex shedding mode to change and implementing a turbulence model in the Vortex Method, hence increasing the maximum Re that it is possible to simulate in the computer.

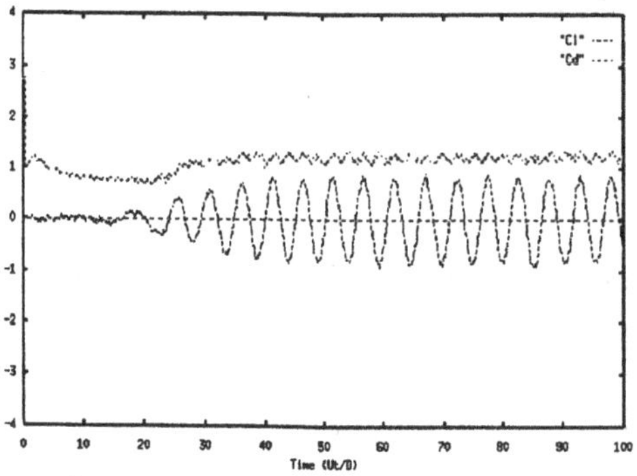

Figure 1 - Force coefficients for *Re*=200

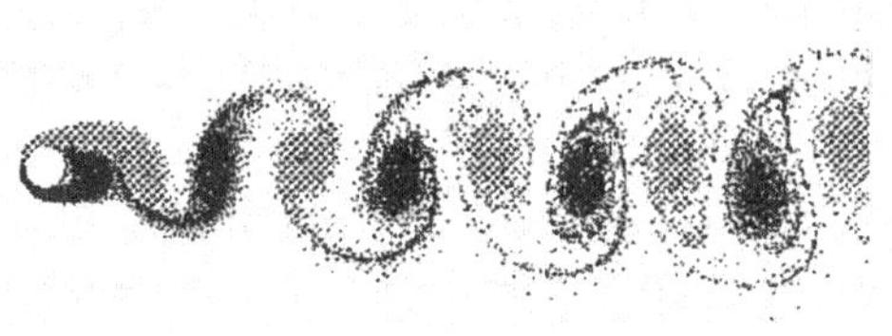

Figure 2- Wake structure for *Re*=200

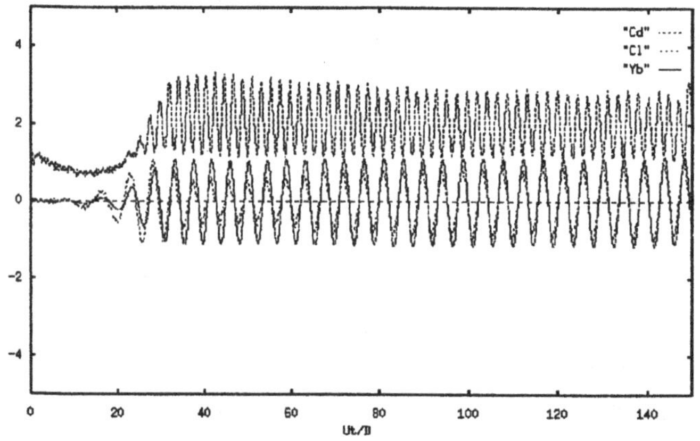

Figure 3 - Results of *Cl, Cd* and *Yb* for $UT_n/D = 5{\cdot}0$

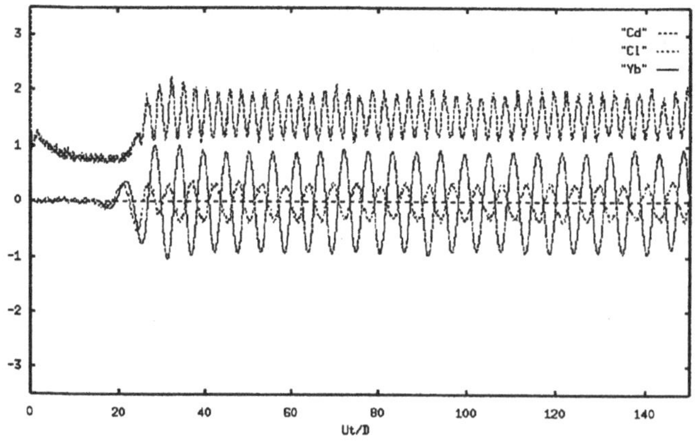

Figure 4 - Results of *Cl, Cd* and *Yb* for $UT_n/D = 7{\cdot}5$

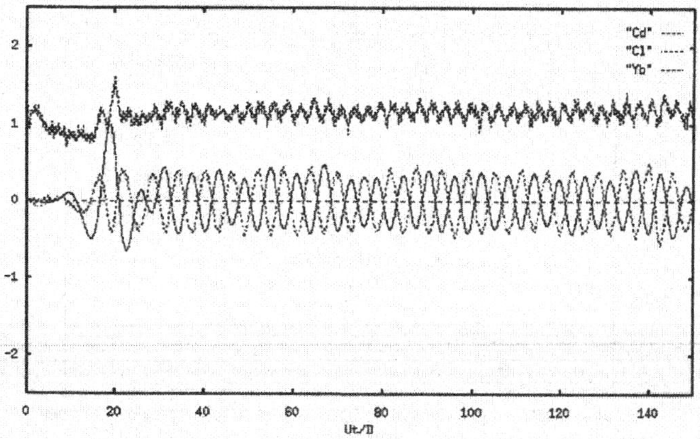

Figure 5 - Results of Cl, Cd and Yb for $UT_n/D = 12{\cdot}5$

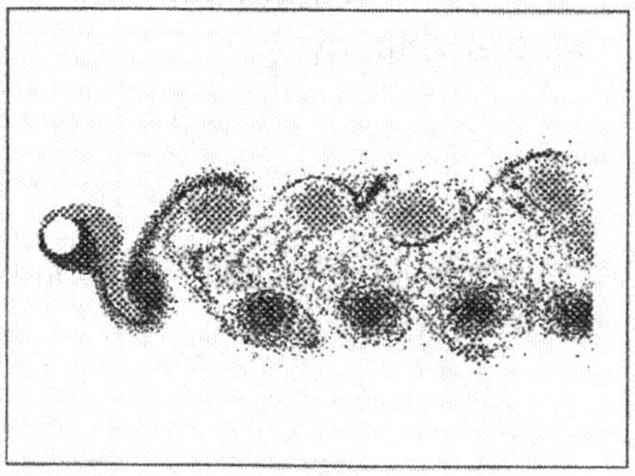

Figure 6 - Wake structure for $UT_n/D=5{\cdot}0$

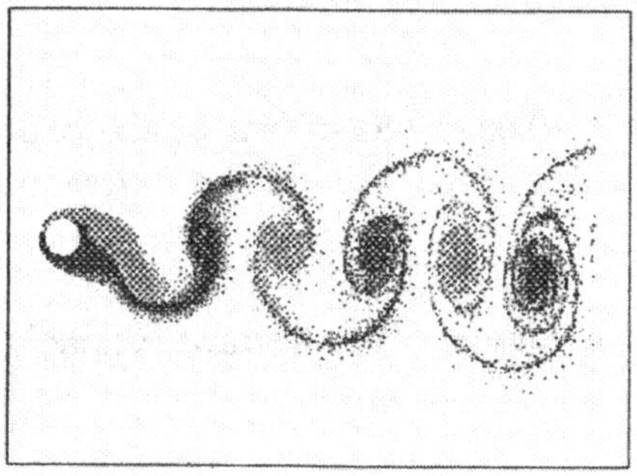

Figure 7 - Wake structure for $UT_n/D=7{\cdot}5$

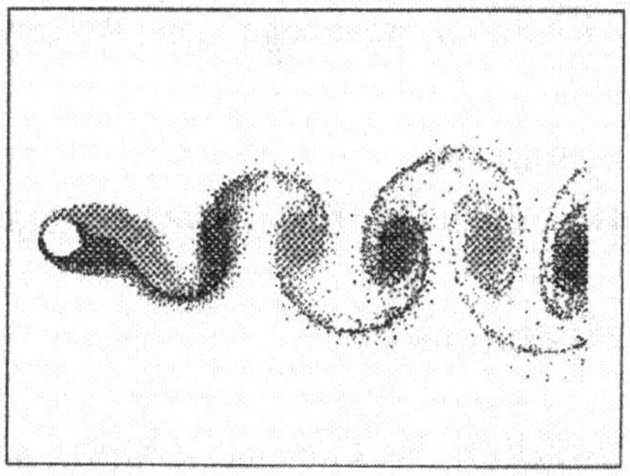

Figure 8 - Wake structure for $UT_n/D=12{\cdot}5$

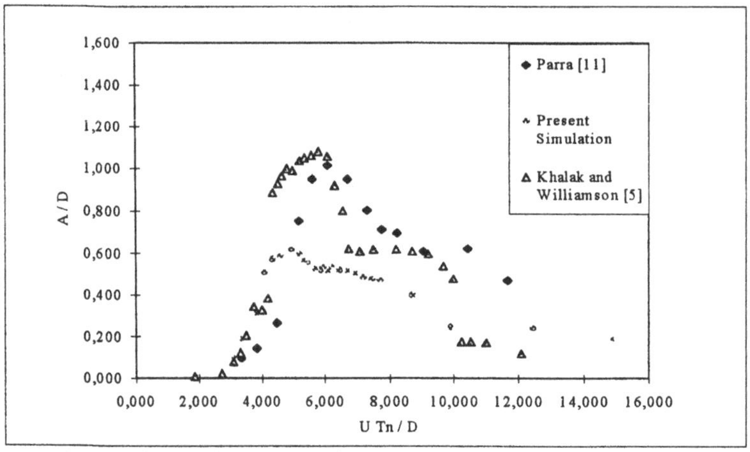

Figure 9 - Non-dimensional amplitude of the circular cylinder transverse oscillation as a function of the reduced velocity

Acknowledgements

The authors (JRM and FS) are grateful to FAPESP (grant 94/3057-0 and 94/3528-3) and CNPq for providing them a Research Grant for this project.

Keywords

vortex shedding, flow induced vibration, computational fluid dynamics

References

1. Arkell, R. H., Graham, J. M. R., and Zhou, C. Y. The effects of waves and mean flow on the hydrodynamic forces on circular cylinders, *Proc. 6th BOSS Conference,* London, 1992.

2. Blevins R. D., *Flow Induced Vibration,* Second edition, Van Nostrand Reinhold, New York, 1990.

3. Brika, D., Laneville, A., Vortex-induced vibration of a long flexible circular cylinder, *Journal of Fluid Mechanics,* Vol. 259, pp. 481-508, 1993.

4. Graham, J. M. R., Computation of viscous separated flow using a particle method, *Numerical Methods in Fluid Mechanics,* Vol.3, pp. 310-317,Ed. K.W. Morton, Oxford University Press, 1988.

5. Khalak, A., and Williamson, C. H. K., Dynamics of a hydroelastic cylinder with very low mass and damping, accepted for publication in the *Journal of Fluids and Structures.*

6. Meneghini, J. R., and Bearman, P. W., Numerical simulation of control of bluff body flow using a discrete vortex method incorporating viscous diffusion, *Proc.*

IUTAM Symposium on Bluff-Body Wakes, Dynamics and Instabilities, pp. 257-262, Ed. H. Eckelmann, Springer-Verlag, 1992.

7. Meneghini, J. R., and Bearman, P. W., Numerical simulation of high amplitude oscillatory flow about a circular cylinder, *Journal Fluids and Structures*, Vol. 9, pp. 435-455, 1995.

8. Meneghini, J. R., and Bearman, P. W., Numerical simulation of the interaction between the shedding of vortices and square-wave oscillations applied to a circular cylinder, *Proc. Second International Conference on Hydrodynamics*, Vol. 2, pp.785-790, Ed. Chwang, Lee & Leung, Balkema, Hong Kong, 1996.

9. Meneghini, J. R., Numerical Simulation of Bluff Body Flow Control Using a Discrete Vortex Method, *PhD thesis*, University of London, UK, 1993.

10. Parkinson, G., Phenomena and modelling of flow-induced vibrations of bluff bodies, *Prog. Aerospace Sci.*, Vol. 26, pp. 169-224, 1989.

11. Parra, P. H. C. C., Mod. Semi-empírico de Vib. Ind. por Vortex Shedding - Análise Teórica e Experimental, *MSc Thesis*, University of São Paulo, Brazil, 1996.

12. Sarpkaya, T., Computational methods with vortices - The 1988 Freeman Scholar Lecture, *J. Fluids Engineering*, Vol. 111, pp. 5-52, 1989.

13. Williamson, C.H.K. and Roshko, A., Vortex formation in the wake of an oscillating cylinder, *J. Fluids and Structures*, Vol. 2, pp 355-381, 1988.

NUMERICAL STUDY ON INTERACTION PROCESSES AND SEQUENCES IN BISTABLE FLOW REGIME OF TWO CIRCULAR CYLINDERS

C. W. Ng

Ove Arup & Partners Hong Kong Limited

56/F Hopewell Centre, 183 Queen's Road East, Hong Kong, China / Email:william.ng@arup.com

Norman W. M. Ko

Department of Mechanical Engineering, The University of Hong Kong

Pokfulam Road, Hong Kong, China / Email:nwmko@hkucc.hku.hk

ABSTRACT

The flow past two circular cylinders normal to stream in side-by-side arrangement was investigated numerically by a two-dimensional discrete-vortex model to examine the interaction processes and sequences in the bistable flow regime. Different types of perturbation and perturbation levels were used to perturb the flow to investigate the interaction sequences. The results obtained are discussed and analyzed.

1. INTRODUCTION

A recent investigation (1), based on two-dimensional discrete vortex method and flow visualization, has established four interaction processes and sequences of vortex cycle, vortex amalgamation, vortex pairing and vortex dipole (see Fig. 1) as process mechanism in the bistable flow regime (2,3) of two circular cylinders in side-by-side arrangement. The initiation and occurrence of the interaction processes depend on the deflections and conditions of gap shear layers. The occurrence of pairing of two vortices of the same sense of rotation and of the amalgamation of three vortices is affected by the condition of the third vortex having opposite circulation. Pairing of vortices is observed when the circulation of the third vortex is comparatively small, while amalgamation is found when the circulations of the vortices are comparable.

Perturbation study (1) has been carried out to indicate the importance on the condition of one of the gap vortices in the initiation of these interaction processes. As an extension of the earlier study, for better understanding of the processes and sequences in the bistable regime, the flow processes under other types of perturbation and perturbation levels were examined. It is hoped that better insight on the flow features and mechanisms involved could be obtained.

2. THE DISCRETE VORTEX MODEL

The mathematical formulation of the discrete-vortex model and the numerical method adopted in this study are given in the earlier work (1). Briefly, the governing velocity equation of incompressible flow was formulated from the summation of the two-dimensional irrotational potential flow around the cylinders, and the velocity induced by vorticity:

$$W = Uz + \sum_{n=0}^{\infty}\left(\frac{\alpha_n}{\zeta - f_n} + \frac{\beta_n}{\xi - g_n}\right) + \sum_{k=1}^{N} W_{\Gamma_k},$$

$$W_{\Gamma_k} = \frac{i\Gamma_k}{2\pi}\left\{-\log(z - z_k) + \sum_{n=0}^{\infty}\left(\frac{\xi - l_{2n}}{\varsigma - h_{2n+1}}\frac{\varsigma - h_{2n}}{\xi - l_{2n+1}}\right\},$$

where U is the free-stream velocity, $z = x + iy$, $\zeta = z - c_l$,

$\xi = z - c_2$, c_1 and c_2 the coordinates of the cylinders centres, $f_0 = 0$, $g_0 = 0$, $\alpha_0 = Ua^2$, $\beta_0 = Ub^2$, a and b the radii of the cylinders (Fig. 2). The terms α_n, β_n, f_n, g_n, l_n and h_n are generated by the method of successive images and are given in Ng *et al.* (1).

The generation of vorticity near solid boundary was modelled in the form of discrete vortices (1,4) and they were subsequently discharged into downstream wake region. The generation and evolution of discrete vortices were then calculated to obtain a time-dependent physical picture of flow in vorticity space for analysis.

A circulation reduction scheme was introduced to allow for the turbulent decay characteristics of the primary or Strouhal vortices in real flow (1,5). The decay parameter was based on the experimental decay rate of Re = 1.4×10^5, as determined by Cantwell and Coles in a fluid dynamic study (6). A first-order scheme was employed in the calculations and details are given in References 5 and 7.

3. RESULTS AND DISCUSSION

With the intention of generating asymmetric flow patterns and to establish the four interaction processes under other types of perturbations, series of perturbations were attempted. The first one was to downstream shift a streamwise distance ΔX_{C1} of the first (upper) cylinder. The second one was to transverse shift ΔY_{C2} of the second (lower) cylinder. The third one was to downstream shift $\Delta X_{V2,V4}$ of the upper inner and lower outer vortices. In the third perturbation type, higher perturbation levels were attempted to better understand the effect. The sequences of the four interaction processes, thus obtained, are tabulated in Table 1.

The flow features obtained under these perturbations are in analogy to the perturbation results of last study (1). The gap flow flip-flops and switches at irregular intervals. Within the flip-flopping phase, the four processes of cycle, amalgamation, pairing and dipole are identified (Fig. 1). Their distributions of vortices are essentially the same as those of last study (1).

With the introduced asymmetry by shifting the first (upper) cylinder, with increasing perturbation level ΔX_{C1} there are decreasing numbers of cycle of quasi-antiphase vortices from five cycles to one, before the initiation and start of other interaction processes (Table 1a). There are five cycles of quasi-antiphase vortices in the first perturbation, while only one distorted cycle is found in the third perturbation. The number of cycle in the second and third perturbations before the initiation of other interactions is the same. However, the first distorted cycle in the third perturbation is immediately followed by amalgamation/pairing, rather than amalgamation in the second perturbation (Table 1a). This suggests that provided the perturbation level in this type is high enough, amalgamation/pairing is initiated and formed directly after the first distorted cycle, bypassing the amalgamation process.

With the introduced asymmetry by shifting the upper inner and lower outer vortices, there are five cycles of quasi-antiphase vortices before the occurrence of first amalgamation in the lowest perturbation level (Table 1c, $\Delta X_{V2,V4} = +0.05$), while no cycle is found in the higher perturbation levels. It is also observed that earlier occurrence of the first vortex dipole is found for the highest perturbation level in compared with the intermediate perturbation level. These results further indicate the importance of phase difference of vortices in triggering other interaction processes.

The occurrence of amalgamation or amalgamation/pairing or pairing-dipole depends on the presence of fluids of opposite vorticity of the third vortex. The quantity or strength of its vorticity plays an important role in the type of interaction which may occur. Table 2 summarizes the mean circulation ratios of the individual vortices of different interaction processes under different perturbation types. In pairing process the circulation of the third vortex of opposite vorticity is significantly lower than those of the pairing vortices, giving a higher circulation ratio. The very high ratio 7.23 shown in Table 2 (ΔX_{C1} = + 0.1) is due to the exceptionally low circulation of the third vortex. For amalgamation/pairing the mean circulation ratio of approximately 1.55 is in between the amalgamation and pairing processes. For amalgamation and dipole processes, the mean circulation ratios are 1.20 and 1.28 respectively, i.e. they are nearly the same.

Other computation at wider cylinder spacing T/d = 2, under different perturbations, were conducted. Results obtained also yield the occurrences of cycles of quasi-antiphase vortices, amalgamation and pairing-dipole. The interaction sequences for two perturbations are tabulated in Table 3.

In this cylinder configuration of T/d = 2, the mean circulation ratio of the inner vortex pair to the outer vortex pair of the cycle process is 1.27. The mean circulation ratio of the paired vortices with the third vortex of opposite vorticity is 2.27. The circulation ratio of amalgamating vortices is 1.17. Thus, no significant difference of the mean circulation ratios was observed, in compared with the last cylinder configuration of T/d = 1.75.

4. CONCLUSIONS

The interaction processes and sequences in the bistable flow regime of two circular cylinders in side-by-side arrangement under various perturbation types and levels were examined by a two-dimensional discrete-vortex model. Interaction processes and sequence of cycle of quasi-antiphase vortices, amalgamation/pairing, pairing and vortex dipole are identified within the near wakes of the cylinders. The mean circulation ratios of vortices for the four interaction processes are calculated. In pairing process, the circulation of the paired vortices is significantly higher than the third vortex of opposite vorticity. This is different from other interaction processes. Results obtained also indicate that the sequence of interaction processes can be altered by the perturbation levels and the types of perturbation applied.

REFERENCES

(1) Ng, C.W., Cheng, V.S.Y. and Ko, N.W.M. (1997) "Numerical study of vortex interactions behind two circular cylinders in bistable flow regime", Fluid Dynamics Research, Vol.19, p.379-409.

(2) Bearman, P.W. and Wadcock, A.J. (1973) "The interaction between a pair of circular cylinders normal to a stream", Journal of Fluid Mechanics, Vol.61, p.499-511.

(3) Kim, J.H. and Durbin, P.A. (1988) "Investigation of the flow between a pair of circular cylinders in the flopping regime", Journal of Fluid Mechanics, Vol.196, p.431-448.

(4) Stansby, P.K. (1981) "A numerical study of vortex shedding from one and two circular cylinders", Aero. Quart., Vol.32, p.48-71.

(5) Kiya, M., Sasaki, K. and Arie, M. (1982) "Discrete-vortex simulation of a turbulent separation bubbles", Journal of Fluid Mechanics, Vol.120, p.219-244.

(6) Cantwell, B. and Coles, D. (1983) "An experimental study of entrainment and transport in the turbulent near wake of a circular cylinder", Journal of Fluid Mechanics, Vol.136, p.321-374.

(7) Sarpkaya, T. and Schoaff, R.L. (1979) "Inviscid model of two-dimensional vortex shedding by a circular cylinder", AIAA Journal, Vol.17, p.1193-1200.

Table 1. Sequences of interaction processes at different perturbation methods. T/d = 1.75.

(a) Downstream shift of first cylinder at $7<\tau<17$												
ΔX_{C1}	+ 0.05	C*	C	C	C	C	A	P	D	C	P	D
	+ 0.1	C*	A	A	A/P	P	D	C	A	A/P	D	A/P
	+ 0.2	C*	A/P	A/P	P	D	C	A/P	D	A/P	A/P	D
(b) Transverse shift of second cylinder at $7<\tau<17$												
ΔY_{C2}	+ 0.05	C*	C	A	P	D	A	A/P	A/P	D	A/P	D
	– 0.05	C*	C	C	C	A	A	P	D	C	C	C
(c) Downstream streamwise shift of upper inner and lower outer vortices at $\tau=12$												
$\Delta X_{V2,V4}$	+ 0.05	C*	C	C	C	C	A	P	D	C	A	P
	+ 0.5	A*	A	A/P	D	A/P	P	D	A/P	D	A/P	P
	+ 1.0	A*	P	D	A/P	A/P	A/P	D	A/P	P	D	C

Note:
* Perturbation
C: cycle of quasi-antiphase vortices; A: amalgamation; A/P: amalgamation/pairing; P: pairing; D: vortex dipole.

Table 2. Summary of circulation ratios of vortices.

		Interaction Process				
		Cycle(1)	Amalgamation(2)	Amalgamation/ Pairing(2, 3)	Pairing(3)	Dipole(4)
ΔX_{C1}	+ 0.05	1.30	1.14	1.41	2.18	1.23
	+ 0.1	1.26	1.19	1.64	7.23	1.33
	+ 0.2	1.27	-	1.52	2.61	1.33
ΔY_{C2}	+ 0.05	1.27	1.22	1.69	2.38	1.26
	– 0.05	1.36	1.26	1.54	2.34	1.28
$\Delta X_{V2,V4}$	+ 0.05	1.29	1.16	1.59	3.26	1.22
	+ 0.5	1.31	1.20	1.48	2.31	1.30
	+ 1.0	1.24	1.25	1.50	3.08	1.25
Mean Ratio		**1.29**	**1.20**	**1.55**	**3.17**	**1.28**

Note:
(1) Mean circulation of gap vortices divided by outer vortices.
(2) Mean circulation of amalgamated vortices of same sense of rotation divided by third vortex of opposite vorticity.
(3) Mean circulation of paired vortices divided by third vortex of opposite vorticity.
(4) Mean circulation of vortices within the vortex dipole.

Table 3. Sequences of interaction processes at different perturbation methods. T/d = 2.

(a) Downstream shift of first cylinder at $7<\tau<17$												
ΔX_{C1}	+ 0.2	C*	C	C	A	P	D	A/P	D	A/P	D	A/P
(b) Transverse shift of second cylinder at $7<\tau<17$												
ΔY_{C2}	+ 0.2	C*	A	A/P	A/P	D	A	P	D	C	C	C

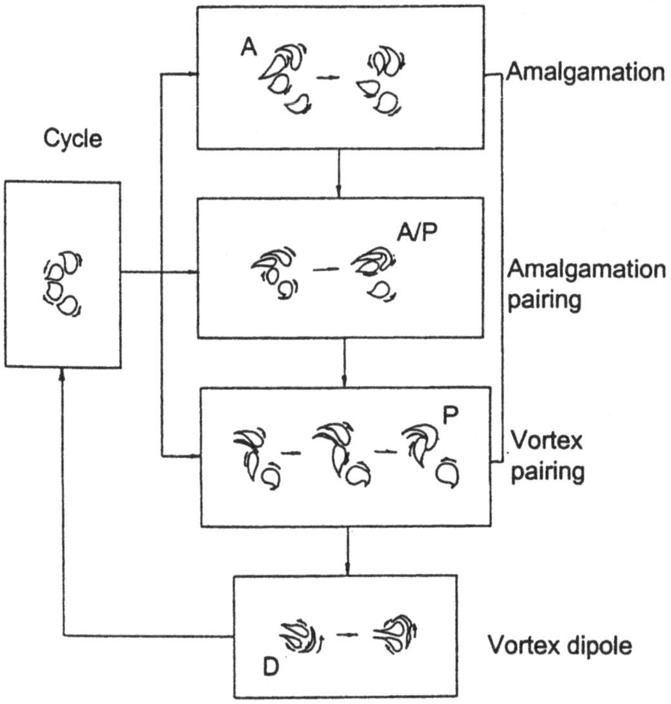

Fig. 1 Schematic diagram of interaction processes and sequences (1).

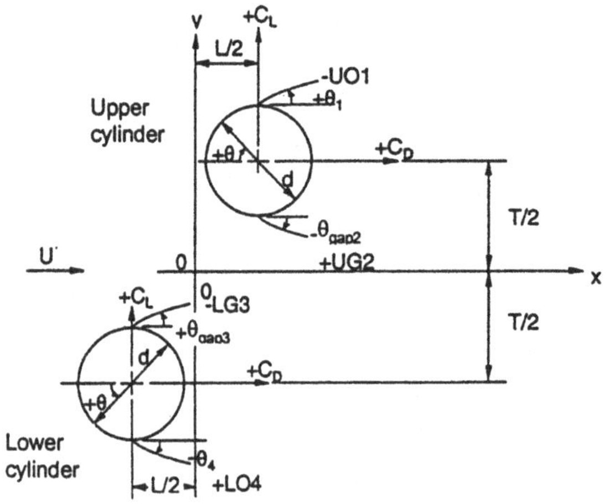

Fig. 2 Flow configuration.

VORTEX METHOD ANALYSIS OF TURBULENT FLOWS

Peter S. Bernard
Department of Mechanical Engineering, University of Maryland
College Park, Maryland, 20742 / Email: bernard@eng.umd.edu

Athanassios A. Dimas and Isaac Lottati
Krispin Technologies, Inc.
1370 Piccard Dr., Rockville, Maryland 20850 / Email: adimas@krispintech.com

ABSTRACT

This paper describes our recent efforts in developing a vortex method suitable for turbulent flow prediction. In this approach, vortex tubes are used as the principal computational element. They are supplemented adjacent to solid boundaries by an unstructured array of triangular vortex sheet prisms in several layers. The tubes are meant to model actual flow features; the sheets provide enhanced efficiency in resolving the turbulent viscous sublayer. A fully adaptive, parallel implementation of the Fast Multipole Method is used to determine velocities. Following a discussion of some of the principal features of the method, we present preliminary results of computations of several complex engineering flows.

1 INTRODUCTION

A primary motivation for the introduction of vortex methods almost three decades ago [1,2] was the belief that they should have unique advantages for the simulation and modeling of turbulent flows at high Reynolds number. In particular, unlike methods which use either a diffusive subgrid model or numerical diffusion to model turbulence as an equivalent, low and variable Ryenolds number flow, they can potentially represent the full range of high Reynolds number physics.

Subsequent development of the methodology revealed that successful application of vortex methods to complex, high Reynolds number, three-dimensional engineering flows depends critically on their resolution and speed. A "reasonable" accounting of small scales is necessary to fairly model turbulence, and an efficient solution to the difficult N-body problem embedded in the Biot-Savart calculation of velocities must be found.

Only relatively recently has the development of adaptive, parallel versions of the fast multipole method (FMM) [3,4,5] opened up the door to the kind of resolution needed to treat engineering flows. Less certain has been how to orchestrate a 3D turbulent flow calculation using vortex elements which both adheres to the physics of turbulence yet limits the numbers of vortex elements to the point where with the help of the parallel FMM, practical turbulent flow predictions can be made.

This paper will describe a vortex method which has been specifically designed to achieve these ends. It is meant for turbulent flow predictions and we hope it may be regarded as a useful alternative to traditional Reynolds-Averaged Navier-Stokes (RANS) and grid-based LES models. In this approach vortex tubes are used to represent the vorticity field everywhere except close to the boundary, where a fixed, thin, unstructured triangular mesh of finite thickness vortex sheets is used. The sheets are stacked normal to the surface several layers deep, with a half-thickness sheet adjacent to the boundary, and are meant to cover the viscous sublayer of turbulent wall flows.

After a description of the methodology we present examples from a number of recent calculations showing the effectiveness of the approach. These include calculations of an engine intake flow, the flow past a prolate spheroid and a minivan. The latter two calculations have been made using the VORCAT^{TM} code, which has been developed through U.S. Department of Energy sponsorship. This represents a commercial implementation of the scheme designed for general engineering applications (see www.krispintech.com).

2 THE VORTEX METHOD

2.1 Vortex tube elements

To reasonably well represent the dynamics of turbulent flow, a vortex method needs to resolve the energy containing scales of turbulence down to, perhaps, the inertial range, as well as the dynamical processes leading to the generation of new coherent vortices near boundaries. In both these instances it can be expected that tube-like vortical elements are a realistic model for the actual structure found in flows. For this reason there is likely to be advantage in utilizing vortex filaments as the primary computational element in the numerical scheme. The vortex tubes are composed of a number of straight-line segments, or "vortons." Note that we impart a quasi-physical significance to our vortex tubes, i.e. we do not view them as purely computational elements. From this perspective, the small scale structures of turbulent flow do not generally need multiple vortex elements for their representation. This is one means for making more effective use of vortices in the calculation.

As described at length by Chorin [6], a field of vortex tubes tend to stretch and fold while bringing energy to dissipation scales. Such natural adaptivity is a real physical effect in turbulence, but it is one which must be controlled lest vast numbers of vortex elements appear to describe the dynamics of the dissipation process in excruciating detail. An effective solution to this problem was pioneered by Chorin [7]: remove small scale hairpin shapes that form in the course of the folding process, as shown in Fig. 1. The rationale for hairpin removal is that the velocity field associated with hairpins is mostly local. Moreover, further calculation of the hairpin will only track the movement of the localized energy to dissipation scales. Thus, rather than invest in computing the small scale folding process, a simple estimate of the energy loss can be achieved by merely cutting away the hairpin and reattaching the ends of the tube. The smallest resolvable scale is determined by the minimum length of the vortex segments composing a filament. This can be adjusted depending on many criteria ranging from physical to computational considerations. Clearly, the smaller the allowed resolution, the more accurate the model.

The dynamics of the vortex tubes are governed by the 3D vorticity equation

$$\underbrace{\frac{\partial \mathbf{\Omega}}{\partial t} + (\nabla \mathbf{\Omega})\mathbf{u} = (\nabla \mathbf{u})\mathbf{\Omega}}_{convect\ filament\ endpoints} + \underbrace{\frac{1}{R_e}\nabla^2 \mathbf{\Omega}}_{decay\ model} , \quad (1)$$

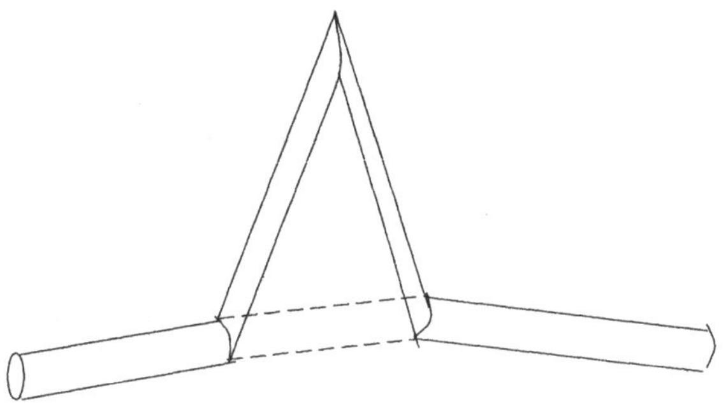

Figure 1. Hairpin removal via Chorin's algorithm.

where $\mathbf{u}$ is the velocity, $\mathbf{\Omega}$ is the vorticity and R_e is the Reynolds number. This is solved as indicated: convection, vortex stretching and reorientation are accommodated by moving the end points of the vortons. As they lengthen, they are divided into smaller segments. A decay model, described in the next section, is used to account for long time diffusive effects on the vortex tubes. In particular, we may assume that viscous dissipation plays some role in the demise of coherent vortices in a turbulent flow. This is especially a concern in internal flows, such as periodic channel flow, where vortices, once present, will persist indefinitely unless steps are taken to model their aging process. In contrast, for external flows, viscous decay of vortices can be assumed to take place downstream of the region of interest and hence is of less of a concern. Apart from the decay model, Eq. (1) is a standard approach for filament calculations [7, 8].

2.2 Vortex tube creation and destruction

Two different models of the vortex destruction process have been developed in order to help restrict vortices to those which are most dynamically important. The intent is also to mimic the finite lifespan of coherent vortices in real turbulent flow. Future research will look into optimal ways for accommodating the destruction process, though it should be noted – as suggested above – that this is not a great concern for external flows.

Preliminary calculations of channel flow [9] have shown that long lived vortices often have many segments, and tend to be without favored orientation. In this state it is safe to imagine that a vortex is "dissipated" and may be removed from the calculation without undue harm. In practical terms this proves to be an effective means for preventing runaway growth in the number of vortices.

An alternative approach is to consider the accumulated effect of vortex stretching versus viscous spreading on each individual tube. The well known

Gaussian core solution [2] establishes an equilibrium radius for tubes, r_e, in which vortex stretching is in equilibrium with diffusion. For a tube whose length increases from Δs to $\Delta s'$ in time Δt it may be shown that

$$r_e = 2\sqrt{\frac{\Delta t}{R_e(\Delta s'/\Delta s - 1)}}. \qquad (2)$$

Generally, the radius of a vortex is either above or below the value of r_e associated with local stretching. If $r > r_e$, the vortex tends to get thinner, if $r < r_e$, it tends to widen. While it may be possible to model the process by which vortices change core size [10], such an approach is unsuited to the present purposes, since it fails to provide for retiring of old structures. The latter objective can be fulfilled, however, by providing for loss of circulation from the tubes whenever $r < r_e$. The amount is determined by the extent to which vorticity would diffuse beyond r due to the imbalance of vortex stretching and diffusion. The necessary relation is

$$\Gamma^{n+1} = \Gamma^n \left(1 - \frac{4\Delta t}{R_e}\left(\frac{1}{r^2} - \frac{1}{r_e^2}\right)\right), \qquad (3)$$

where Γ^n is the circulation at time n. If $r > r_e$, the circulation may be left the same, i.e.

$$\Gamma^{n+1} = \Gamma^n, \qquad (4)$$

and if the tube faces a net contraction, i.e. $\Delta s' < \Delta s$, then the maximum circulation loss can be assessed, namely

$$\Gamma^{n+1} = \Gamma^n \left(1 - \frac{4\Delta t}{R_e}\frac{1}{r^2}\right). \qquad (5)$$

For tubes composed of multiple segments, the average of Δs and $\Delta s'$ over all segments can be used.

As the scheme is presently constituted, the same radius is given to all vortices. Since high shear near the walls leads to high stretching rates and hence small r_e, the influence of (3) or (5) in reducing Γ should be most pronounced away from the wall where vortices are less organized and experience slower rates of stretching. Whenever the circulation of a vortex drops below a threshold, it is dropped from the calculation, thus providing a second means for eliminating vortices.

Providing for each new generation of vortices as they are produced in a turbulent flow is also a critical aspect of the numerical scheme. The algorithm must at the same time be both sensitive to the physical process by which new structures appear, yet not so unconstrained as to allow for the formation of impossibly large numbers of new vortices. One means for accommodating these conditions has been previously developed [9, 11] and has been adapted for use in VORCAT. In this, vorticity arriving at the outer sheet layer is formed into new tubes whenever it surpasses a threshold. This will happen in regions of significant turbulence ejection and where the flow is separating. To further restrict the number of vortons, new ones are allowed to acquire vorticity over several consecutive time steps until they have convected away from their initial location.

2.3 Vortex sheet elements

Once vorticity is generated on solid surfaces, its evolution within the vortex sheet mesh is followed by solving the vorticity equation in conservative form:

$$\frac{\partial \mathbf{\Omega}}{\partial t} = \nabla \times \mathbf{F} + \frac{1}{R_e}\nabla^2 \mathbf{\Omega}, \qquad (6)$$

where $\mathbf{F} = \mathbf{u} \times \mathbf{\Omega}$. An explicit forward scheme is used for the temporal discretization and an unstructured, finite-volume, upwind scheme for the spatial discretization on the right-hand side of (6).

The derivatives of $\mathbf{F}$ at grid points are computed using the divergence theorem,

$$\frac{\partial F_i}{\partial x_j} = \frac{1}{V}\int_S F_i n_j dS, \qquad (7)$$

where V is the volume of the surrounding prisms and S is their bounding surface. An upwind scheme is used for the flux computation on the surface. The surface of each prismatic volume consists of several triangular or quadrilateral areas. Each of these is bounded by two triangular prisms: the "R" prism is the one towards which the surface unit normal vector, $\mathbf{n}$, is directed, while the other is denoted as the "L" prism. For every surface, the value of F_i is chosen based on the value of the normal component of the velocity field, i.e., the projection along vector $\mathbf{n}$. To this purpose, u_R is defined as the normal component of the velocity at the centroid of the "R" prism, while u_L is defined at the centroid of the "L" prism. Then, the appropriate F_i value is chosen based on the following criterion:

$$F_i = \begin{cases} F_i^L \text{ if } u_L \geq 0 \text{ and } u_L + u_R \geq 0 \\ F_i^C \text{ if } u_L < 0 \text{ and } u_R > 0 \\ F_i^R \text{ otherwise} \end{cases} \qquad (8)$$

where F_i^R is the value of F_i at the centroid of prism "R", F_i^C is the value of F_i at the surface itself, and F_i^L is the value of F_i at the centroid of prism "L".

A similar approach may be used for the diffusion term, although for high Reynolds number flows only the wall normal diffusion term is retained and is computed by a finite-difference scheme.

3 VELOCITY FIELD

In a vortex method the velocity field is recovered from the vorticity field via

$$\mathbf{u}(\mathbf{x},t) = \int_{\Re^3} K(\mathbf{x}-\mathbf{x}')\mathbf{\Omega}(\mathbf{x}',t)d\mathbf{x}' + \mathbf{u}_p(\mathbf{x},t), \quad (9)$$

where

$$K(x,y,z) = -\frac{1}{4\pi \mid \mathbf{x} \mid^3} \begin{pmatrix} 0 & -z & y \\ z & 0 & -x \\ -y & x & 0 \end{pmatrix}, \quad (10)$$

is the Biot-Savart kernel and $\mathbf{u}_p$ is a potential flow added to enforce the non-penetration boundary condition at solid surfaces. Following standard practice [7,8], that part of the integral in (9) due to the ith vorton is approximated via

$$\int_{\Re^3} K(\mathbf{x}-\mathbf{x}')\mathbf{\Omega}(\mathbf{x}',t)d\mathbf{x}' \approx -\frac{\Gamma_i}{4\pi} \frac{\mathbf{r}_i \times \mathbf{s}_i}{\mid \mathbf{r}_i \mid^3} \phi(r/\sigma) \quad (11)$$

where

$$\phi(r) = \left(1 - \left(1 - \frac{3}{2}r^3\right)\right) e^{-r^3}$$

is a high-order smoothing function, Γ_i is the circulation, $\mathbf{r}_i = \mathbf{x} - \mathbf{x}_i$, $\mathbf{s}_i$ is an axial vector along the length of the tube and σ is a smoothing parameter.

To evaluate the contribution to the velocity from the triangular-prism sheets, we take advantage of their being thin to convert the volume integral in (9) to an area integral times the sheet thickness. For enhanced accuracy the vorticity field is taken to be piecewise linear over the triangles. In this case, analytic evaluation of the integral in Eq. (9) is done where possible, and where not, it is evaluated numerically using Simpson's rule. Note that after integration $K(\mathbf{x} - \mathbf{x}_i)$ is for the most part desingularized, though in some cases, particularly near the edge of triangles, a finite velocity is only the end result of a cancelation of singular terms. These special cases are handled in the code by forcing analytic cancelation.

The potential flow necessary to ensure non-penetration is derived from a collection of source panels covering the same unstructured triangular surface mesh as used in computing the sheet vorticity field. A piecewise linear distribution of the source strength field, q, is assumed. q is determined from numerical solution of the defining equation

$$\frac{1}{4\pi} \int_S \frac{q(\mathbf{n} \cdot \mathbf{r})}{r^3} dS' = -\mathbf{n} \cdot \mathbf{u_v} \quad (12)$$

where $\mathbf{r} = \mathbf{x} - \mathbf{x}'$. The left-hand side is the surface normal velocity induced on the surface point $\mathbf{x}$ by q and the right-hand side is the opposite of the surface normal velocity induced by the vortex elements. By enforcing Eq. (12) at the node points of the triangularized surface, a linear system of equations for nodal point source strengths results, namely,

$$A_{ij} q_j = -\mathbf{n} \cdot \mathbf{u}_{\mathbf{v}_i} \quad (13)$$

where each element of A_{ij} is the normal velocity induced at surface node point i by a piecewise linear source distribution with unit strength at node point j and zero elsewhere. The matrix A_{ij} is computed in the precomputation stage of the code and need not be repeated. At each time step the right-hand side of Eq. (13) changes and a new set of values for q_i must be determined.

The evaluation of $\mathbf{u}_p$ from the source panels involves integrals identical to those appearing in Eq. (9), so the same integral evaluations can be used for both purposes.

For those vortons lying within the sheet region adjacent to the boundary, the velocity is found using three-dimensional linear interpolation over the nodal points of the triangular prisms. This is found to enhance the accuracy of vorton motions and most importantly prevent the passage of vortons into solid bodies.

Vorticity in the half-thickness vortex sheets touching solid surfaces is determined by enforcing the non-slip boundary condition via finite difference approximation to their definition in terms of velocities. The vorticity in the wall half-sheets does not contribute to the velocity elsewhere in the flow since they are imagined to be matched with half-thickness image vortex sheets which exactly cancel their induced wall-normal velocity at the boundary surface. In other words, a wall sheet and its image induce a velocity tangential to the wall in the region between them: one which balances the velocity due to all other sheets and vortons.

Outside the region between the sheets, their contribution to velocities exactly cancel [12].

4 FAST MULTIPOLE METHOD

A fully adaptive, parallel implementation of the FMM has been developed and is now part of the VORCAT code. In the adaptive FMM, the domain is partitioned into nested cubic boxes such that no box has more than a predetermined number of vortons. Regions with a high concentration of vortons are covered by many small boxes, whereas, regions with few vortons have a small number of large boxes. The partitioning results in an oct-tree structure with the computational domain at the root of the tree, and the shape of the tree determined by the distribution of the vortons. The tree structure provides the hierarchy necessary to achieve an operation count of $O(N_v \log N_v)$ instead of the nominal $O(N_v^2)$ for N_v vortons.

Efficiency in the FMM is gained by combining the vector potential induced by vortons in the root boxes into truncated expansions in spherical harmonics about the centers of the boxes. These expansions are combined into expansions over parent boxes at higher levels in the tree. Expansions at the highest level are then shifted down a separate tree of field points formed from boxes containing the points where the velocity needs to be evaluated. The cumulative effect of many vortons is thus brought down to the evaluation of a single expansion at the root level of the field tree. Precautions have to be taken to make sure that vortons in adjacent boxes contribute to each others' velocity outside of the multipole expansions, in order to preserve the accuracy of the computation. This adaptive fast multipole method is based on the algorithm of Strickland and Baty [4].

Parallelization of the adaptive FMM was helped by assuming that computational costs far exceeds memory requirements even for $O(10^7)$ vortons. Consequently, each processor was given its own copy of the tree, and even though each processor does not need the source expansion coefficients for every node in the tree, a total exchange of coefficients in the source expansion was done anyway. These assumptions make it significantly easier to parallelize the adaptive FMM algorithm.

The parallel algorithm follows the Bulk Synchronous Parallel (BSP) model for scalable computing [13]. BSP is a parallel programming model for designing efficient, portable, and predictable scalable code. A BSP computation consists of a sequence of parallel supersteps. Each superstep consists of computations on values held locally plus any necessary communications followed by a barrier synchronization. After the barrier, any remote memory accesses are guaranteed to have taken effect.

The adaptive FMM computation has 2 supersteps: one for the source expansions, and one for the field expansions and velocities. The tree is partitioned in such a way that the resulting sub-trees have, as close as possible, the same number of points. These sub-trees are used to distribute the work in the 2 supersteps. Details of the algorithm including timings on an Origin 2000 have been reported by Collins, et al. [5].

5 RESULTS

The capability of VORCAT in predicting turbulent flow is currently being assessed. The validation process takes many directions including applying VORCAT to the prediction of a variety of well documented turbulent flows. Included in this is an exploration of its sensitivity to parameters such as σ, maximum vorton length, hairpin removal angle, vorton creation threshold, sheet density and thickness, number of sheet levels as well as parameters associated with the decay model, the FMM and other aspects of the velocity computation. We concentrate here on presenting preliminary results from three recent applications of VORCAT. The first is a calculation showing the interaction between flow entering an idling aircraft engine and the ground plane. This was done with a prototype of VORCAT. The second two applications study the flow past a 6:1 prolate spheroid at angle of attack and a minivan. Each of these were computed using the most recent, fully adaptive, VORCAT code.

5.1 Engine intake flow

The airflow between the ground and the intake of a stationary or taxiing aircraft jet engine at high thrust levels, may lead to strong vorticity generation [14]. In this, small vortices generated on the ground are drawn into the engine inlet. Under conditions which are not well understood, the stream of rotational fluid strengthens to form discernible vortical structures extending from the ground up to the engine. Such vortices are a potential hazard since they can create suction forces strong enough to disrupt the flow ahead of the compressor blades possibly causing compressor stall, or even pick up foreign objects from the runway and inject them into the engine.

Current RANS solvers cannot correctly predict vorticity generation by the separated ground flow induced by suction of the aircraft engine. To investigate the capabilities of VORCAT in this complex flow, we considered an idealized problem wherein the jet engine is modeled as a cylinder with the same length/diameter and height/diameter ratios (3 and 1.5, respectively) as an actual engine for which this phenomenon is known to occur. Pictures of the vorton distribution for the case with a cross-wind at 5% of the engine intake velocity and $R_e = 5,000,000$ are shown in the sequence of front views in Figs. 2 - 7 in intervals of 22 time units since impulsive startup of the engine. It is seen that vortices generated on the ground are drawn into the engine intake. With time, these strengthen and appear to combine into several substantial vortices. A sideview at the last time step is given in Fig. 8 and a top view in Fig. 9. It is interesting to note the tendency for larger scale vortical structures to form throughout the turbulent boundary layer on the ground plane. We also see that the effect of the crosswind is to blow the intake vortices to the side; in other calculations without crosswind, illustrated through the streamline plot in Fig. 10, the intake flow is centrally aligned. This flow differs markedly from the equivalent potential flow. Future calculations will take into account swirl in the engine intake plane.

5.2 Prolate spheroid

The flow past a 6:1 prolate spheroid under a variety of conditions has been the object of considerable attention in physical experiments [15,16] and thus provides a convenient venue with which to examine the effectiveness of our vortex method in predicting high Reynolds number flows. In the case studied here, the Reynolds number based on free stream velocity and axial length is $R_e = 4,200,000$. The spheroid has 4752 triangles covering its surface and there are 6 sheet levels, each of thickness 0.0017. Figs. 11 - 13 show the vorton distribution from the side, front and back, respectively, at approximately $t = 1.5$ (1500 time steps) after impulsive startup. There are approximately 200,000 vortons in the image, and it should be noted there is also considerable vorticity (not depicted) in the vortex sheets adjacent to the surface.

The figure shows massive separation off the surface in the form of a rolled up vortex. Pressure contours, computed using the integral formulation of Uhlman [17] and shown in Fig. 14, also reflect this behavior. From the pressure and surface vorticity, estimates of the drag and lift forces have been made. For the time step shown in the figure, the drag and lift coefficients are, respectively, $C_D = 1.06$ and $C_L = .53$, values well within the range predicted in physical experiments [15,16], which run from $0.9 - 1.2$ for C_D. The force coefficients in the axial (x) and normal (y) directions are $P_x = 0.64$, $P_y = 0.98$ for the pressure and $V_x = 0.01$ and $V_y = 0.01$ for the viscous forces. When the forces are calculated over a period of time they show a distinct periodicity suggesting that alternate shedding of the large vortex is likely to occur. Further extension of the current calculations in time hope to verify the occurrence of this behavior.

5.3 Minivan

The minivan which is being investigated is depicted in Fig. 15 in the form of the surface mesh triangularization used by VORCAT. In this study there is no ground plane. The computed vorton distribution at approximately $t = 2$ after impulsive start is shown in Fig. 16. At this time a prominent feature of the wake flow is a single large vortex. Symmetry breaking was not forced in the calculation: evidently, rotation of the opposite sign is possible depending on the particular details of the simulation. On the side shown in the figure, the flow separates off the hatch back of the van and curls up into the large wake vortex. Note the system of parallel vortices formed on the side in the direction of flow. This study is ongoing and will be extended in the future to include comprehensive force predictions, the addition of a ground plane and other changes in boundary conditions.

6 CONCLUSION

The vortex method for turbulent flow simulation described herein has made a promising start toward achieving the capability of efficiently modeling complex, high Reynolds number turbulent flows. Reasonable force and pressure predictions have been made; flow and wake structure developing on the prolate spheroid and minivan show many realistic features. Furthermore, the method has been demonstrated to have some capability for modeling the difficult engine intake flow. Further work will investigate the sensitivity of the approach to the various simplifications used in creating an efficient scheme, and attempt to demonstrate a close connection between the modeled and experimental predictions for both the flows presented here as well as a range of other important flows.

ACKNOWLEDGMENT

This work was supported by the Department of Energy under SBIR Award No. DE-FG02-97ER82413 and resulted in a software package that is pending a US patent. Computer time on the Origin2000 was provided by the CFD Group of SGI, Eagan, Minnesota.

REFERENCES

[1] Chorin, A. J. (1973), "Numerical study of slightly viscous flow," J. Fluid Mechanics, Vol. 57, p. 785 - 796.

[2] Leonard, A. (1975), "Numerical simulation of interacting, three-dimensional vortex filaments. *Lec. Notes in Phys.*, Vol. 35, p. 245 - 250.

[3] Greengard, L. and Rokhlin, V. (1987), "A fast algorithm for particle simulations," J. Comput. Phys., Vol. 73, p. 325-348.

[4] Strickland, J. H. and Baty, R. S. (1993), "A three dimensional fast solver for arbitrary vorton distributions," *Technical Report SAND93-1641*, Sandia National Laboratories.

[5] Collins, J. P., Dimas, A., and Bernard, P. S. (1999), "A parallel adaptive fast multipole method for high performance vortex method based simulations," Proc. ASME IMECE'99, Nashville, Tenn.

[6] Chorin, A. J. (1982), "The evolution of a turbulent vortex," Comm. Math. Physics, Vol. 83, p. 517 - 535.

[7] Chorin, A. J. (1993), "Hairpin removal in vortex interactions II," J. Comput. Phys., Vol. 107, p. 1-9.

[8] Puckett, E. G. (1993), "Vortex methods: an introduction and survey of selected research topics," in *Incompressible Computational Fluid Dynamics: Trends and Advances*, M. D. Gunzburger and R. A. Nicolaides, ed., Cambridge University Press, p. 335-407.

[9] Bernard, P. S. (1999), "Toward a Vortex Method Simulation of Non-Equilibrium Turbulent Flows," in *Modeling Complex Turbulent Flows*, M.D. Salas et al., Kluwer Academic Pub., p. 161 - 181.

[10] Rossi, L. (1995), "Resurrecting core spreading vortex methods: a new scheme that is both deterministic and convergent," SIAM J. Sci. Comp., Vol. 17, p. 370.

[11] Bernard, P. S., A. Dimas, and P. Collins, (1999), "Turbulent flow modeling using a fast, parallel, vortex tube and sheet method," *ESAIM Proceedings, Third International Workshop on Vortex Flow and Related Numerical Methods*, Vol. 7 Sept 1999 Editors: A. Giovannini, et al. http://www.emath.fr/Maths/Proc/Vol.7/index.html, p. 46 - 55.

[12] Bernard, P. S. (1995), "A deterministic vortex sheet method for boundary layer flow," J. Comput. Phys., Vol. 117, p. 132-145.

[13] McColl, W. F (1995), "Scalable computing," *Lecture Notes in Computer Science*, Vol. 1000, p. 8, Springer-Verlag.

[14] Campbell, J. F. and Chambers, J. R. (1994) "Patterns in the sky: natural visualization of aircraft flow fields," NASA SP-514.

[15] Chesnakas, C. J., and Simpson, R. L. (1994) "Full three-dimensional measurements of the cross-flow separation region of a 6:1 prolate spheroid," Experiments in Fluids, Vol. 17, p. 68-74.

[16] Wetzel, T. G., Simpson, R. L. and Chesnakas, C. J. (1998), "Measurement of three- dimensional crossflow separation," AIAA Journal, Vol. 36, p. 557-564.

[17] Uhlman Jr., J. S. (1992), "An integral equation formulation of the equations of motion of an incompressible fluid," NUWC-NPT Technical Report 10,086, Naval Undersea Warfare Center Division, Newport, Rhode Island.

Figure 2. Front view of engine flow at t=33.

Figure 3. Front view of engine flow at t=55.

Figure 4. Front view of engine flow at t=77.

Figure 5. Front view of engine flow at t=99.

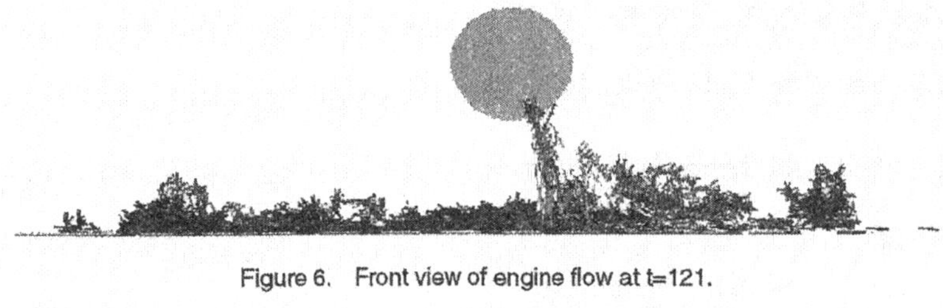

Figure 6. Front view of engine flow at t=121.

Figure 7. Front view of engine flow at t=143.

Figure 8. Side view of engine flow at t=143.

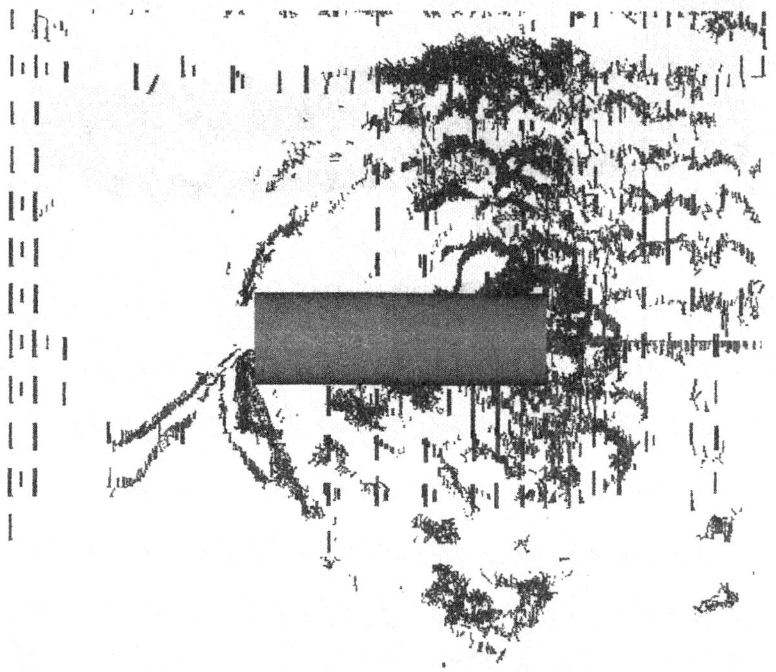

Figure 9. Top view of engine flow at t=143.

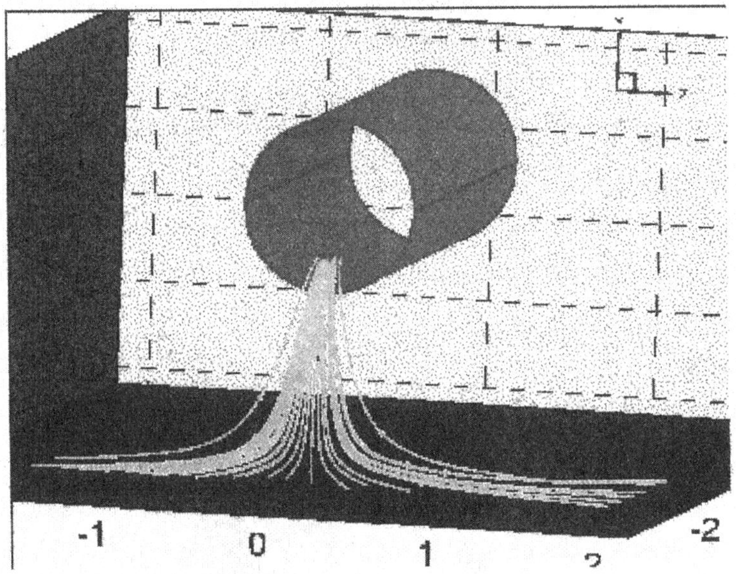

Figure 10. Streamlines without cross flow.

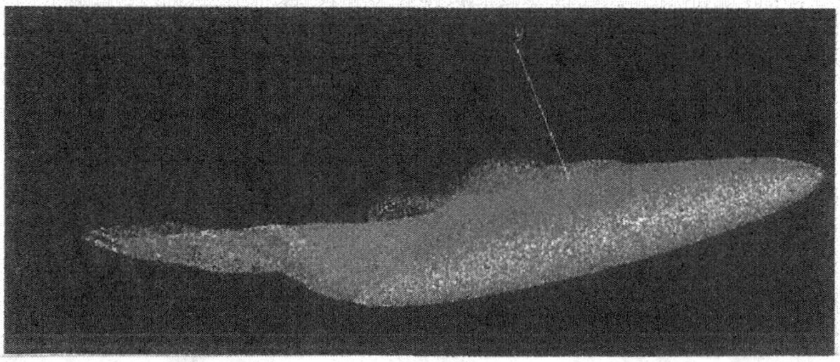

Figure 11. Side view of prolate spheroid at 30° angle of attack.

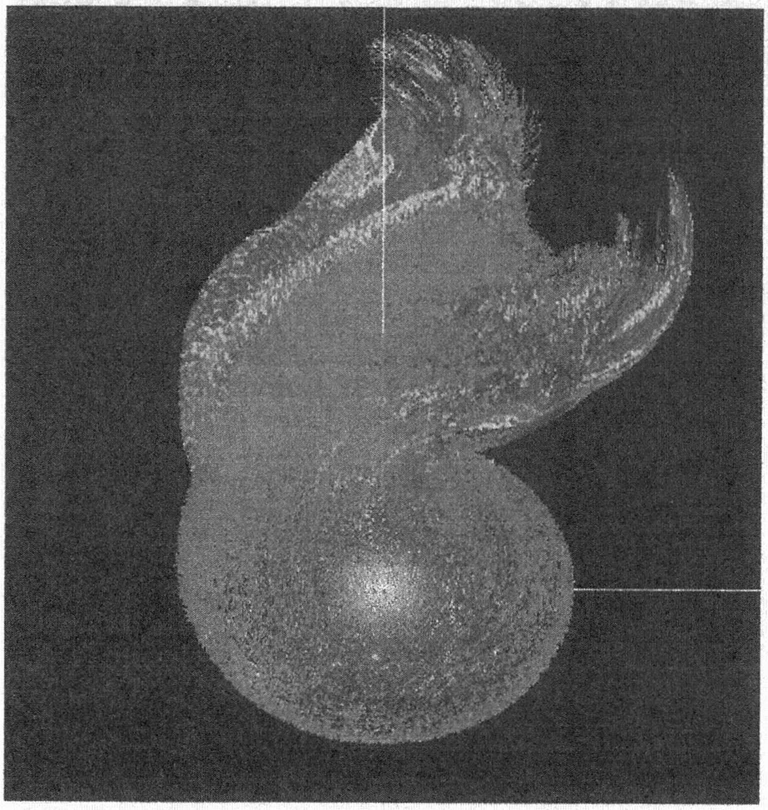

Figure 12. Front view of prolate spheroid at 30° angle of attack.

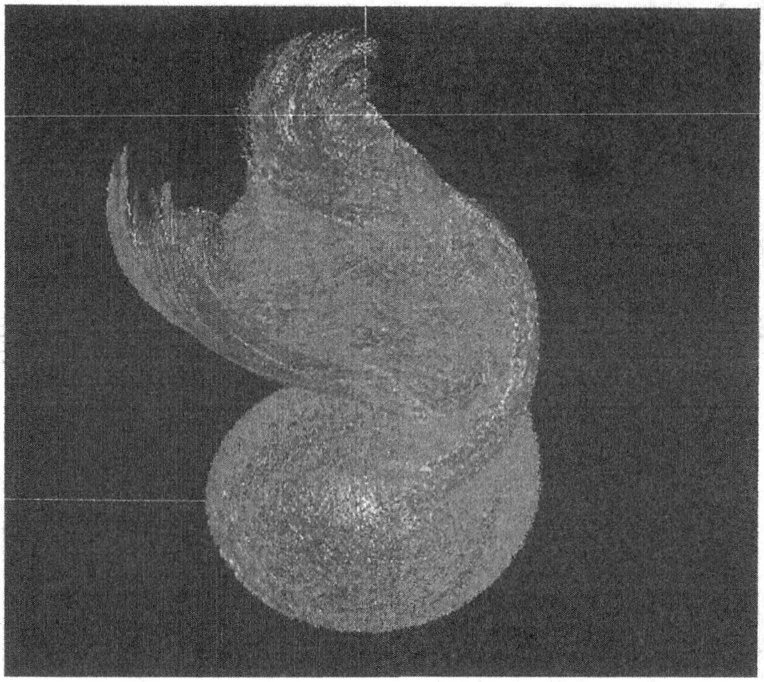

Figure 13. Rear view of prolate spheroid at 30° angle of attack.

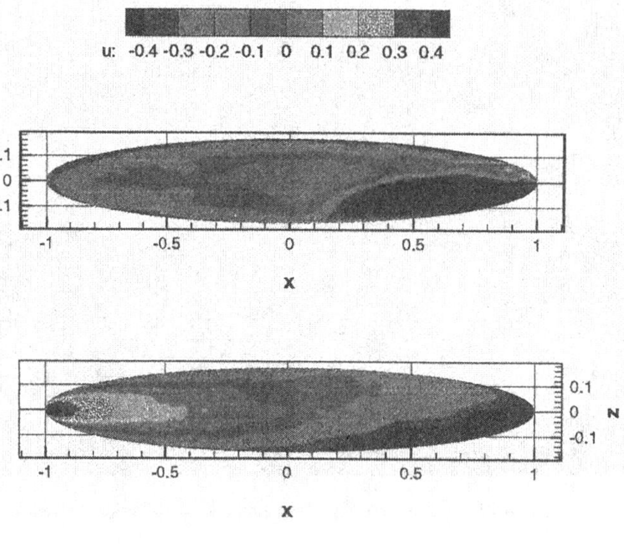

Figure 14. Pressure contours, top and bottom view.

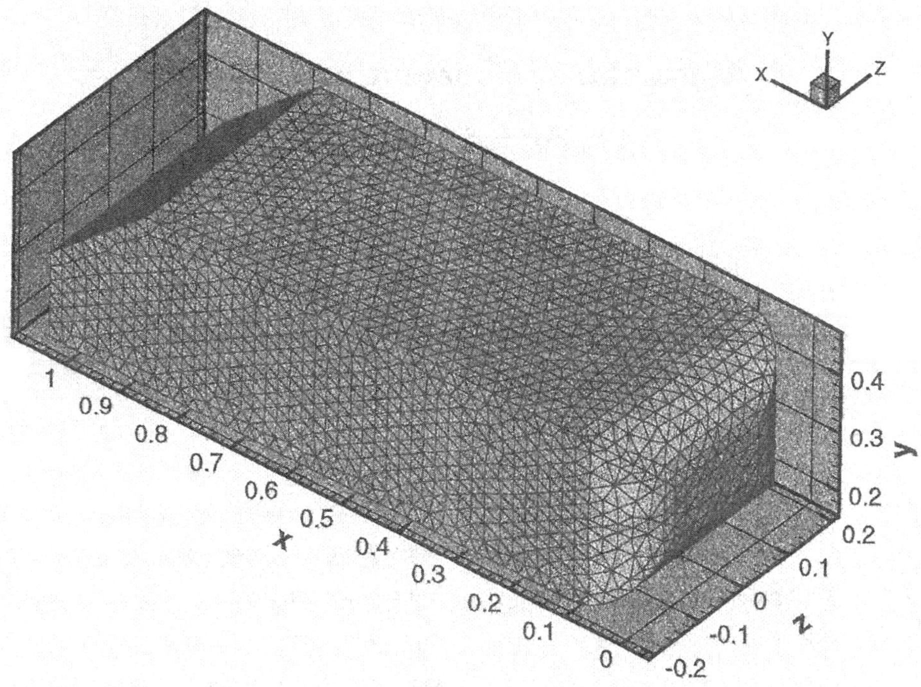

Figure 15. Minivan sheet mesh.

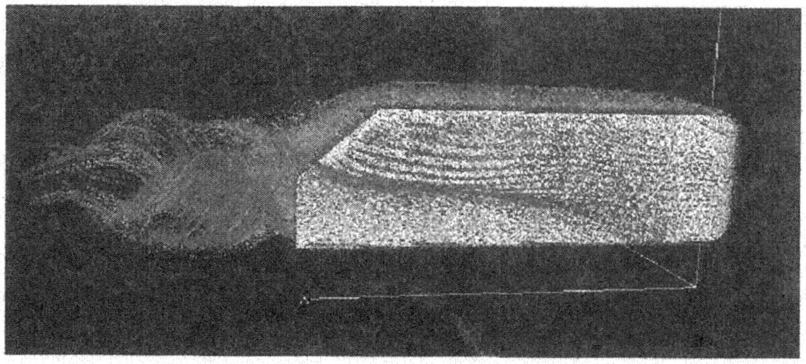

Figure 16. Minivan, side view.

Dynamics of Coherent Structures in a Forced Round Jet

S. Izawa, H. Ishikawa, and M. Kiya
Division of Mechanical Science, Hokkaido University
Sapporo, 060-8628, Japan / izawa@ring-me.eng.hokudai.ac.jp

ABSTRACT

Three-dimensional vortex method simulations of forced, incompressible, spatially developing round jets at moderate Reynolds numbers are made to investigate dynamics of vortical structures based on resolvable or grid-scale vorticity motions. The jet is sinusoidally forced by helical or multiple disturbances to clarify evolution of the vortical structures. The multiple disturbance is a combination of two helical disturbances of the same mode and amplitude, rotating in the opposite directions. The viscous diffusion of vorticity is represented by core spreading which approximates the growth of the rectilinear viscous vortex tube. Compact and highly organized, coherent structures are observed by isosurfaces of vorticity field and shown to be synonymous with initial disturbance modes.

1. INTRODUCTION

This paper discusses effects of initial disturbances on the development of an impulsively started round jet. A round jet experiences a dramatically high rate of spreading by appropriate combinations of frequencies of axisymmetric and orbital disturbances (Reynolds & Porekh [1]); the jet acquires an elliptic or square cross-section by a proper combination of two helical disturbances rotating in the counter directions (Cohen & Wygnanski [2]; Long & Petersen [3]). Large-scale coherent structures (CS) characterized by organized vorticity distributions in the developing region of a jet are intrinsic features of mixing layers at high Reynolds number. CS also play a crucial role in its growth rate, in production of Reynolds stresses and jet noise, and in the entrainment and mixing of a jet with its surroundings. Thus, if we are familiar with the dynamics and topology of CS, wide applications in engineering are expected by active control of a jet. Evolution of CS in the forced jet, however, is not fully understood because of experimental difficulties.

Danaila et al. [4] showed by direct numerical simulations that CS appear to be synonymous with instability modes in a naturally evolving axisymmetric jet at low Reynolds number ($Re \leq 500$). Grinstein et al [5] investigated the near field of an azimuthally excited, reacting, round jet in a combined computational/experimental study. In the present work, we study the active control of jet development. We focus on how CS in the impulsively started round jet are modified and the entrainment rate - rate at which fluid from the jet and from its surroundings become entangled and mixed as they join at the mixing layers, is changed by initial disturbances. The control is an open-loop control in which the jet is forced by axisymmetric or helical disturbances. The numerical simulation is made by a three-dimensional vortex blob method in which the viscous diffusion of vorticity is represented by core spreading (Leonard [6]).

2. NUMERICAL METHOD

2.1. Vortex Method

In a vortex method, the evolution of a vortical region is represented by a collection of vortex blobs with overlapping cores. A vortex blob α is defined by the position $\boldsymbol{x}^\alpha$, vorticity $\boldsymbol{\omega}^\alpha$, volume $d^3\boldsymbol{x}^\alpha$. Vorticity field $\boldsymbol{\omega}$ is given by

$$\boldsymbol{\omega}(\boldsymbol{x},t) = \sum_\alpha \boldsymbol{\omega}^\alpha(t) d^3\boldsymbol{x}^\alpha f_\delta(\boldsymbol{x} - \boldsymbol{x}^\alpha(t)) \tag{1}$$

where f_δ is the spherical core function with the cut-off radius σ_α. The core function f_δ is taken as [7] [8]

$$f_\delta(\boldsymbol{x}) = \frac{1}{\sigma_\alpha^3} f\left(\frac{|\boldsymbol{x}|}{\sigma_\alpha}\right), \tag{2}$$

with the normalization

$$4\pi \int_0^\infty f(\xi)\xi^2 d\xi = 1. \tag{3}$$

To maintain the accuracy of the representation in Eq. (1), the value of σ_α should be larger than the distance between the center of neighboring blobs. We use the high-order algebraic smoothing [8]

$$f(\xi) = \frac{15}{8\pi}\frac{1}{(\xi^2+1)^{7/2}}. \tag{4}$$

The evolution of position $\boldsymbol{x}^\alpha$ and vorticity $\boldsymbol{\omega}^\alpha$ is described by the Biot-Savart law and the vorticity equation in Lagrangian form, as follows

$$\frac{d\boldsymbol{x}^\alpha}{dt} = -\frac{1}{4\pi}\sum_\beta \frac{r_{\alpha\beta}^2 + (5/2)\sigma_\beta^2}{\left(r_{\alpha\beta}^2+\sigma_\beta^2\right)^{5/2}} \boldsymbol{r}^{\alpha\beta} \times \boldsymbol{\gamma}^\beta, \tag{5}$$

$$\frac{d\boldsymbol{\gamma}^\alpha}{dt} = (\boldsymbol{\gamma}^\alpha \cdot \nabla)\boldsymbol{u}^\alpha, \tag{6}$$

where $\boldsymbol{r}^{\alpha\beta} = \boldsymbol{x}^\alpha - \boldsymbol{x}^\beta$, $r_{\alpha\beta} = |\boldsymbol{r}^{\alpha\beta}|$, and $\boldsymbol{\gamma}^\beta = \boldsymbol{\omega}^\beta d^3\boldsymbol{x}^\beta$ is the strength.

We use the vorticity equation without the viscous diffusion of vorticity, Eq. (6). The viscous diffusion is approximated by a core spreading model, in which the cut-off radius increases with time t in the form

$$\frac{d\sigma_\alpha^2}{dt} = 4\nu, \tag{7}$$

where ν is the kinematic viscosity [6].

2.2. Flow Configuration

Calculations are performed for an impulsively started round jet issuing with velocity U into the same fluid at rest from a straight nozzle of radius R. The xyz-coordinate system is defined such that the x-axis is along the axis of the nozzle, the y- and z-axes normal to the x-axis, and the origin is at the center of the nozzle exit.

The surface of the nozzle is constructed by 96 vortex blob panels. Vorticity of the vortex blobs is determined so as to satisfy the zero cross-flow velocity at the center of gravity of the panels. The jet flow is produced by a source disk located in a yz-plane at x = -0.262R, which consists of 3791 sources arranged in 30 circular shells of the same thickness. The image nozzle is placed on the opposite side of the above-mentioned yz-plane, together with the image flow of the jet. Thus the physical flow in the region $x \geq$ -0.262R is the jet issuing from the nozzle with length 0.262R attached to a non-slip plane.

2.3. Numerical Procedure

Nascent vortex blobs are introduced into the flow at 0.061R downstream of the edge of the nozzle. To satisfy Kelvin's law, the circulation of the nascent vortex blob $\Gamma_{\rm nj}(t)$ at time t after its introduction is determined by

$$\Gamma_{\rm nj}(t) = \Gamma_{\rm pj}(t) - \Gamma_{\rm pj}(t - \Delta t_{\rm n}) \tag{8}$$

with $\Gamma_{\rm nj}(0) = \Gamma_{\rm pj}(0)$, where $\Gamma_{\rm pj}$ is the circulation of the vortex blobs belonging to the j-th panel at the edge of the nozzle, and $\Delta t_{\rm n}$ is the time interval of introducing the nascent vortex blobs. The nascent vortex blobs are advanced by the velocity induced at their center to simulate the shedding of vorticity from the edge.

Vorticity of the vortex blobs generally increases with time owing to the vortex stretching. Thus, to maintain the spatial resolution, a vortex blob α of length l_α is divided into two vortex blobs of the same vorticity $\boldsymbol{\omega}^\alpha$, cut-off radius σ_α, and length $l_\alpha/2$, located at $\boldsymbol{x}^\alpha$ $\pm(l_\alpha/2)\boldsymbol{\omega}^\alpha/\,|\boldsymbol{\omega}^\alpha|$, when the length l_α becomes greater than twice its initial value.

The time interval of advancing the vortex blob Δt is $0.03R/U$, while the time interval of introducing the vortex blobs is $\Delta t_n = 2\Delta t$. The cut-off radius of the nascent vortex blobs is chosen as $0.196R$ to realize the core overlapping. The same cut-off radius is also used for the vortex blobs of the panels. The time advancement is made by the second-order Adams-Bashforth scheme.

2.4. Mode of Forcing

A disturbance for forcing is introduced by a sinusoidal oscillation of source strength in the outermost two shells of the source disk. This induces a sinusoidal streamwise velocity perturbation of the form $u_F = a\cos(m\phi - 2\pi ft)$, where a, m and f are the amplitude, mode number, and frequency of forcing, respectively, and ϕ is the azimuthal angle with respect to the z-axis. In this paper, an axisymmetric disturbance m = 0, helical disturbance m = 1 or 2, and combined disturbances $m = \pm 1$ or ± 2 are studied; $m = \pm k$, k being an integer, implies combination of two helical disturbances $m = k$ and $m = -k$ of the same amplitude. The combined disturbance can be written as $u_F = 2a\cos(m\phi)\cos(2\pi ft)$, so that the maximum disturbance is fixed at particular azimuthal angles.

The forcing frequency is chosen as $f = 0.56U/R$, which was found by preliminary calculations to be the fundamental frequency of instability in the initial shear layer of the unforced jet. The amplitude is changed in a range, $a = 0.015U - 0.31U$. Note that $a = 0.025U$ is near the upper limit of linear growth of infinitesimal disturbances in free shear layers.

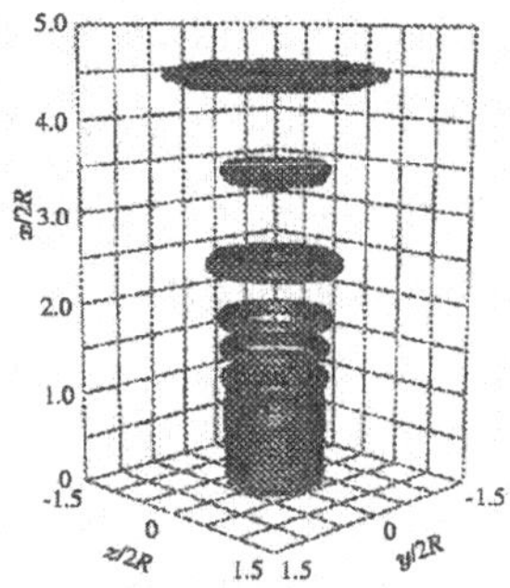

Fig.1: Isosurfaces of magnitude of vorticity ω/ω_0 = 0.60 in unforced jet (Re = 2,000) at $Ut/2R$ = 11.16. ω_0 is the average magnitude of vorticity of nascent vortex blobs.

Numerical simulations are performed at Reynolds numbers Re (= $2RU/\nu$) of 2,000 and 20,000.

3. RESULTS AND DISCUSSION

Results will be presented for the initial disturbance mode effects in Sections 3.1. Those for the forcing amplitude effects will be presented in Section 3.2.

3.1. Vortex Dynamics in Forced Jets

Isosurfaces of vorticity $\omega = |\boldsymbol{\omega}|$ is shown in figure 1, at $Ut/2R$ = 11.16 after the start of flow, where the level of vorticity is represented with reference to the average vorticity of nascent vortex blobs ω_0. The shear layer rolls up into vortices about $x/2R$ = 1.2; rolled-up vortices merge to generate larger vortices which are fairly axisymmetric at the above mentioned time. Note that the forcing by the axisymmetric mode m = 0 produces no significant change in the structure at the same time.

The vortical structures appear to be synonymous with initial disturbance modes. Figure 2 shows the process of merging vortex spiral produced by the forcing m = 1. The spiral has negative streamwise vorticity (grey regions in figure 2 (d), (f)), while the positive vorticity (black regions) is produced by local merging of the vortex spiral, which occurs downstream of $x/2R$ > 1.5.

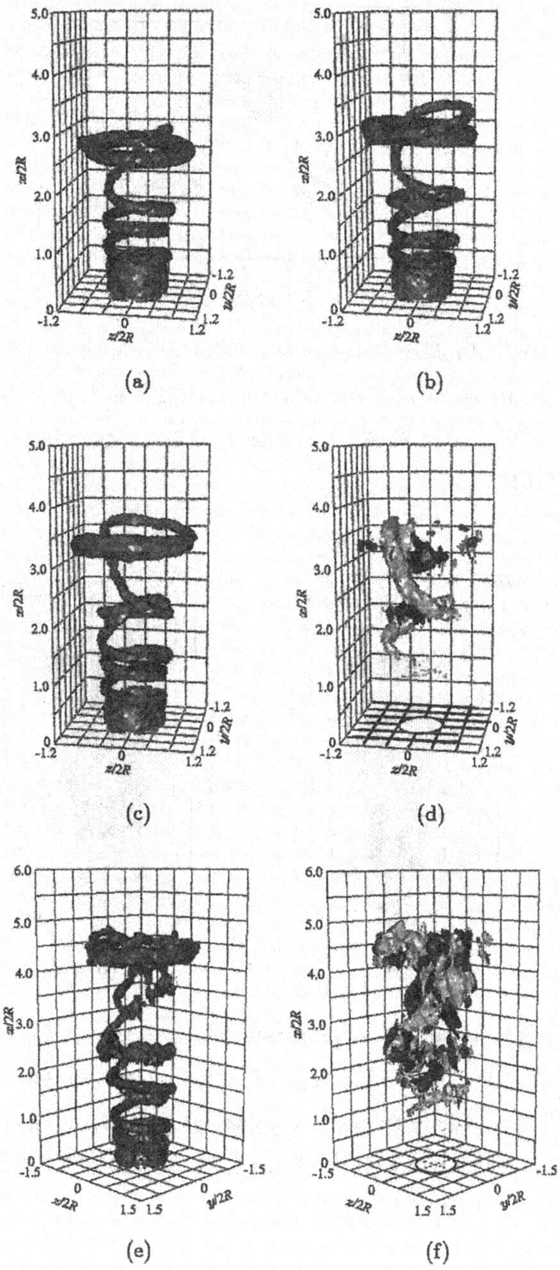

Fig.2: Process of merging of vortex spiral in forced jet ($m = 1$, Re = 2,000) represented by isosurfaces of ω at (a) $Ut/2R = 7.20$, (b) 7.92, (c) 8.64, (e) 11.16, and by isosurfaces of ω_x at (d) $Ut/2R = 8.64$, (f) 11.16 (black regions for positive ω_x and grey regions for negative ω_x). $a/U = 0.015$.

Here, the streamwise vorticity refers to the component of streamwise vorticity ω_x. The local merging is the

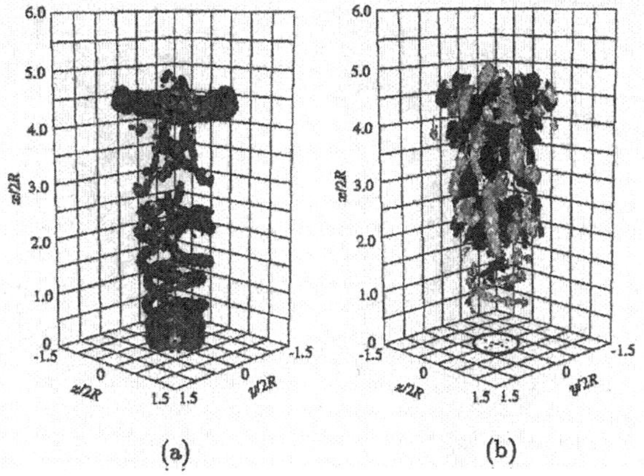

Fig.3: Isosurfaces of (a) magnitude of vorticity $\omega/\omega_0 =$ 0.60 and (b) x-component of vorticity $\omega_x/\omega_0 = \pm 0.16$ (black regions for positive ω_x and grey regions for negative ω_x) in forced jet ($m = 2$, Re = 2,000) at $Ut/2R$ = 11.16. $a/U = 0.015$.

merging of two windings of the spiral. This process generates a part of the spiral tilted with respect to the x-axis in such a way to generate the positive streamwise vorticity and eventually a pair of counter rotating streamwise vortices. The vortex spirals are much distorted in the region of $x/2R > 3$ by the effect of the topmost large vortex ring, which has engulfed the vortex spirals to generate streamwise vorticity of both signs. This mechanism is basically true also for $m = 2$, in which the process of vortex merging is much more complex (figure 3(b)). In this case, pairs of streamwise vortices are more clearly observed.

Figure 4 shows the vortical structures for the combined disturbances $m = \pm 1$. This forcing generates the maximum amplitude of disturbance in the xz-plane (where $\phi = 0$ and π), where the rolling-up of the shear layer is enhanced to have a staggered arrangement in the same plane (in frame (c)). This rolling-up can be interpreted as merging of the two spirals associated with the disturbances or a double-pairing interaction of inclined vortex rings. In terms of the double-pairing interaction, a vortex ring generated by the disturbance is tilted, and merges with the down-

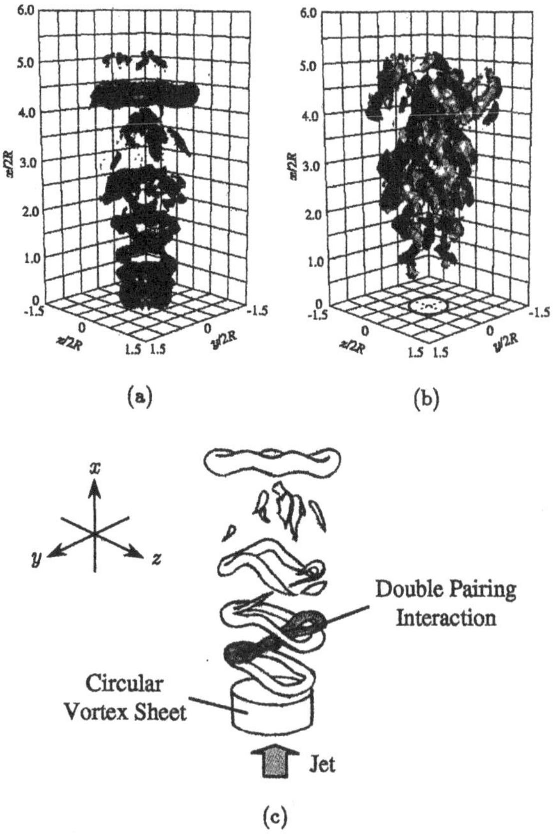

Fig.4: Isosurfaces of (a) magnitude of vorticity $\omega/\omega_0 = 0.60$, (b) x-component of vorticity $\omega_x/\omega_0 = \pm 0.16$ (black regions for positive ω_x and grey regions for negative ω_x) in forced jet ($m = \pm 1$, $Re = 2{,}000$) at $Ut/2R = 11.16$, and (c) topological interaction model for coherent structures in (a). $a/U = 0.015$.

stream vortex ring in parts and the other side pairs with the upstream vortex ring. The merging parts are connected by other parts of the vortex spirals or rings; these merging parts are tilted in such a way that pairs of negative and positive streamwise vortices are generated. Further merging of the vortices makes the vorticity distributions much complicated. Although the jet contains many small-scale structures produced by disturbances, the velocity profiles have no high wave-number components (figure 5). This is because the velocity distributions are calculated from the volume integral of vorticity distributions (Biot-Savart law).

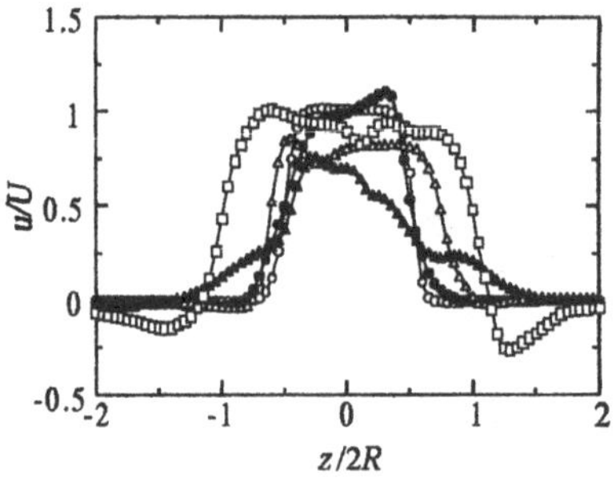

Fig.5: Profiles of streamwise velocity u in xz-section for $m = \pm 1$ at $Re = 2{,}000$ and $Ut/2R = 11.16$. ○, $x/2R = 0.5$; ●, 1.5; △, 2.5; ▲, 3.5; □, 4.5. $a/U = 0.015$.

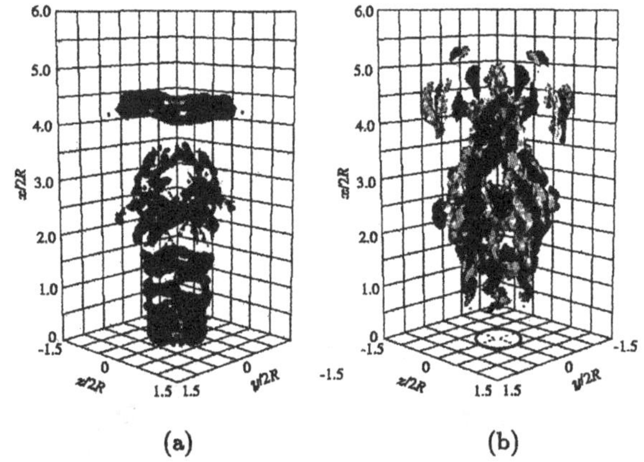

Fig.6: Isosurfaces of (a) magnitude of vorticity $\omega/\omega_0 = 0.60$ and (b) x-component of vorticity $\omega_x/\omega_0 = \pm 0.16$ (black regions for positive ω_x and grey regions for negative ω_x) in forced jet ($m = \pm 2$, $Re = 2{,}000$) at $Ut/2R = 11.16$. $a/U = 0.015$.

Figure 6 shows isosurfaces of ω and ω_x for the combined disturbances $m = \pm 2$. The vorticity distributions extend more outwards than those for any other mode, especially in the region centered around $x/2R \simeq 2.0$. This is caused by rapid outward motion of rolling-up vortices generated at four positions of maximum forcing amplitude. At the same time the rolling-up vortices effectively generate small-scale structures in the region $x/2R \simeq$

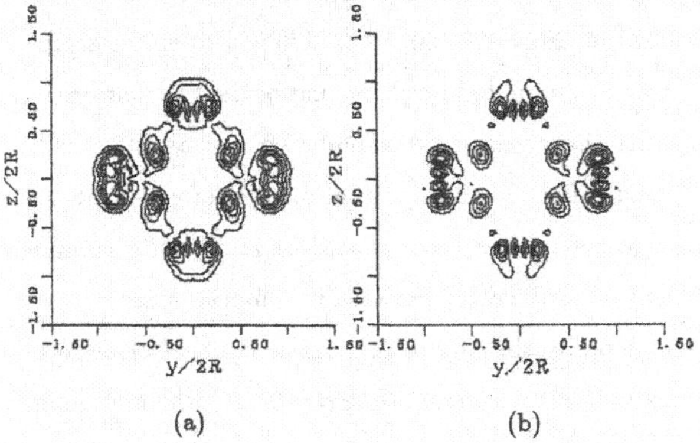

Fig.7: Contours of (a) ω and (b) ω_x in yz-section at $x/2R = 2.0$ in forced jet ($m = \pm 2$, Re = 2,000) at $Ut/2R = 11.16$. $a/U = 0.015$.

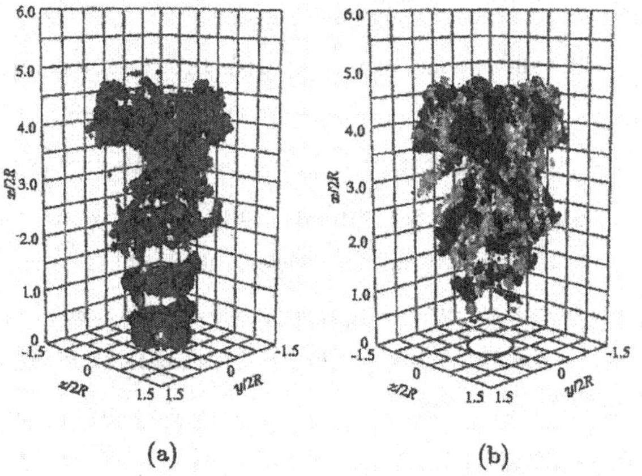

Fig.8: Isosurfaces of (a) magnitude of vorticity $\omega/\omega_0 = 0.60$ and (b) x-component of vorticity $\omega_x/\omega_0 = \pm 0.16$ (black regions for positive ω_x and grey regions for negative ω_x) in forced jet ($m = \pm 1$, Re = 20,000) at $Ut/2R = 10.68$. $a/U = 0.015$.

1.5 − 3.0. These structures take highly organized distributions in yz-planes as shown in figure 7.

The vortical structure for the combined disturbance $m = \pm 1$ at Re = 20,000 and $Ut/2R = 10.7$ are illustrated in figure 8. Compared with the results of Re = 2,000, the streamwise vorticity is generated closer to the nozzle, and high vorticity regions increase, especially in the topmost vortex.

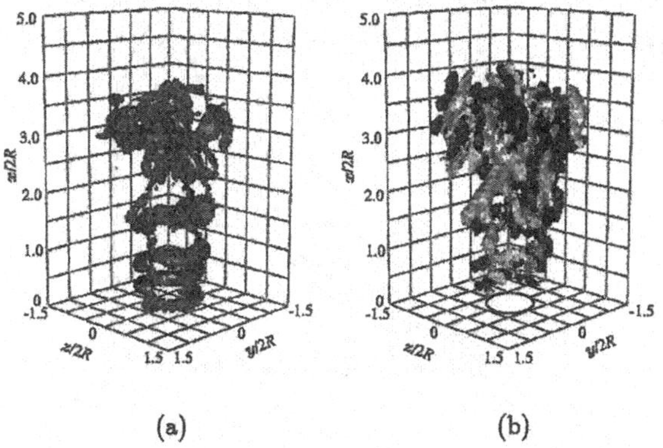

Fig.9: Isosurfaces of (a) magnitude of vorticity $\omega/\omega_0 = 0.60$ and (b) x-component of vorticity $\omega_x/\omega_0 = \pm 0.16$ (black regions for positive ω_x and grey regions for negative ω_x) in forced jet ($m = \pm 1$, $a = 0.31U$, Re = 2,000) at $Ut/2R = 9.0$.

3.2. Forcing Amplitude Effects on Jet Development for $m = \pm 1$

As the forcing amplitude increases, the vortical structures in the forced jet become much more complicated (figure 9). Vortex rings roll up just behind the nozzle exit and deform three-dimensionally by double pairing interaction much earlier than the results of lower forcing amplitude.

Figure 10 shows the non-dimensional entrainment ratio $Q(x,t)/Q_0$ of the forced jet at selected time, where Q is defined by

$$Q(x,t) = \int\int \rho u dy dz \qquad (9)$$

and Q_0 is the initial reference flux at $x = 0$ in the unforced jet. ρ is the density. Note that the comparisons are made at different times for each forcing condition because the number of vortex blobs increases rapidly as time advances. The rate of entrainment of fluid from the surroundings into the jet, $dQ(x,t)/dx$, first becomes significantly larger than that for the unforced jet. This is a common feature for the forced jets. In particular, the variation Q for a = 0.31U is about 10% larger than that for the unforced jet in the region

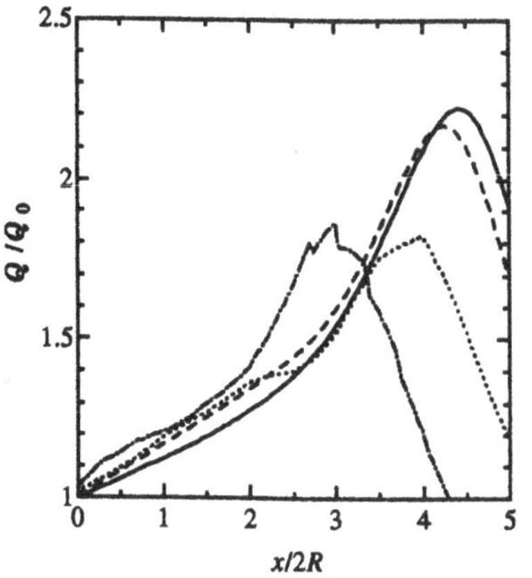

Fig.10: Flow rate Q as a function of streamwise distance. —, unforced jet ($Ut/2R$ = 11.16); – – –, forced jet ($m = \pm 1$, Re = 2,000, $Ut/2R$ = 11.16, a/U = 0.15); ·····, forced jet ($m = \pm 1$, Re = 20,000, $Ut/2R$ = 10.68, a/U = 0.15); —·—, forced jet ($m = \pm 1$, Re = 2,000, $Ut/2R$ = 9.0, a/U = 0.31).

$x/2R < 2.0$–2.5.

4. ERROR ESTIMATION

Greengard [9] argues that the solution of the viscous core spreading model will not converge to the exact solution of Navier-Stokes equations even in the limit of infinitely many vortex elements. However, the core spreading model is expected to yield a fairly good approximation to the solution within a finite time after the start of flow (Nakajima & Kida [10]). Although an error estimation of vortex methods was carried out only for two-dimensional flows, we try to estimate the errors of the present simulations based on this two-dimensional estimation. When the Reynolds number is sufficiently high, the error for convection of vorticity is of the order of $(\Delta t)^2$ and that for diffusion is of the order of $4(U\Delta t/2R)/Re^{1/2}$. If Δt is sufficiently small, the vortex methods are expected to give reliable results within a finite time after the introduction of the vortex blobs into the flow. In the present simulations, $U\Delta t/2R$ is equal to 0.03 so that the convection error per time step is of the order of 0.001, while the diffusion error is of the order of 0.008 at Re = 2,000 and 0.003 at Re = 20,000. It is difficult to estimate the error accumulated until the end of the calculations.

On the other hand, circulation of nascent vortex blobs varies in the azimuthal direction for helical disturbances. Thus there should be streamwise component of vorticity ω_x between neighboring nascent vortex blobs at the edge of the nozzle. For the helical forcing $m = \pm 1$ with amplitude $a = 0.31U$, the maximum change of circulation was found to be about 77% at most, while the change in circulation between neighboring vortex blobs was less than 10%. Therefore, in the present simulation, the vorticity ω_x was not introduced from the edge. The streamwise component was generated by tilting and stretching of rolling-up vortices in the streamwise direction.

To validate the vortex method, the results should be compared with results of DNS or experiments. Moreover, the statistical properties of the flow such as Reynolds stresses and time-mean velocity are required. However rapid increase in the number of vortex blobs as time advances prevented the computation time extending to obtain the statistical properties. These are the issues to be studied in the future.

5. CONCLUSIONS

The vortex method simulation serves to understand the dynamics and topology of coherent structures in the forced jets under helical disturbances. The structure of the jet is sensitive to initial disturbance modes. The vortex spirals produced by helical disturbances lead to the pair of negative and positive streamwise vortices. The entrainment rate was found to be significantly larger than that for a unforced jet in the near field, especially at high Reynolds number or high forcing amplitude.

RERENCES

[1] Reynolds, W. C. & Parekh, D. E. (1987), Control of Structure in Turbulent Flows, ***AFOSR-TR-88-***

0036, Thermosciences Division, Department of Mechanical Engineering, Stanford University.

[2] Cohen, J. & Wygnanski, I. (1987), The evolution of Instabilities in the Axisymmetric Jet, Part 1. The Linear Growth of Disturbances near the Nozzle, ***Journal of Fluid Mechanics***, Vol. 176, pp. 191-235.

[3] Long, T. A. & Petersen, R. A. (1992), Controlled Interactions in a Forced Axisymmetric Jet, Part 1. The Distortion of the Mean Flow, ***Journal of Fluid Mechanics***, Vol. 245, pp. 37-55.

[4] Danaila, I., Dušek, J. & Anselment, F. (1997), Coherent Structures in a Round, Spatially Evolving, Unforced, Homogeneous Jet at Low Reynolds Numbers, ***Physics of Fluids***, Vol. 9, pp. 3323-3342.

[5] Grinstein, F. F., Gutmark, E. J. & Obeysekare, U. (1996), Streamwise, and Spanwise Vortex Interaction in an Axisymmetric Jet. A Computational and Experimental Study, ***Physics of Fluids***, Vol. 8, pp. 1515-1524.

[6] Leonard, A. (1980), Vortex Methods for Flow Simulation, ***Joournal of Computational Physics***, Vol 37, pp. 289-335.

[7] Knio, O. M. & Ghoniem, F. (1990), Numerical Study of a Three-Dimensional Vortex Method, ***Joournal of Computational Physics***, Vol 86, pp. 75-106.

[8] Winckelmans, G. S. & Leonard, A. (1993), Contributions to Vortex Particle Methods for Computation of Three-dimensional Incompressible Unsteady Flows, ***Joournal of Computational Physics***, Vol. 109, pp. 247-273.

[9] Greengard, C., (1985), The Core Spreading Vortex Method Approximates the Wrong Equation, ***J. Comp. Phys.***, Vol. 61, pp. 345-348.

[10] Nakajima, T. & Kida, T., (1990), A Remark of Discrete Vortex Method (Derivation from Navier-Stokes Equation), ***Trans. Jpn. Soc. Mech. Eng.***, (in Japanese), Vol. 56, No. 531, B, pp. 3284-3291.

APPLICATION OF DISCRETE VORTEX METHOD AS TURBULENT MODELLING TOOL FOR LOCAL EDDY GENERATION IN COASTAL WATERS

Keita Furukawa, Tadashi Hibino and Munehiro Nomura

Marine Environment Division, Port and Harbour Research Institute, MOT, Japan,
3-1-1, Nagase, Yokosuka 239-0826, Japan
E-mail: furukawa_k@cc.phri.go.jp

ABSTRACT

To model local eddy generation in coastal waters, a discrete vortex method (DVM) has been developed. Because eddies are generated as a result of boundary layer separation from a headland, thel modelling of detailed boundary shapes using boundary elements and complex vortex fields observed near headlands using a vortex tracking method is ideal.

For the DVM to accommodate flow simulations in coastal waters, the model considers shallow water effects. To achieve the effects, a depth averaged two-dimensional Navier-Stokes equation is used as the governing equation. It adds a bottom sheer stress term to the DVM.

Three case studies are presented to demonstrate the feasibility of using the DVM in simulations on coastal water flows with eddy generation.

INTRODUCTION

In coastal waters, local eddies may be generated by tidal current separation in the lee of islands, headlands and narrow channels (Wolanski et. al. (1); Ingram and Chu (2); Signel and Geyer (3)). Since this phenomenon is unsteady and localized in space and time, it presents two major difficulties when studied. One is the difficulty of numerical modelling due to limited computer resources and time constraints. To track vortex fields near separation points, a fine mesh system is required. The other is the difficulty in observing complex vortex fields. This requires that velocity fields be sampled continuously and simultaneously in widely spread areas.

This paper reveals features of local eddy generation and its role in coastal waters using a discrete vortex method (DVM). To assess the method, observation results are also presented as case studies. The results include velocity field data observed by a high-frequency radar system (HF Radar), an acoustic Doppler current profiler (ADCP), and other oceanographic observation methods.

DISCRETE VORTEX METHOD

A depth averaged two-dimensional Navier-Stokes (NS) equation is used as the governing equation for the DVM (Furukawa and Wolanski (4)). To discretize the governing equation, a discrete vortex having total circulation Γ and area S is defined. In a coordinate system that moves with circulation Γ, the NS equation can be simplified to:

$$\frac{d\Gamma}{dt} = -\gamma_b \nabla \times \left(\frac{\mathbf{u}|\mathbf{u}|}{H} \right) S + K_x \nabla^2 \Gamma \qquad (1)$$

where $\boldsymbol{u}$ is the velocity vector, H is the water depth, K_x is the horizontal eddy viscosity coefficient, γ_b is the bottom friction coefficient and t is the time.

In Equation (1), the first term of the right hand side denotes the generation of an eddy by bottom friction, which represents a depth effect in the model. The second term on the right hand side represents the vorticity diffusion by the eddy viscosity. The eddy viscosity is determined by a one-equation model (Furukawa and Wolanski (4)).

We choose a discrete vortex to have a core; it has a Gaussian vorticity distribution (after Chorin (5); Leonard (6); Furukawa and Hosokawa (7)). The tangential velocity u_r induced by the discrete vortex is:

$$u_r = \frac{\Gamma}{2\pi r}\left[1 - \exp\left(-\frac{r^2}{4K_x t}\right)\right] \tag{2}$$

where r is the distance from the center of the vortex.

A number of discrete vortices are introduced in the boundary layer of land-sea boundaries (a boundary vortex as a boundary element). Each vortex induces the tangential velocity by its own circulation intensity Γ_r. Each Γ_i is determined by the conditions of 'zero normal velocity at a land-sea boundary' as an integration of the induced velocity at the control points from all discrete vortices mapped in the calculation domain.

CASE STUDIES

Case 1: Water mixing at the mouth of a bay

Eddy generation processes at the mouth of a bay will affect the long-term water exchange as they are seen in the changes of residual flow patterns. A series of predicted tidal residual patterns at the mouth of Tokyo Bay, Japan by an FEM method has shown different flow patterns at successive (a) neap tide, (b) spring tide and (c) neap tide (**Figure 1**). At spring tide, a strong clockwise circulation always persistsat the north of Cape Futtsu. In contrast, no significant standing circulations can be observed at neap tide.

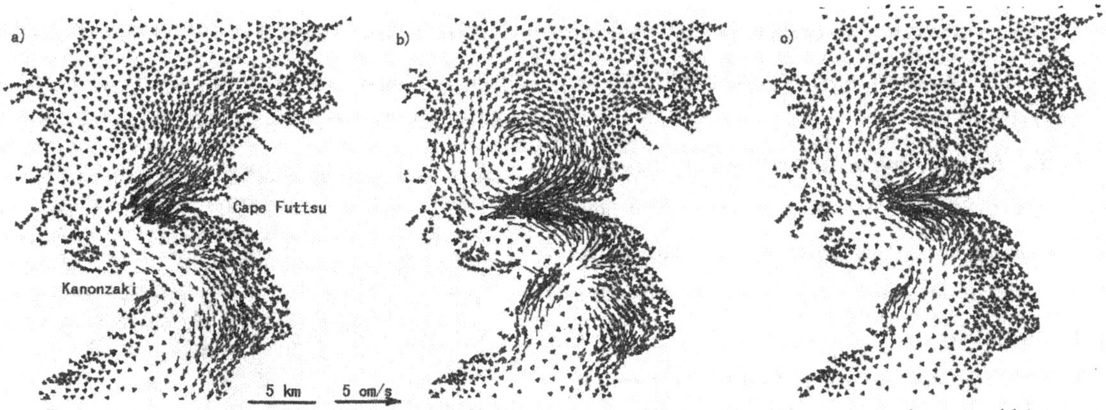

Figure 1: Successive residual flow predicted by FEM (a: neap tide, b: spring tide, c: successive neap tide)

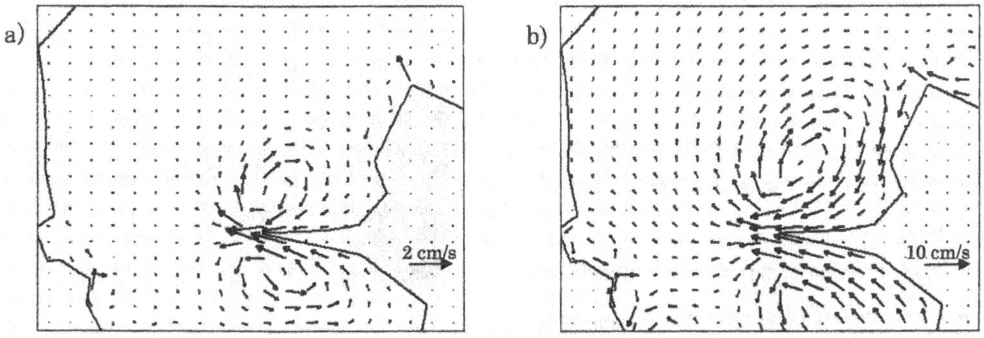

Figure 2: Residual flow pattern predicted by DVM (a: neap tide, b: spring tide)

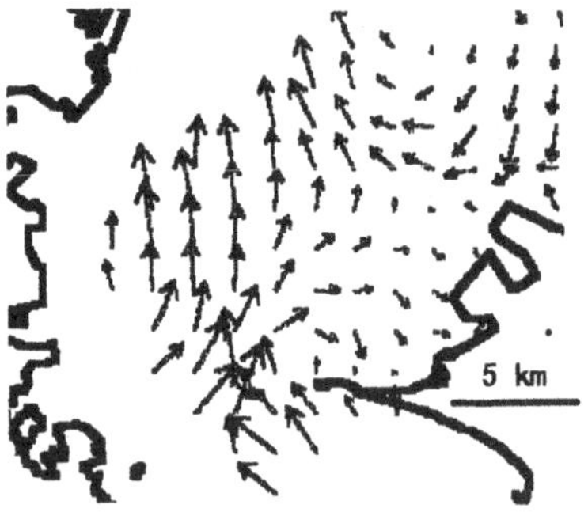

Figure3: Instantaneous view of circulation observed by HF Radar on 15:00, 4 Dec., 1998

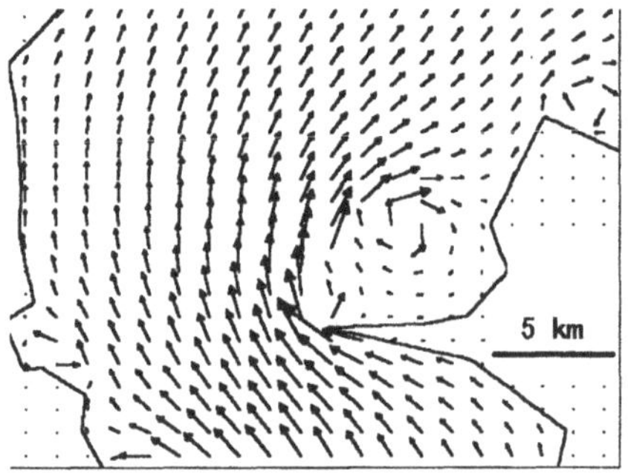

Figure 4: DVM calculation result correspond to the condition with Fig.2.

CTD cast observations have also supported the existence of two different patterns of saline distributions at different tidal phases.

Vertical mixing is enhanced near Cape Futtsu at spring tide, and horizontally well mixed patterns are observed at neap tide. Accordingly, the circulation in this case can be a controller for a bay water exchange mechanism. The origin of the residual circulation can be considered as vortex generation by Cape Futtsu, Coriori's force and an estuarine circulation enhanced by a density current.

To test the first hypothesis, the DVM is employed. As shown in **Figure 2**, the change of the residual flow field according to the tidal phase is reproduced by the DVM, but its location and size do not quite agree with the FEM predictions.

To get a more accurate comparison for the temporal propagation of local eddy generation, a flow pattern determined by the HF radar set is shown in **Figure 3**. The figure shows a small adjacent circulation north of Cape Futtsu caused by vortex separation from the cape. Nevertheless, another clockwise circulation also exists in the middle of Tokyo Bay. It can be considered a part of the estuarine circulation.

Figure 4 shows a corresponding flow field predicted by the DVM. The method results in a good agreement with the observed local eddy formation around Cape Futtsu. It is clear that the standing residual circulation is not only the results of local eddy generation or an estuarine circulation.

Case 2: Saline water patch generation by manmade headland

Patches in coastal waters are known to be of various sizes and to have varying degrees of importance in natural processes. Some of these patches are induced by local eddies generated by headlands (Geyer and Signell (8); Signell and Geyer (3)) and islands (Wolanski and Hamner (9); Furukawa and Wolanski (4)).

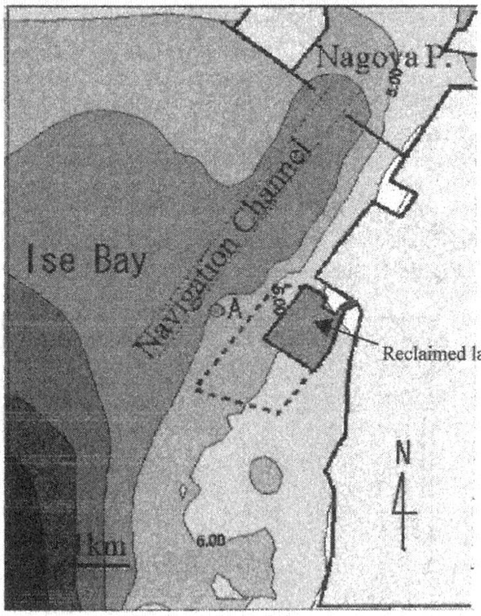

Figure 5: Reclaimed land No.5 in top of Ise Bay.

Small-scale water mass structures (patches) play an important role in the fate of dissolved and suspended matter in coastal waters (Hamner and Hauri (10)). The dissolution and suspension are key processes for coastal environments.

Reclaimed land No. 5 is located at the inner part of Ise Bay, Japan (**Figure 5**). Ise Bay is known as one of the most "enclosed" bays in Japan (Unoki (11)). The Kiso river discharges into the top of the bay and is a source of low saline water patches in the bay. The depth around the reclaimed land is c.a. 5 m, and there is a 10 m depth navigation channel for Nagoya Port to the west. The tidal currents in the channel are north-eastward at flood tide, and south-westward at ebb tide. The tip of reclaimed land No. 5 forms a man-made headland that protrudes into oscillating tidal currents.

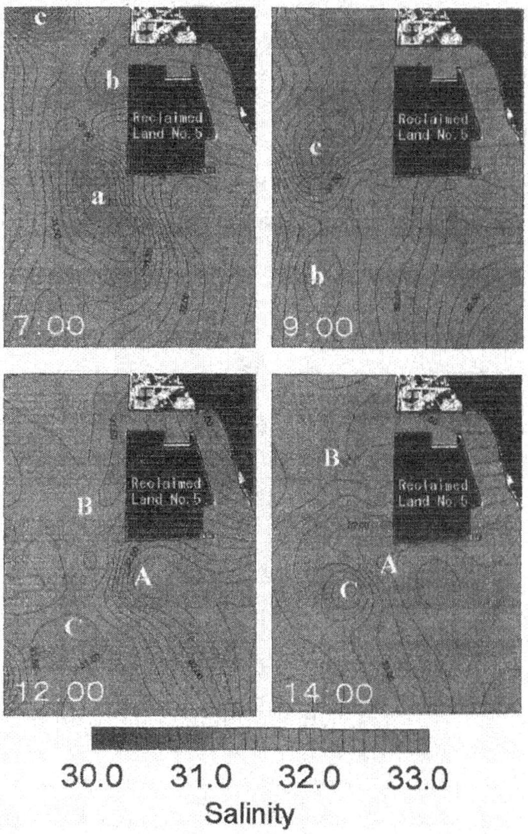

Figure 6: Surface saline distribution on 30/10/97 from CTD castings at 42 points.

At different tidal phases (ebb and flood), patchs of saline water have been detected (**Figure 6**). The patches from the saline distributions are clearly marked as a-c for ebb tide and A-C for flood tide. At ebb tide, some of the patches with a relatively high saline water mass are convected from north to south in response to 0.2 m s^{-1} southward tidal currents. At flood tide, low saline water masses (B and C in **Figure 6**) are divided into two parts due to the high saline water mass (A in **Figure 6**) existing during northward transit at the tidal currents. This process has been found to create saline patches, or in other words, to generate eddies at the headland creating these patches.

An experiment was carried out to model such eddy generation. A large-scale hydraulic tank in the Ise-Bay Experiment Facility at the 5th Port Construction Office, MOT was used for the experiment. The model basin of Ise Bay has a 1/2000 scale for the horizontal axis and a 1/159 scale for the vertical axis. The general circulation in Ise Bay was reproduced with simple M2 tides. **Figure 7** shows the flow patterns of small floating tracers in the hydraulic model test. The tracers were trapped southward at ebb tide, and trapped northward at flood tide. The circulation broke up into patches, and the size of the circulation at ebb tide became the same as the width of the headland. The eddy size at flood tide was 1/3 smaller than at ebb tide.

First, the model was run for a constant south-westward current (ebb tide). The flow was visualized through an instantaneous current vector distribution. It showed the

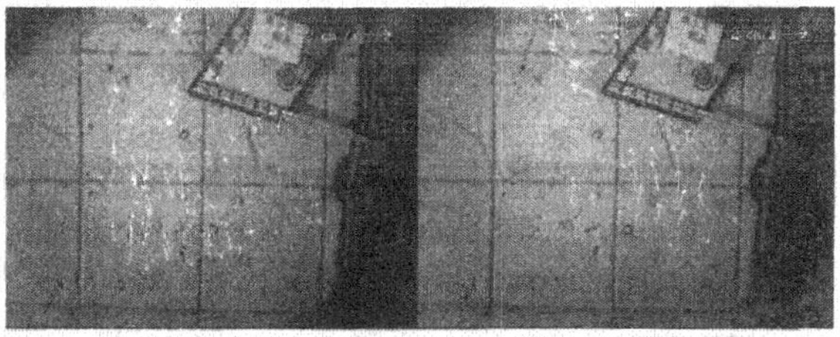

Figure 7: Current around reclaimed land visualized by floating tracers in hydraulic model (left: ebb tide, right: flood tide).

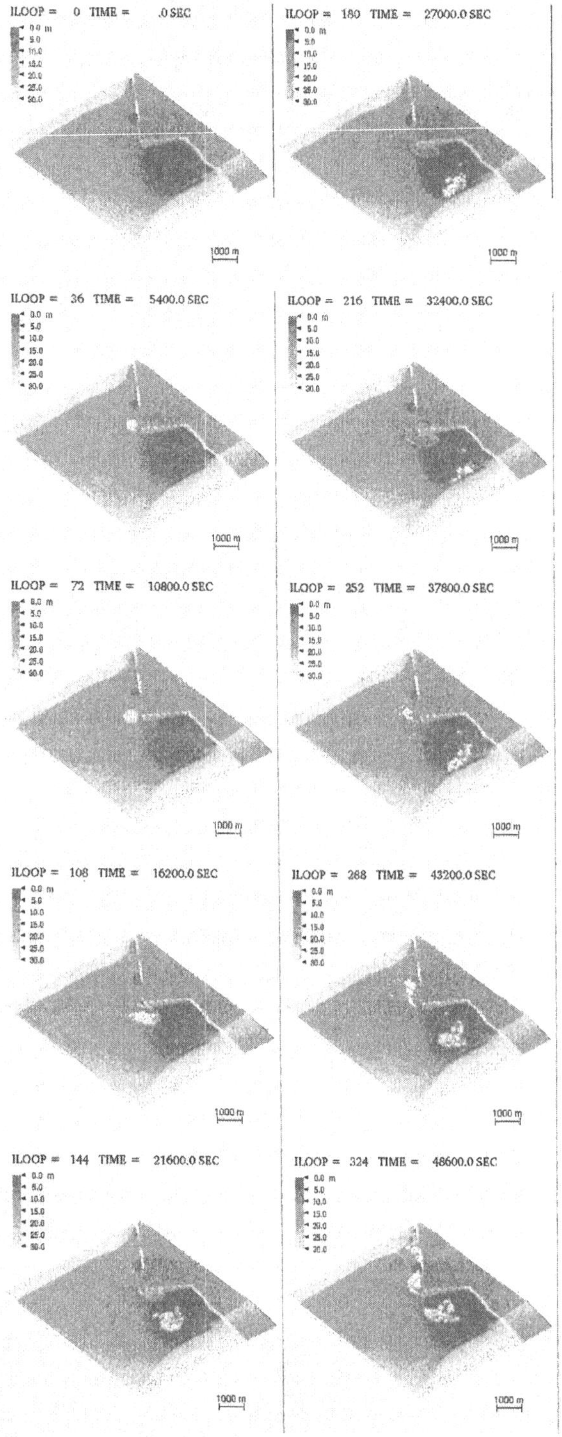

Figure 8: Flow pattern around manmade headland. Tracers were released from tip of headland. The color of tracer is white for tracer released at flood tide, and gray for tracer released at ebb tide. Time elapsed from top right to bottom right, then top left to bottom left.

development of a complex circulation in the lee of the headland. No separation of the eddy was observed in this scenario.

Next, the model was run for an oscillating current simulating an M2 tide. The flow was visualized through an instantaneous tracer distribution injected from the corner of the reclaimed land (**Figure 8**).

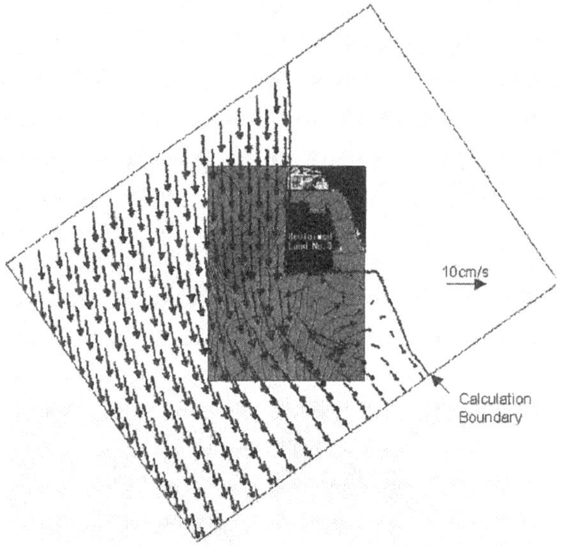

Figure 9: Overlapped image of calculated velocity vector and observed saline distribution at ebb tide.

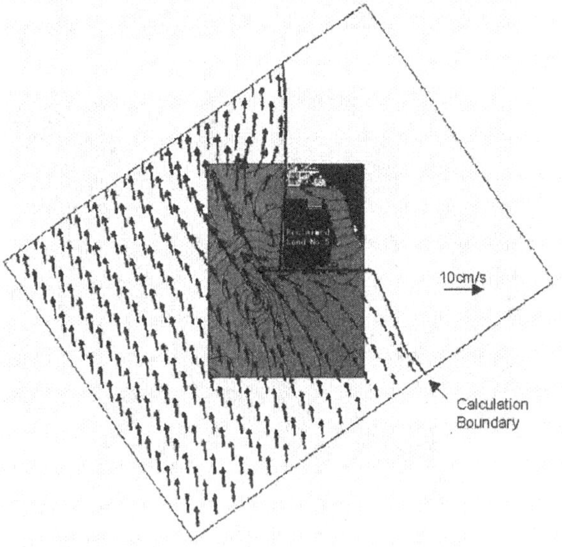

Figure 10: Overlapped image of calculated velocity vector and observed saline distribution at flood tide.

Figure 8 shows eddies generated by reversing the tidal currents at both sides of the headland; their formation was phase-locked with the tide.

Overlapped images of numerical calculations and observed saline distributions at ebb tide and flood tide are shown in **Figure 9** and **Figure 10**, respectively. A close examination of the images suggests that the cause of the observed saline patches is the local eddies generated at the headland by reversing the tidal currents. At ebb tide, the low saline water from the top of the bay is transported from north to south. The separation of the local eddies from the headland assists in spreading the low saline water mass to the middle part of the bay. At flood tide, a high saline water mass travels along the coast from south to north. From the headland, the high saline water mass penetrates into the low saline water mass, so the coastal waters become patchy.

Case 3: Vortex generation at a small strait between islands

A vortex pair associated with a tidal jet through a narrow passage between reef islands of the Palm Islands Group, Australia has been observed (**Figure 11**). The nearly oscillating flows northward at ebb tide, and southward at flood tide at a maximum speed of 1 m/s. It gives a significant driving force to separate the boundary layer, and makes complex eddies interact.

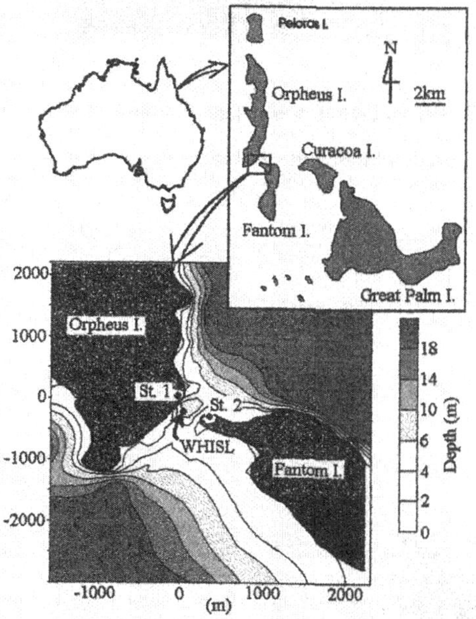

Figure 11: Palm Islands group and observation stations for VHF radar (St.1,2) and Current monitoring points (WHISL)

In visual observations, a mushroom-shaped jet has been seen at flood tide in the north of the passage, and a complex front structure has been seen just after a slack high tide (**Figures 12 and 13**).

Figure 12: View from St.1 facing North East ward at ebb tide. Thick lines on picture indicate front of sea surface. Arrow indicate its movement.

Figure 13: View from St.1 facing South East ward at just after low slack tide. North ward flow and South ward returning flow makes strong shier layer.

A VHF radar system (McLaren (12)) can be used to clearly detect eddy formations (**Figures 14 and 15**). His system gives direct vortex propagation images because of its high spatial resolution (150 m).

The predictions by the DVM are in good agreement with the velocity fields observed by the VHF radar (**Figures 16-18**). Using the model in this case, numerical passive tracers were released in the line

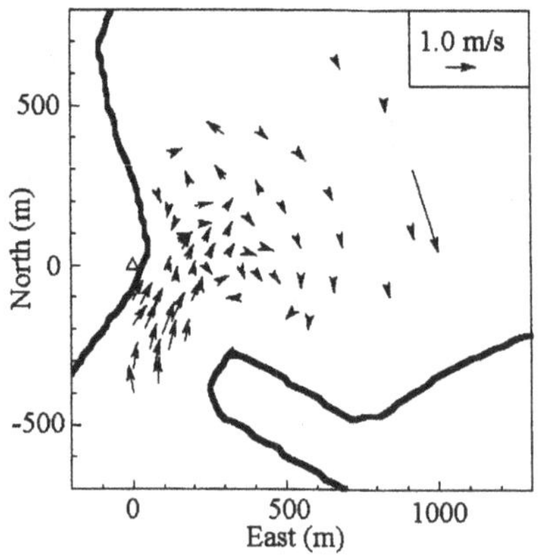

Figure 14: Ebb tide current observed by VHF radar system.

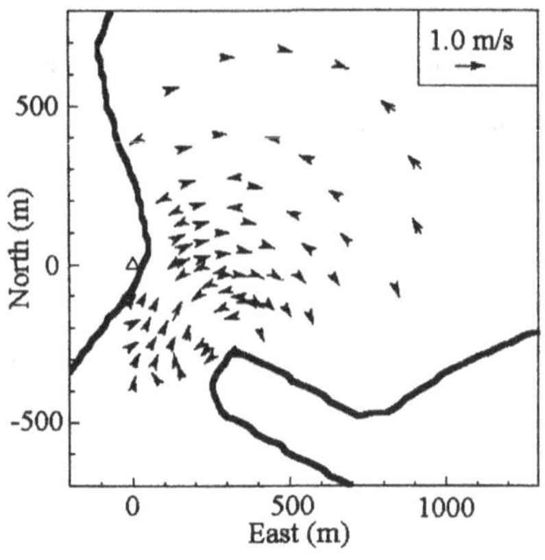

Figure 15: Flood tide current observed by VHF radar system.

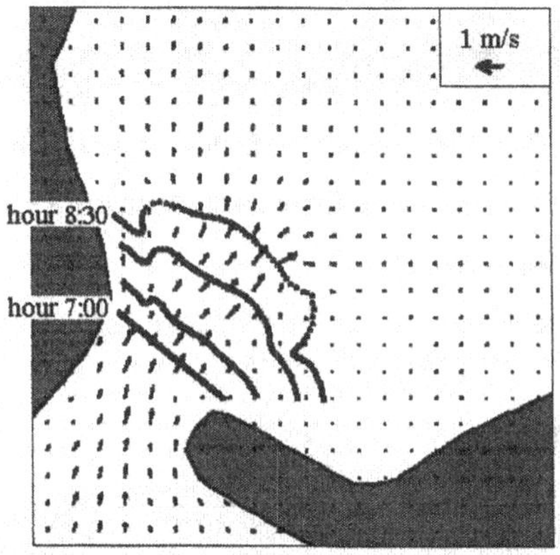

Figure 16: Ebb tide current and tracer distribution whitch were released at start of ebb tide.

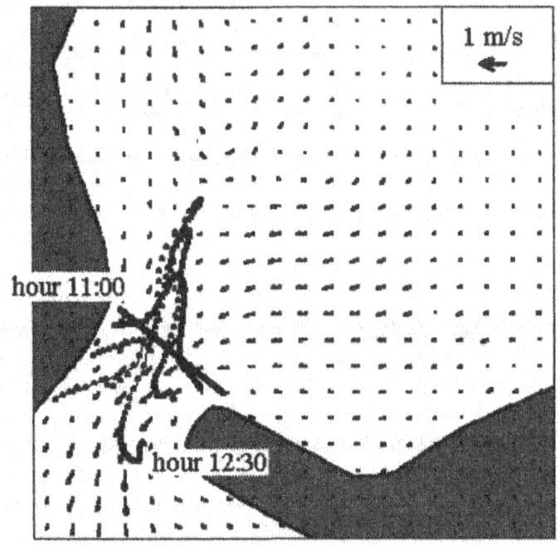

Figure 17: Slack tide to flood tide current and tracer released at slack tide.

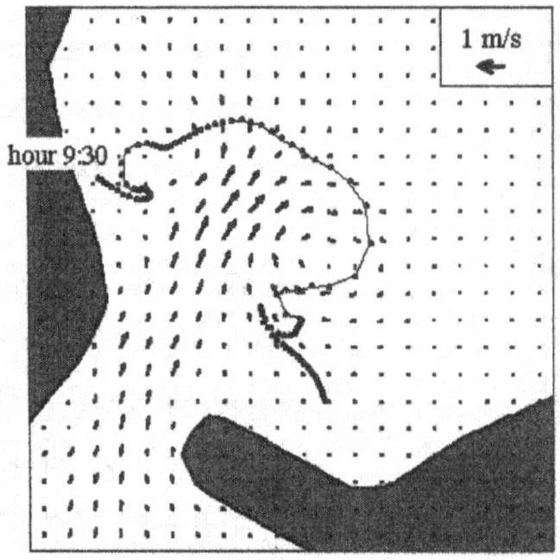

Figure 18: Final stage of ebb tide current correspond to Figure 12, 14

indicated as hour 7:00 in **Figure 16**. They showed similar patterns to formations of floating materials observed in the field. The rough bottom surfaces of coral reefs provided extra friction for the flow through the passage, thereby distorting the vortex pair. Accordingly, the vortex pair induced a northward flow. This flow was indicated as the remaining core flow in the center of the passage as shown in **Figure 19**. As a result, the strong residual flow in the center of the passage was forced north. This residual flow might be a conveyer that brings nutrients and suspended matters to the north of the passage.

CONCLUSION

The proposed discrete vortex method was found to be a valuable tool for studying local eddy generation when used appropriately. The results obtained here showed a good agreement with field observations and FEM predictions in areas where eddy generation was significant.

ACKNOWLEDGMENTS

This study was supported by a Program for Promoting Fundamental Transport Technology Research from the Corporation for Advanced Transport and Technology (CATT), the Japan Research Development Corporation (JRDC), and James Cook University.

REFERENCES

(1) Wolanski, E., Imberger, J. and Heron, M.L. (1984) Island wakes in shallow coastal waters. *Journal of Geophysical Research*, **89**(C6), 10553-10569.

(2) Ingram, R.G. and Chu, V.H. (1987) Flow around islands in Rupert Bay. *Journal of Geophysical Research*, **92**(C13), 14521-14533.

(3) Signel, R.P. and Geyer, W.R. (1991) Transient eddy formation around headlands. *Journal of Geophysical Research*, **96**(C2), 2561-2575.

(4) Furukawa, K. and E. Wolanski (1998) Shaloow-water frictional effects in island wakes. *Estuarine, Coastal and Shelf Science*, **46**: 599-608.

(5) Chorin, A.J. (1973) Numerical study of slightly viscous fluid. *Journal of Fluid Mechanics*, **57**(4): 785-796.

(6) Leonard, A. (1980) Vortex methods for flow simulation. *Journal of Computational Physics*, **37**: 289-335.

(7) Furukawa, K. and Y. Hosokawa (1996) Parameter fitting on a discrete vortex method with wake simulation behind an inclined flat plate. *Computational Fluid Dynamics Journal*, **5**(1): 89-105.

(8) Geyer, W.R. & Signell R. 1990 Measurement of tidal flow around a headland with a shipboard acoustic Doppler current profiler. *Journal of Geophysical Research*, **95**, 3189-3197.

(9) Wolanski, E. & Hamner W.H. 1988 Topographically controlled fronts and their biological influence. *Science*, **241**, 177-181.

(10) Hamner, W.M. & Hauri, J.R. 1977 Fine-scale currents in the Whitsunday islands, Queensland, Australia: Effect of tide and topography. *Australian Journal of Marine and Freshwater Research*, **28**, 333-359

(11) Unoki, S. 1985 Ise and Mikawa bays. In *Coastal Oceanography of Japanese Islands.* ed. by The Coastal Oceanography Committee, Tokai University Press, Tokyo (in Japanese).

(12) McLaren, N.R. (1995) VHF ocean radar with applications in coastal monitoring. PhD. Thesis, James Cook Univ. of North Queensland, Australia, 139 p.

UNSTEADY AERODYNAMIC FORCES ACTING ON A LARGE FLAT FLOATING STRUCTURE

Hiroki Tanaka, Kazuhiro Tanaka and Fumio Shimizu
Department of Mechanical System Engineering, Kyushu Institute of Technology
680-4 Kawazu Iizuka-city 820-8502 Japan
e-mail hiroki@mse.kyutech.ac.jp

Youhachirou Watanabe
Nagasaki R&D Center, Mitsubishi Heavy Industries Co. LTD
Nagasaki, Japan

ABSTRACT

Large flat floating structures such as floating heliport are examined and one of the problems is the flow induced vibration in strong wind. The unsteady aerodynamic forces induced by wind vibrate the structure a little and the vibration produces new unsteady aerodynamic forces, so the vibration some times increases until large amplitude. As the reason, it is very important to get the aerodynamic forces acting on the vibrating structures and examine the possibility of flutter on the floating structures.

Vortex method was used to calculate the aerodynamic forces on the vibrating structure, because the method is easy and useful to deal with the conditions of vibrating boundary and open boundary. Sink and source are also used to satisfy the change of structure volume. Unsteady Bernoulli's equation was used to calculate the pressure on the surface of structure.

Flutter characteristics were also calculated with the unsteady aerodynamic forces obtained by the vortex method. Flutters are strongly influenced by the mass of structure, frequency and damping force. However, the damping force on water surface is so great that it is found that the flutter will hardly occur on the floating structures.

1. NOMENCLATURE

n	: direction cosine
p	: pressure
t	: time
u, v	: velocity in x and y directions
x, y	: x and y coordinates
Γ	: circulation
ϕ	: velocity potential
ρ	: density of fl

2. CALCULATION PROCEDURE

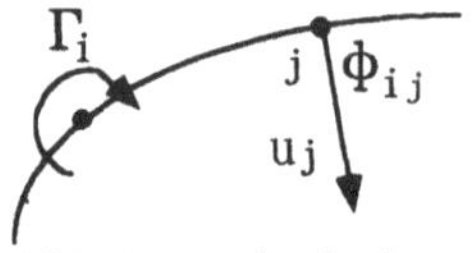

Figure 1 Vortex and velocity potential

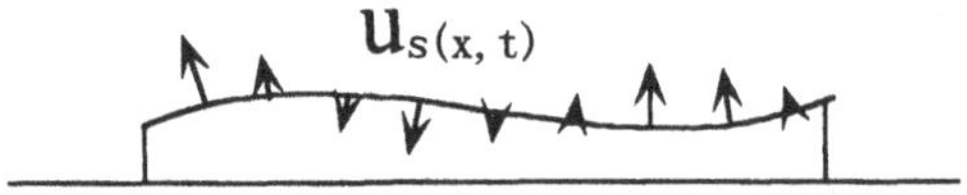

Figure 2 Vibrating surface of structure

2.1 Velocity potential

The vortex method is a common method and has been closely explained for example in reference (1). However, there are some different points used in this paper, so a brief explanation is done in this section.

As shown in figure 1, velocity potential of point I induced by the vortex of point j is expressed as follows,

$$\phi_{ij}(t)=\frac{-\Gamma_j(t)}{2\pi}\tan^{-1}\frac{y_i(t)-y_j(t)}{x_i(t)-x_j(t)}=A_{\phi ij}\Gamma_j(t) \qquad (1)$$

These values are a function of time, but to make simple, the letter 't' is neglected from here. In the case where there are many vortices, the velocity potential can be expressed as follows,

$$\phi_i=\sum_{j=1}^{N}A_{\phi ij}\Gamma_j=\sum_{j=1}^{N}\frac{-1}{2\pi}\tan^{-1}\frac{y_i-y_j}{x_i-x_j}\Gamma_j=\{A_{\phi ij}\}^T\{\Gamma_j\} \qquad (2)$$

The equation can be expressed by vector and matrix as follows,

$$\{\phi_i\}=[A_{\phi ij}]\{\Gamma_j\} \qquad (3)$$

There are two kinds of vortices. One is the vortices fixed on the surface of structure and is called bound vortex because they do not move. Another is free vortex that is generated from the end of the surface and move with the flow velocity. So, the vortices are distinguished by putting subscript 'b' and 'f'.

1.2 Induced Velocity by Vortex

As shown in figure 1, velocity at the point i induced by vortex j can be expressed as followings,

$$u_{ij} = \frac{\partial \phi_{ij}}{\partial x} = \frac{\partial A_{\phi ij}\Gamma_j}{\partial x} = A_{uij}\Gamma_j \tag{4}$$

$$v_{ij} = \frac{\partial \phi_{ij}}{\partial y} = \frac{\partial A_{\phi ij}}{\partial y} = A_{vij}\Gamma_j \tag{5}$$

Total of the induced velocity must be summing up the influence of all vortices and so they can be expressed as follows,

$$u_i = \sum_{j=1}^{N_b} A_{uij}\Gamma_{bj} + \sum_{k=1}^{N_f} A_{uik}\Gamma_{fk} = [A_{uij}]\{\Gamma_{bj}\} + [A_{uik}]\{\Gamma_{fk}\} \tag{6}$$

$$v_i = \sum_{j=1}^{N_b} A_{vij}\Gamma_{bj} + \sum_{k=1}^{N_f} A_{vik}\Gamma_{fk} = [A_{vij}]\{\Gamma_{bj}\} + [A_{vik}]\{\Gamma_{fk}\} \tag{7}$$

2.2 Boundary Condition

Surface of the structure vibrates and the vibrating velocity changes with the position and time as shown in figure 2. As wind flow can not go through the structure, the flow velocity in normal direction must be the same as the moving velocity of the surface.

Normal component of the surface at point i can be expressed as follows;

$$u_{nsi} = \dot{x}_{si} n_{xi} + \dot{y}_{si} n_{yi} \tag{8}$$

The air flow velocity in normal component on the surface can also be expressed as follows;

$$u_{ani} = (u_{bi} + u_{fi} + u_{wi})n_{xi} + (v_{bi} + v_{fi} + v_{wi})n_{yi} \tag{9}$$

where the suffix 'w' means the wind velocity. These values are the function of time. The normal component of airflow must be the vibrating velocity of the boundary, so the boundary condition is expressed as follows,

$$u_{sni} = u_{ani}$$

Therefore,

$$(u_{bi} + u_{fi} + u_{wi} - \dot{x}_{si})n_{xi} + (v_{bi} + v_{fi} + v_{wi} - \dot{y}_{si})n_{yi} = 0 \tag{10}$$

2.3 Generation of free vortex

Flow goes away form the surface of structure at a separation point. In the case of floating structures, if the amplitude of vibration is not great, the separation point can be considered to be the trailing edge of the deck as shown in figure 3. The air just above the boundary layer of the deck end moves with the velocity u_t and after that it becomes free vortex. In real flow, the velocity on deck surface must be zero, so at the deck end, the vortex made in the boundary layer should move with half of the flow velocity. Then, The free vortex generated in time Δt can be expressed as follows,

$$\Gamma_{fK} = \frac{1}{2} u_t^2 \Delta t_K \tag{11}$$

The free vortex moves freely with the neighboring air, so moving velocity of the free vortex can be expressed as

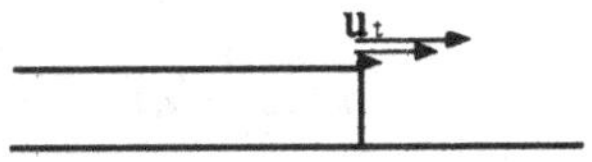

Figure 3 Flow at the deck end

Figure 4 Plate model

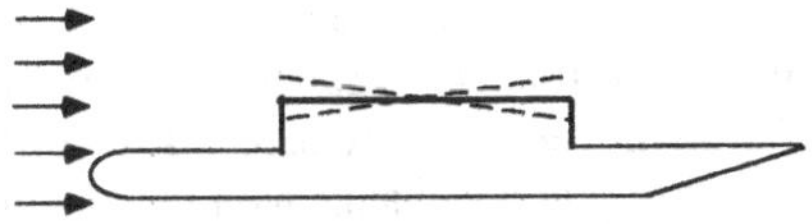

Figure 5 Wing model

Figure 6 Symmetric model

follows,

$$u_{fK} = [A_{uKj}]\{\Gamma_{bj}\} + [A_{uKk}]\{\Gamma_{jk}\} + u_{wK} \tag{12}$$

$$v_{fK} = [A_{vKj}]\{\Gamma_{bj}\} + [A_{vKk}]\{\Gamma_{jk}\} + v_{wK} \tag{13}$$

The position of free vortex can be obtained by integrating the above velocities with time after its separation.

2.4 Determination of Bound Vortex

The wind velocity and vibration of the deck surface must be assumed before the calculation. The free vortex can be obtained from the velocity of deck end and so, the bound vortex only remains as unknown values.

Rearranging the equation(10) using equation (6), (7), the following equation is obtained,

$$\left([n_{xi}][A_{uij}] + [n_{yi}][A_{vij}]\right)\{\Gamma_{bj}\}$$
$$= [n_{xi}]\{-u_{fi} - u_{wi} + \dot{x}_{si}\} + [n_{yi}]\{-v_{fi} - v_{wi} + \dot{y}_{si}\} \tag{14}$$

If we introduce the number of control point equal with the number of bound vortices, we can get the values of bound vortex.

3. UNSTEADY PRESSURE

Unsteady Bernoulli's equation is expressed as follows,

$$p = p_0 - \frac{1}{2}\rho(u^2 + v^2) - \rho\frac{\partial \phi}{\partial t} \tag{15}$$

Unsteady term is the third term in left-hand side. If the boundary is fixed, then the partial differential with respect to time is easy. However, when the boundary is vibrating, the control points are also moving but the differential must be done on the stationary point.

Using the equation (2), the following equation is

obtained,

$$\frac{\partial}{\partial t}\phi_i = \sum_{j=1}^{N} A_{\phi ij}\frac{\partial}{\partial t}(\Gamma_j) + \sum_{j=1}^{N}\frac{\partial}{\partial t}(A_{\phi ij})\Gamma_j \tag{16}$$

The first term is due to the change of vortex itself and the second term is due to the movement of vortex. As shown in the equation (2), $A_{\phi ij}$ is a function of the position i and j, so the movement of vortex produces the unsteady pressure.

4. SPECIAL PROBLEM ON FLOATING BODY

There is a water level around the floating structure and we must consider that the water level is also the boundary. So we tried to calculate the pressures on the model shown in figure 4. Pressure distributions on the steady structures did not change with the length of water levels spread front and rear if the length is greater than the width of floating structure. However, unsteady aerodynamic pressure changed greatly with the water level length, even though the length is much greater unit. The model shown in figure 5 showed the same characteristics as the model in figure 4.

Usually structures in water or air have pressures on both side and so, for example, lift force on a wing can be obtained by subtracting the upper pressure from lower pressure. In this case, the absolute values of velocity potential does not produce any difference, because the absolute values are canceled each other on upper and lower surface. However, in the case of floating structure, we cannot subtract the pressure so we must calculate the correct value of velocity potential including absolute values.

So the next, the symmetric model was introduced as shown in figure 6. In this case, the model corresponds to an infinite water level nevertheless there are not any vortices on water surface. However, in some vibration mode, inner volume of the symmetric model changes. The vortex method can not be applied to the volume change, so sink and source are introduced to satisfy the boundary of the structure. Vortex was used to calculate the effect of free vortex only.

5. EVALUATION OF CALCULATION

5.1 Steady aerodynamic force on a box

Sample of the floating structure used in this paper is two dimensions. Width is 1.0 and standard height above water is 0.05 of width. At first, the steady aerodynamic force on stationary structure is calculated. The calculated pressure distribution was compared to the experiment of similar shape. The experiment was done with the model which consists of a deck and lots of columns to support the deck, so the shape is not the fully same as that for calculation. The calculated and measured values are shown in figure 7. Features of the pressure distribution are in good agreement with the experiment except around at the deck end.

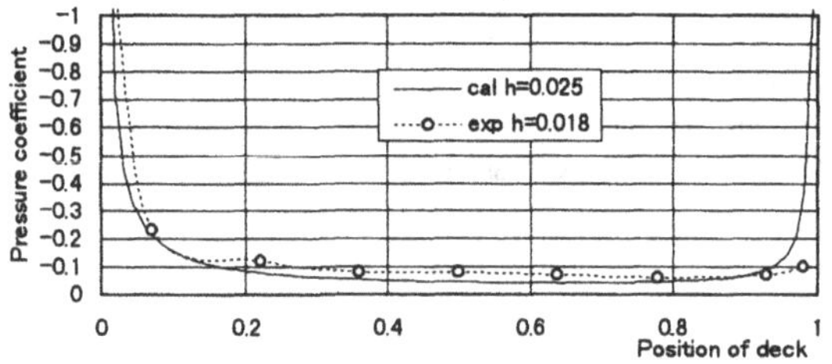

Figure 7 Comparison of pressure coefficients on the deck between calculation and experiment.

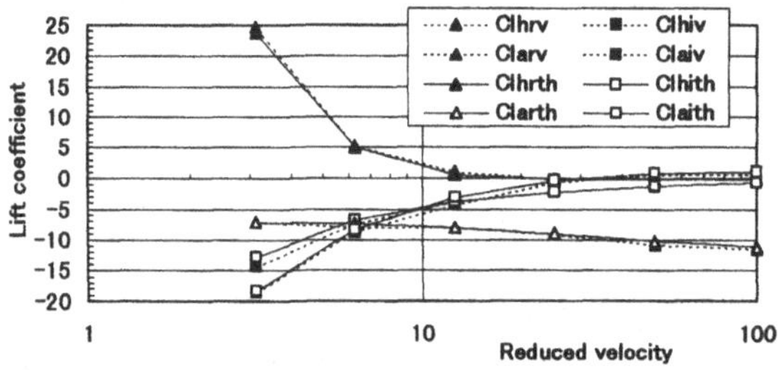

Figure 8 Lift coefficient of flat plate, comparison of calculated by vortex method and analytical method

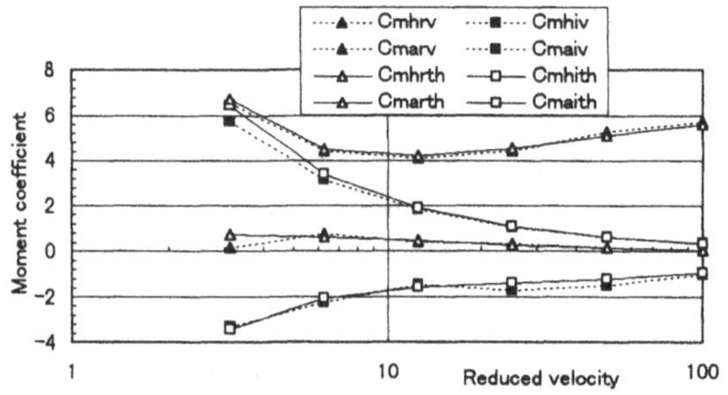

Figure 9 Moment coefficient of flat plate, comparison of calculation by vortex method and analytical method.

5.2 Unsteady Aerodynamic forces on a flat plate

Calculation of unsteady aerodynamic forces on a flat plate was conducted to confirm the accuracy of calculation. The flat plate is vibrating in heaving and pitching motion. The amplitude of heaving motion is 0.025 of length and pitching motion is 0.05 radian. Calculated aerodynamic forces are shown in figure 8 and 9. Real part means the force component in phase with the vibration of plate and imaginary part is the component of 1.5708 radian go ahead of the vibration.

As seen in the figures, the calculated values by vortex method are in good agreement with the analytical values in wide range of reduced velocity.

6. EQUATION OF FLUTTER

The unsteady aerodynamic force must be arranged to be able to calculate the flutter velocity of structure. At first, we must obtain the vibration mode in which the structure will vibrate in the wind but it is not able to know the flutter mode beforehand. So we assume that the flutter mode is made up of combination of the modes of natural frequency. We introduce four modes corresponding to the first to forth natural frequency. The natural frequencies and modes are expressed as follows,

$$f_1,\ \psi_1(x),\ f_2,\ \psi_2(x),\ f_3,\ \psi_3(x),\ f_4,\ \psi_4(x),$$

The equation of motion of mass-spring system can be expressed as follows,

$$m\ddot{y} + D\dot{y} + ky = 0 \tag{17}$$

In the case of two-dimensional flat plate, the displacement y is the function of position x. So the displacement is expressed as follows,

Displacement

$$y(x,t) = \phi(t)\sum y_j = \phi(t)\sum a_j\psi_j(x) \tag{18}$$

where

a_j : Amplitude of displacement

$\psi_j(x)$: Spatial mode of displacement

$\phi(t)$: Time mode

Forces produced by the vibration are expressed as follows,

Inertia force

$$F_m = m\ddot{y} = m\ddot{\phi}(t)\sum a_j\psi_j(x) \tag{19}$$

Damping force

$$F_D = D\dot{y} = D\dot{\phi}(t)\sum a_j\psi_j(x) \tag{20}$$

Stiffness force

$$F_K = Ky = K\phi(t)\sum a_j\psi_j(x) \tag{21}$$

The aerodynamic forces on the deck are also functions of time and point as shown in figure 10.
The pressure expressed as follows,

$$p(x,t) = \sum(p_j(x,t)) \tag{22}$$

Total force on the plate is expressed as follow,

$$F_T = F_m + F_D + F_K - P = 0 \tag{23}$$

Applying the virtual work method, the following equation is obtained,

$$\ddot{\phi}\int m\left(\psi_i\sum a_j\psi_j\right)dx + \dot{\phi}\int D\left(\psi_i\sum a_j\psi_j\right)dx$$
$$+\phi\int K\left(\psi_i\sum a_j\psi_j\right)dx - \int p(x,t)\psi_i dx = 0 \tag{24}$$

The fourth term of equation (24) is the aerodynamic term, so the integration of pressure is necessary to calculate the flutter. It is considered that the pressure is generated by each vibration modes, so the following equation is obtained,

$$p_j(x,t) = a_j p_{cj}(x,t) \tag{25}$$

Integrating the pressure multiplied by vibration mode, the following equation is obtained;

$$\int p(x,t)\psi_i dx = \int \psi_i \sum a_j p_{cj}(x,t)dx \tag{26}$$

Now, freezing the time, the instantaneous pressure can be expressed as follows,

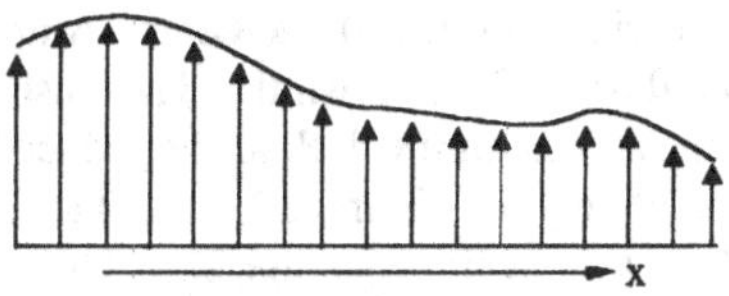

Figure 10 Instantaneous pressure distribution on plate

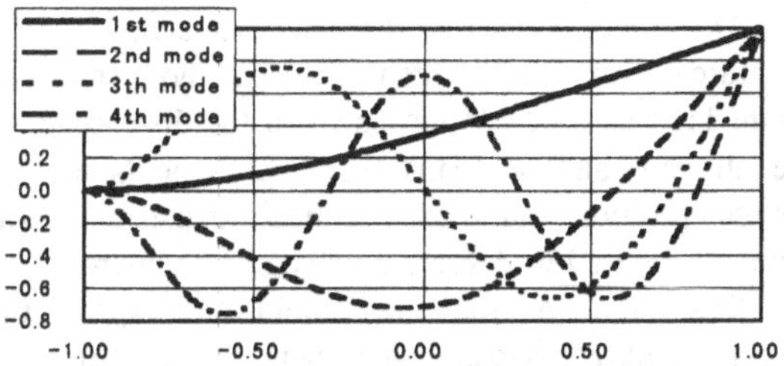

Figure 11 Mode of natural frequency of flat plate clamped one side

$$\psi_{pij}(t) = \int \psi_i p_{cj}(x,t)dx \tag{27}$$

When the structure vibrates with a steady sinusoidal motion, the vibration can be expressed as follows,

$$\phi(t) = \cos(\omega t) \tag{28}$$

Then, component of pressure in phase with the vibration is

$$\psi_{pRij} = \frac{1}{\pi}\int_0^{2\pi}\psi_{pij}(t)\cos(\omega t)dt \tag{29}$$

Component of out of phase is

$$\psi_{pIij} = \frac{-1}{\pi}\int_0^{2\pi}\psi_{pij}\sin(\omega t)dt \tag{30}$$

Then, the pressure can be expressed as follows,

$$\psi_{pij}(t) = \psi_{cpij}\phi(t) = \left(\psi_{pRij} + i\psi_{pIij}\right)\phi(t) \tag{31}$$

If the pressure distribution at arbitrary time can be expressed by the summation of equation (31), the flutter equation can be calculated and the solution is obtained. So, it is necessary to reduced(24) the wind force in the form of equation(31).

7. NATURAL MODE OF FLOATING STRUCTURE

It was shown that the vortex method with sink and source is able to calculate reasonable unsteady aerodynamic forces, so the forces on floating structure were calculated. At first, we assumed the vibration mode in which the structure vibrates in the wind will consists of the modes of natural frequencies.

Four natural modes of flat plate clamped at one side were used in this calculation. Mode shapes corresponding to first to forth natural frequencies are shown in figure 11.

8. AERODYNAMIC FORCE ON FLOATING STRUCTURE

8.1 Pressure Distribution

length was divided in to 80 elements. Amplitude of vibration is 0.025 of deck width. The instantaneous pressure distributions on the deck at the reduced velocity of 10.0 are shown in figure 12, 13, 14, 15 which correspond to the p_j in equation(22). To make non-dimensional value, the pressure was divided by dynamic presser and vibration amplitude of the floating structure.

The time 0 means that displacements of the structure due to vibration are the same as those shown in figure 11. The time $\pi/2$ means when the displacements are all zero and π means the displacements are all opposed positions and so on. The pressure distributions at time π are in 180 degrees revolved as compared with the time zero. At the time π and $3\pi/4$, all the displacements are zero and so, if there is no vibration, all the four pressure distributions must be the same and they are also the same as that shown in figure 7. However, They are different at time π and $3\pi/4$. These characteristics show that the pressures have phase differences and must be expressed by real part and imaginary part as shown in equation(31).

8.2 Aerodynamic force coefficient

Now, four modes are introduced in this paper, so there exists 16 kinds of aerodynamic forces and each force has real part and imaginary part. The aerodynamic force was made non-dimensional values as follows,

$$c_\psi = \psi_{cp} \Big/ \frac{1}{2}\rho u^2 y \qquad (32)$$

where y means the vibration amplitude.

Each coefficients are shown in figure 16, 17 18, 19. In the range where the reduced velocity is below 20, the aerodynamic coefficients change greatly with the reduced velocity. It means that the aerodynamic forces are strongly affected by the vibration of structure. However, on the range of reduced velocity above 20, the forces do not vary so much and imaginary parts reduce to zero. In this region, the frequency of vibration is so small that the phenomena can be considered as quasi-steady.

9. FLUTTER CALCULATION

It is necessary to get structural characteristics of the floating structure. Introducing the natural frequency, damping coefficient, equation(24) can be written as follow,

$$[\psi_{mij}]\{-(2\pi f)^2 a_j\phi_0\} + [i\psi_{mij} g(2\pi f_{ij})^2]\{a_j\phi_0\} + [\psi_{mij}(2\pi f_{ij})^2 - \psi_{cpij}]\{a_j\phi_0\} = 0 \qquad (33)$$

As the equation(33) is homogeneous, determinant of the coefficient matrix must be zero. So, the following equation is obtained,

$$-|\psi_{mij}| + (1+ig)\left|\psi_{mij}\left(\frac{f_{ij}}{f}\right)^2\right| - \frac{1}{(2\pi f)^2}|\psi_{cpij}| = 0 \qquad (34)$$

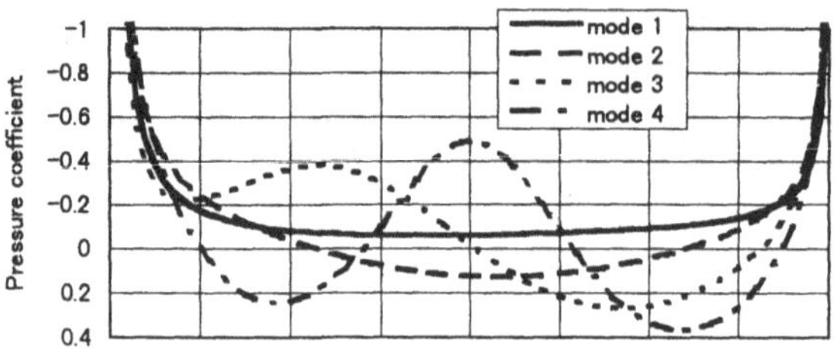

Figure 12 Pressure distribution in four modes at the time =0

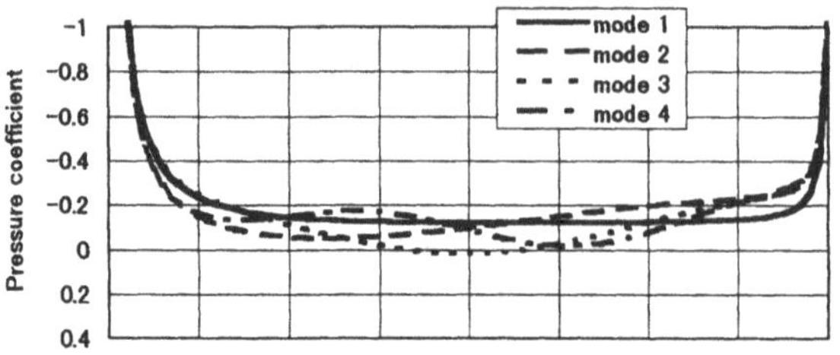

Figure 13 Pressure distribution in four modes at the time = π /

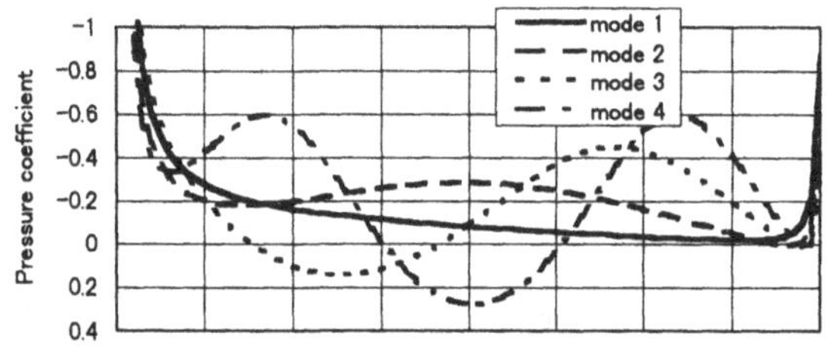

Figure 14 Pressure distribution in four modes at the time = π

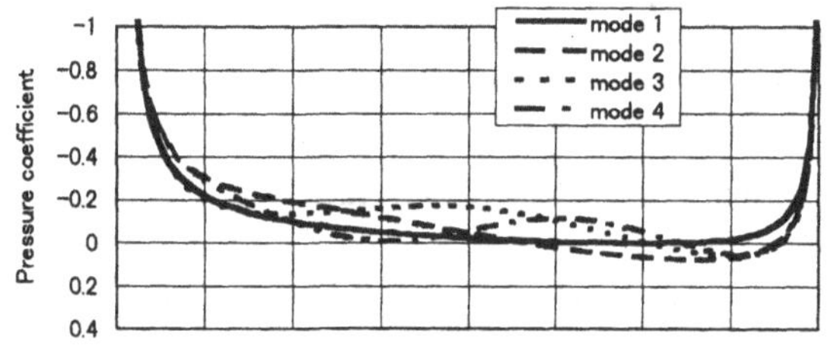

Figure 15 Pressure distribution in four modes at the time =3 π /4

Now, changing the equation into non-dimensional form, the following equation is obtained,

$$-[\psi^*_{mij}] + (1+ig)\left(\frac{f_0}{f}\right)^2\left[\psi^*_{mij}\left(\frac{f_{ij}}{f_0}\right)^2\right] - \frac{1}{2(2\pi)^2}u^{*2}[\psi^*_{pij}] = 0 \qquad (35)$$

The unknown values in the equation(35) are damping coefficient and frequency and so we put the term as follows,

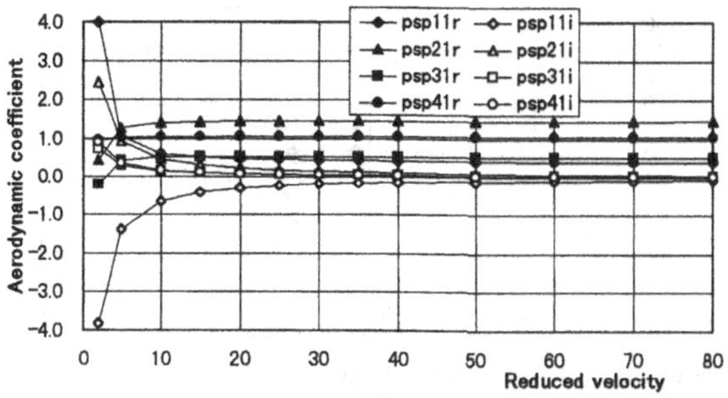

Figure 16 Aerodynamic coefficients vibrating in the first mode

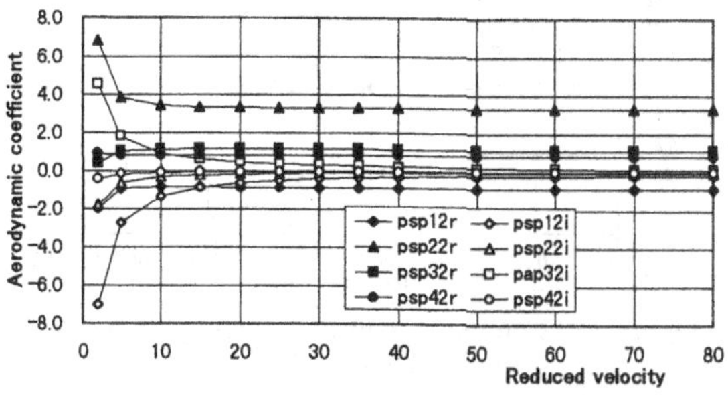

Figure 17 Aerodynamic coefficients vibrating in the second mode

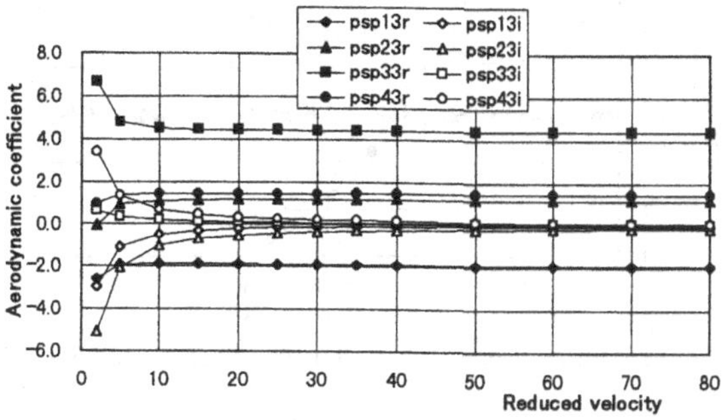

Figure 18 Aerodynamic coefficients vibrating in the third mode

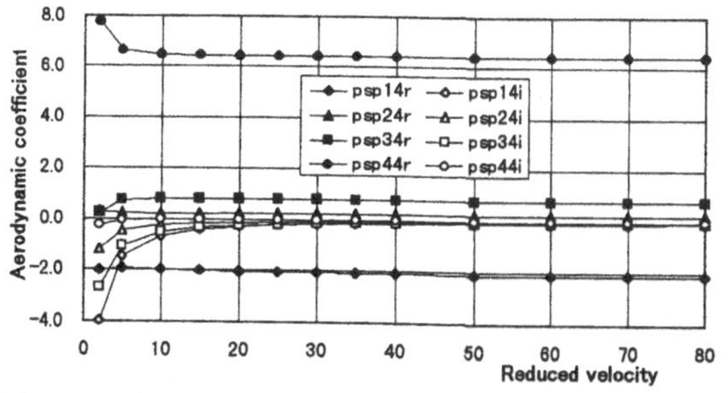

Figure 19 Aerodynamic coefficients vibrating in fourth mode

$$x = (1+ig)\left(\frac{f_0}{f}\right)^2 \tag{36}$$

Substituting this into equation and change equation(35) a little, we can be obtained the following equation,

$$[I]x - \left(\left[\psi^*_{mij}\left(\frac{f_{ij}}{f_0}\right)^2\right]\right)^{-1}\left(\left[\psi^*_{mij}\right] + \frac{u^{*2}}{2(2\pi)^2}\left[\left(\psi^*_{pRij} + i\psi^*_{pIij}\right)\right]\right) = 0 \tag{37}$$

We can get eigenvalues and from the value, exciting force and frequency can be get as follows,

$$x = x_R + x_I = x_R\left(1 + \frac{x_I}{x_R}\right) \tag{38}$$

$$f = f_0 \times \sqrt{x_R} \tag{39}$$

$$g = x_I \times \frac{1}{\sqrt{x_R}} \tag{40}$$

The value of damping g means that the value is necessary to keep the steady vibration and so if the real damping coefficient is less than the calculated value, the flutter will occur and vibrate with large amplitude.

Mass ratios of each mode used here are all 116. Natural frequencies of floating structure are very difficult to calculate, because there is the interaction between structure and water. So the calculations were made by frequencies as a function.

At first, the frequencies of flat plate clamped on one side are used. The results are shown in figure 20. In this case flutter will occur at the reduced velocity over 300. This means the flutter does not occur. In the most serious case when the all frequencies are the same, the results are shown in figure 21. The structural damping is less than 0.1, the flutter will occur at very small velocity. However, the damping coefficients of floating structures are thought to be more than 0.6, so the flutter does not occur.

In the case where the natural frequencies of each mode are 5% different each other, the results are shown in figure 22. The most possible flutter velocity is about 21 but the damping coefficient is less than 0.6 so the flutter does not occur. In the case of 10% difference of natural frequency shown in figure 23, the most possible velocity is about 30 and the damping coefficient is also increased from that of 5% difference. The figure 23, 24, 25 show that as the frequency difference increase, the most possible flutter velocity becomes higher and the damping coefficient also becomes greater.

In the case of this study, the reduced velocity which the structure will encounter on water is less than about 20-30 and so it is clarified that the flutter dose not occur on this structure.

10. DISCUSSION AND CONCLUSION

The vortex method including sink and source was used to get the unsteady aerodynamic force on flexible structure. The static pressure distribution on floating structure is rather in good agreement with the experiment and unsteady aerodynamic forces on vibration flat plate are also in good agreement. So, it is conceivable that this method can be applied to calculate the aerodynamic forces on flexible floating structure.

Flutter calculation was conducted with the calculated unsteady aerodynamic forces and it was clarified that the variety of natural frequency makes the flutter speed and damping force great. However the natural frequency have so great effect on the flutter speed that it is necessary to get correct natural frequencies.

REFERENCE

(1) Lewis, R. I. (1991), "Vortex element methods for fluid dynamic analysis of engineering system" Cambridge University Press UK.

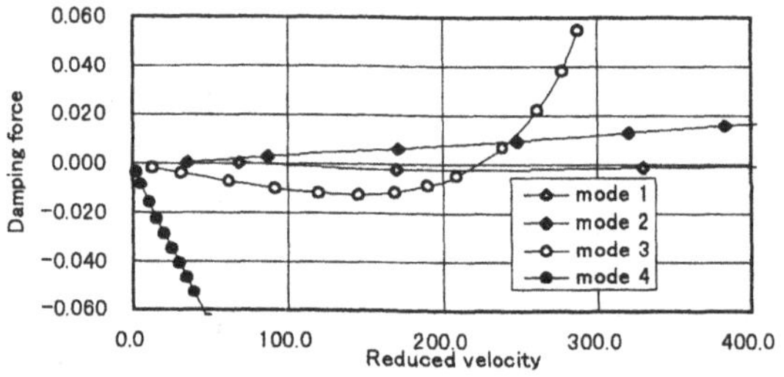

Figure 20 Flutter calculation, frequency =1.0, 6.27, 17.5, 34.5

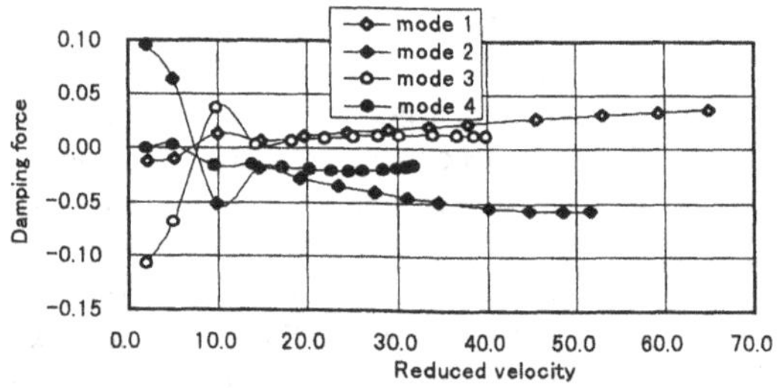

Figure 21 Flutter calculation, frequency = 1.0, 1.0, 1.0, 1.0

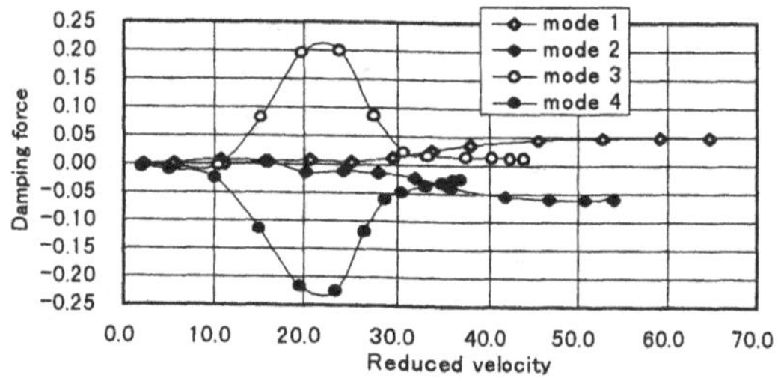

Figure 22 Flutter calculation, frequency = 1.0, 1.05, 1.103,1.058

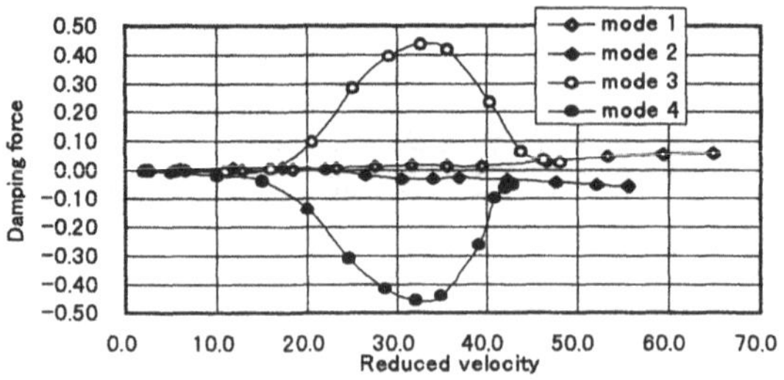

Figure 23 Flutter calculation, frequency =1.0, 1.1, 1.21,1.33

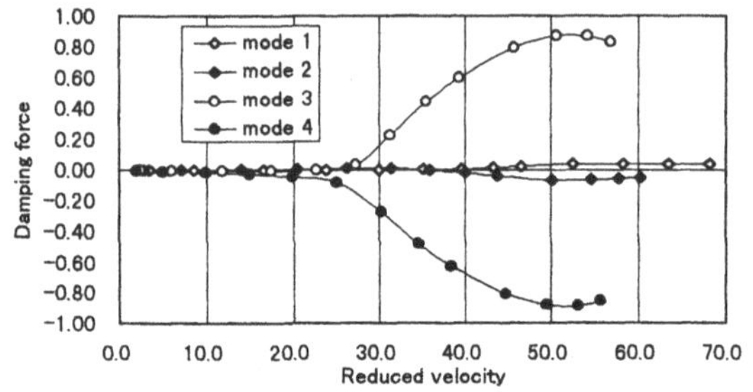

Figure 24 Flutter calculation, frequency =1.0, 1.2, 1.44,1.728

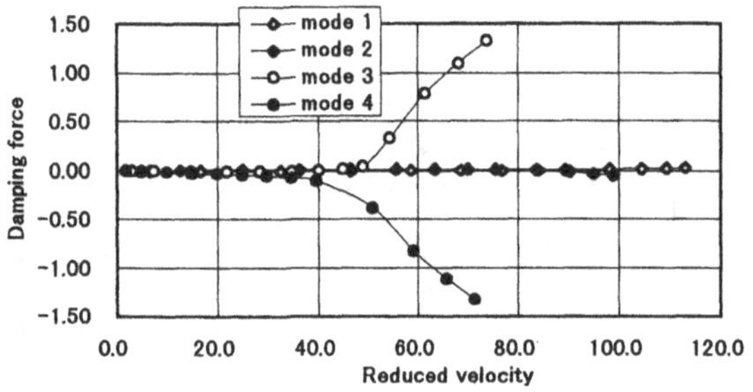

Figure 25, Flutter calculation, frequency = 1.0, 1.5, 2.25,3.375

CONVERGENCE STUDY FOR THE VORTEX METHOD WITH BOUNDARIES

Lung-an Ying
School of Mathematical Sciences
Peking University
Beijing 100871 P.R. China
Email yingla@sxx0.math.pku.edu.cn

ABSTRACT

We present some results on the convergence problems for the vortex method to viscous and inviscid incompressible flow with boundaries. To study the vortex method for the Navier-Stokes equation, one problem is viscous splitting. Within each time step the equation is split into an Euler equation and a Stokes equation, so a vorticity creation operator is defined after each inviscid step to maintain the non-slip condition. We will present some results about the viscous splitting. The Chorin-Marsden formula is one theoretical formulation to the viscous splitting. We will present a convergence result for this formula on convex domains. To study the convergence problem for the vortex blob method, we devote ourselves on the study for the initial-boundary value problems of the Euler equations, which is one step in the numerical algorithm. We will present some convergence results on two dimensional and three dimensional flows.

1 INTRODUCTION

The advantages of the vortex method are: relatively small computer storage is needed, easy to deal with the convection term by using the characteristic method, the mesh is self-adaptive, convenient for high Reynolds number flow, and easy to deal with the boundary condition at infinity. There are a great deal of complex mechanical phenomena in vortex flow. The vortex method is an effective approach to simulate these phenomena, especially those near the boundary of a rigid body. Therefore it is of extremly importance to handle the boundary conditions appropriately. In this respect there are some different ways . We will discuss some mathematical problems arising from the convergence problem of the vortex methods with the emphasis on the presence of boundaries.

One problem is viscous splitting. In the vortex method vortices move along particle trajectories, while the viscosity is treated in different approaches, either by a random walk procedure or a deterministic method. Therefore in the real computation it is a fractional steps method, and viscous splitting is frequently applied. To study the viscous splitting, the presence of boundaries leads to new difficulties. The boundary conditions for the Euler equation and the Stokes equation are different, so the types of boundary conditions is changed frequently, which creates singularities. We will discuss this problem in details.

The other problem is the convergence of the vortex blob method. We will discuss this problem for the Euler equations. Although the flow near boundaries is relatively simple for inviscid flow, the convection step is the main step in the vortex method even if viscous flow is considered. To understand the convergence property it is necessary to study this method applied to the Euler equations. With the presence of boundaries there are some new difficulties. For example the recovery of the velocity from the vorticity is more complicated. Another difficulty is that the vortex blobs may move across the boundaries and it makes no sense if a particle is out of the flow field. We will discuss this problem in details, too.

2 VISCOUS SPLITTING

We consider the following initial value problem:

$$\frac{\partial u}{\partial t} + (u \cdot \nabla)u + \frac{1}{\rho}\nabla p = \nu \triangle u + f,$$

$$\nabla \cdot u = 0,$$

$$u|_{t=0} = u_0,$$

where u is the velocity in two or three dimensional spaces, p is the pressure, f is the body force, ρ is the constant density, and ν is the constant kinetic viscosity.

Let $k > 0$ be the length of temporal steps, and let $\tilde{u}_k$, u_k be the solutions to:

(A)

$$\frac{\partial \tilde{u}_k}{\partial t} + (\tilde{u}_k \cdot \nabla)\tilde{u}_k + \frac{1}{\rho}\nabla \tilde{p}_k = f,$$
$$\nabla \cdot \tilde{u}_k = 0,$$
$$\tilde{u}_k(ik) = u_k(ik - 0),$$
$$\frac{\partial u_k}{\partial t} + \frac{1}{\rho}\nabla p_k = \nu \triangle u_k,$$
$$\nabla \cdot u_k = 0,$$
$$u_k(ik) = \tilde{u}_k((i+1)k - 0),$$

for $t \in [ik, (i + 10k)$, $i = 0, 1, \cdots$, where $u_k(-0) = u_0$.

Or consider a second order scheme

(B)

$$\frac{\partial \tilde{u}_k}{\partial t} + (\tilde{u}_k \cdot \nabla)\tilde{u}_k + \frac{1}{\rho}\nabla \tilde{p}_k = f, \quad t \in [ik, (i+\frac{1}{2})k),$$
$$\nabla \cdot \tilde{u}_k = 0, \quad t \in [ik, (i+\frac{1}{2})k),$$
$$\tilde{u}_k(ik) = \tilde{u}_k(ik - 0),$$
$$\frac{\partial u_k}{\partial t} + \frac{1}{\rho}\nabla p_k = \nu \triangle u_k, \quad t \in [ik, (i+1)k),$$
$$\nabla \cdot u_k = 0, \quad t \in [ik, (i+1)k),$$
$$u_k(ik) = \tilde{u}_k((i+\frac{1}{2})k - 0),$$

$$\frac{\partial \tilde{u}_k}{\partial t} + (\tilde{u}_k \cdot \nabla)\tilde{u}_k + \frac{1}{\rho}\nabla \tilde{p}_k = f, \quad t \in [(i+\frac{1}{2})k, (i+1)k),$$
$$\nabla \cdot \tilde{u}_k = 0, \quad t \in [(i+\frac{1}{2})k, (i+1)k),$$
$$\tilde{u}_k((i+\frac{1}{2})k) = u_k((i+1)k - 0).$$

It is proved by Beale and Majda(1) that (for $f = 0$)

Theorem Let $ik \leq T$, and take an arbitrary $m \geq 0$, then there is a constant $s \geq 0$, such that if

$$||\nabla \wedge u_0|| \leq M_1, \quad ||u_0||_s \leq M_2,$$

then the solution u_k to the scheme (A) satisfies

$$||u_k(ik) - u(ik)||_m \leq C\nu k,$$

and the solution u_k to the scheme (B) satisfies

$$||u_k(ik) - u(ik)||_m \leq C\nu k^2,$$

where u is the exact solution and C depends only on m, M_1, M_2, and T.

Next, let us consider the initial-boundary value problem

$$\frac{\partial u}{\partial t} + (u \cdot \nabla)u + \frac{1}{\rho}\nabla p = \nu \triangle u + f,$$
$$\nabla \cdot u = 0,$$
$$u|_{\partial\Omega} = 0$$
$$u|_{t=0} = u_0,$$

where Ω is a bounded domain with smooth boundary $\partial\Omega$.

Let the approximate solutions $\tilde{u}_k, u_k$ satisfy

$$\frac{\partial \tilde{u}_k}{\partial t} + (\tilde{u}_k \cdot \nabla)\tilde{u}_k + \frac{1}{\rho}\nabla \tilde{p}_k = f,$$
$$\nabla \cdot \tilde{u}_k = 0,$$
$$\tilde{u}_k \cdot n|_{x\in\partial\Omega} = 0,$$
$$\tilde{u}_k(ik) = u_k(ik - 0),$$
$$\frac{\partial u_k}{\partial t} + \frac{1}{\rho}\nabla p_k = \nu \triangle u_k,$$
$$\nabla \cdot u_k = 0,$$
$$u_k|_{x\in\partial\Omega} = 0,$$
$$u_k(ik) = \tilde{u}_k((i+1)k - 0),$$

where n is the unit outward normal vector on $\partial\Omega$.

The convergence problem has been studied by Douglis, Fabes, Alessandrini(2)(3), Zheng, Huang(4), Zhang(5), and Ying(6).

Theorem If $2 \leq s < \frac{5}{2}$, and u, u_0, f are sufficiently smooth, then there exists a constant $k_0 > 0$ such that if $0 < k \leq k_0$, then

$$||\tilde{u}_k(t)||_s \leq M,$$

$$||\tilde{u}_k(t) - u(t)||_r \leq \begin{cases} M'k, & 0 \leq r < \frac{1}{2}, \\ M'k^{\frac{s-r}{s-\theta}}, & \frac{1}{2} \leq r \leq s, \end{cases}$$

where $0 < \theta < 1$.

To prove this assertion, we introduce an auxiliary problem:

$$\frac{\partial \tilde{u}^*}{\partial t} + \frac{1}{\rho}\nabla \tilde{p}^* = f - (u \cdot \nabla)u,$$
$$\nabla \cdot \tilde{u}^* = 0,$$
$$\tilde{u}^* \cdot n|_{x\in\partial\Omega} = 0,$$
$$\tilde{u}^*(ik) = u^*(ik - 0),$$
$$\frac{\partial u^*}{\partial t} + \frac{1}{\rho}\nabla p^* = \nu \triangle u^*,$$
$$\nabla \cdot u^* = 0,$$
$$u^*|_{x\in\partial\Omega} = 0,$$
$$u^*(ik) = \tilde{u}^*((i+1)k - 0).$$

Introducing the Stokes operator $A = -P\Delta$, where P is the Helmholtz operator and Δ is the Laplace operator, the solution u^* can be expressed in terms of

$$u^*(t) = e^{-\nu t A} u_0 + \sum_{i=0}^{[t/k]} \int_{ik}^{(i+1)k} e^{-\nu(t-ik)A} P(f - (u\cdot\nabla)u)(\tau)\, d\tau,$$

then $\tilde{u}^* - u$ can be estimated.

The convergence result relies on the following estimate:

$$\|u_k(t)\|_3 \le C(t-(i+\frac{1}{2})k)^{\frac{s-3}{2}}, \quad 2 \le s < \frac{5}{2},$$

$$\|\tilde{u}_k(t)\|_s \le C(\|u_k(ik-0)\|_s + 1), \quad 2 \le s < \frac{5}{2},$$

$$\|\tilde{u}^*(t) - \tilde{u}_k(t)\|_1 \le Ck.$$

In 1978 Chorin, Hughes, McCracken, and Marsden(7) proposed the following product formula for the initial-boundary value problems:

$$u_k(ik) = (H_k \circ \phi \circ E_k)^i u_0,$$

where E_k is the solver of local Eular flow, H_k is the solver of the heat equation, and ϕ is the "vorticity creation operator". To study the approximate schemes with the additional operator ϕ, we consider the following scheme:

$$\frac{\partial \tilde{u}_k}{\partial t} + (\tilde{u}_k \cdot \nabla)\tilde{u}_k + \frac{1}{\rho}\nabla \tilde{p}_k = f,$$

$$\nabla \cdot \tilde{u}_k = 0,$$

$$\tilde{u}_k \cdot n|_{x\in\partial\Omega} = 0,$$

$$\tilde{u}_k(ik) = u_k(ik-0),$$

$$\frac{\partial u_k}{\partial t} + \frac{1}{\rho}\nabla p_k = \nu \Delta u_k + \frac{1}{k}(I-\phi)\tilde{u}_k((i+1)k-0),$$

$$\nabla \cdot u_k = 0,$$

$$u_k|_{x\in\partial\Omega} = 0,$$

$$u_k(ik) = \phi \tilde{u}_k((i+1)k-0).$$

The nonhomogeneous term at the diffusion step is designed to maintain consistency.

Theorem If $2 < s < \frac{5}{2}$, u, u_0, f are sufficiently smooth, and $\phi : \{u \in (H^1(\Omega))^N; \nabla\cdot u = 0\} \to \{u \in (H_0^1(\Omega))^N; \nabla \cdot u = 0\}$ is bounded in the norm $\|\cdot\|_r$ with $\frac{1}{2} < r \le s$, where N is the dimension of the domain Ω, then there exists a constant $k_0 > 0$ such that if $0 < k \le k_0$, then

$$\|\tilde{u}_k(t)\|_s, \|u_k(t)\|_s \le M,$$

$$\|\tilde{u}_k(t) - u(t)\|_1, \|u_k(t) - u(t)\|_1 \le M'k.$$

In the paper (7) the Chorin-Marsden formula was proved convergent for linear equations, where ϕ is defined as: the velocity field is extended oddly to the exterior to create a vortex sheet, and the heat equation is solved on the whole space without any boundary condition. For nonlinear equations the convergence problem is more difficult, since under this vorticity creation operator solutions are singular, and the estimates are hard to be given. Neither in the convection step nor in the diffusion step the boundary condition $u|_{x\in\partial\Omega} = 0$ is involved in the scheme.

The result by Benfatto and Pulvirenti(8)(9) in 1986 is the following: Let $\Omega = R_+^2 = \{(x_1, x_2); x_2 > 0\}$. In the operator ϕ the tangential component of the velocity is expended oddly, and the normal component of the velocity is extended evenly. It is easy to verify that under this extension the velocity field keeps incompressible on the whole plane. On the other hand the extension by Chorin and Marsden does not satisfy this requirement.

The scheme is the following:

$$F = -\frac{\partial f_1}{\partial x_2} + \frac{\partial f_2}{\partial x_1},$$

$$\omega_k = -\frac{\partial (u_k)_1}{\partial x_2} + \frac{\partial (u_k)_2}{\partial x_1},$$

$$\frac{\omega_k(x,(i+1)k) - \omega_k(x,ik)}{k} + u_k(x,ik)\cdot\nabla\omega_k(x,ik) = F(x,ik),$$

$$-\Delta \psi_k = \omega_k, \quad \psi_k|_{x_2=0} = 0,$$

$$u_k(x,(i+1)k) = \left(\frac{\partial\psi_k}{\partial x_2}, -\frac{\partial\psi_k}{\partial x_1}\right)(x,(i+1)k),$$

$$(u_k)_1(x_1,x_2,(i+1)k) = -(u_k)_1(x_1,-x_2,(i+1)k),$$

$$(u_k)_2(x_1,x_2,(i+1)k) = (u_k)_2(x_1,-x_2,(i+1)k),$$

$$\frac{\partial \tilde{u}_k}{\partial t} = \nu\Delta\tilde{u}_k, \quad x \in R^2,$$

$$\tilde{u}_k(x,ik) = u_k(x,(i+1)k),$$

$$u_k(x,(i+1)k) = \tilde{u}_k(x,(i+1)k-0).$$

Theorem If $u_0 \in W^{2,2}(\Omega)$, $\omega_0 = -\frac{\partial(u_0)_1}{\partial x_2} + \frac{\partial(u_0)_2}{\partial x_1}$,

$$\left\|\frac{\partial^{\alpha_1+\alpha_2}\omega_0}{\partial x_1^{\alpha_1}\partial x_2^{\alpha_2}}\right\| < +\infty, \quad 0 \le \alpha_1 \le 6, \quad 0 \le \alpha_2 \le 2,$$

then for any $\beta \in (0, \frac{1}{4})$ it holds that

$$\lim_{\substack{i\to\infty,\ k\to+0\\ ik=t}} \|\omega(t) - \omega_k(t)\| k^{-\beta} = 0.$$

where

$$\|f\| = \int \sup_{x_2} |\hat{f}(p,x_2)|\, dp,$$

and $\hat{f}$ is the Fourier transform of f in x_1.

We turn now to consider more general domains. Let Ω be a bounded convex domain in R^2 with smooth boundary $\partial\Omega$. The operator ϕ is defined as follows: We take a positive constant d, then consider the set of all straight line segments through $\partial\Omega$ and normal to it, and the length of each segment is d inside and d outside. The union of line segments is a tubular neighborhood of $\partial\Omega$, denoted by S. $\Phi : S \to S$ is the map which reflects across the boundary relative to those line segments. Φ is a smooth mapping. Let $J(x)$ be the Jacobian of Φ at point x. The approximate solutions are solved by the following scheme.

$$\phi\tilde{u}_k(x,ik-0) = \begin{cases} \tilde{u}_k(x,ik-0), & x \in \bar{\Omega}, \\ -|J(x)|\tilde{u}_k(\Phi(x),ik-0), & x \in S\setminus\bar{\Omega}, \\ 0, & x \notin S\cup\Omega, \end{cases}$$

$$\frac{\partial u_k^*}{\partial t} = \nu \Delta u_k^*, \quad x \in R^2,$$

$$u_k^*|_{t=ik} = \phi\tilde{u}_k(x,ik-0),$$

$$\frac{\partial \tilde{u}_k}{\partial t} + (Pu_k^* \cdot \nabla)\tilde{u}_k + \frac{1}{\rho}\nabla\tilde{p}_k = f,$$

$$\nabla \cdot \tilde{u}_k = 0,$$

$$\tilde{u}_k \cdot n|_{x\in\partial\Omega} = 0,$$

$$\tilde{u}_k|_{t=ik} = Pu_k^*(x,(i+1)k-0),$$

Theorem If $u_0 \in X = \{u \in (L^2(\Omega))^2; \nabla\cdot u = 0, u \cdot n|_{\partial\Omega} = 0\}$, $f \in L^2(0,T;X)$, then the weak solution $u \in L^2(0,T;(H_0^1(\Omega))^2) \cap L^\infty(0,T;X)$ is the limit of $\tilde{u}_k, u_k, u_k^*$ in the sense of

$$u_k, \tilde{u}_k \to u(L^\infty(0.T;X), \text{weak}*),$$

$$u_k^* \to u(L^\infty(0.T;(L^2(\Omega))^2), \text{weak}*),$$

$$u_k, u_k^* \to u(L^2(0.T;(H^1(\Omega))^2), \text{weak}),$$

$$\tilde{u}_k \to u(L^2(0.T;(H^\gamma(\Omega))^2), \gamma < 0),$$

$$u_k, u_k^* \to u(L^2(0.T;(H^s(\Omega))^2), s < 1).$$

The proof of this theorem relies on the study of the following problem:

$$\frac{\partial h}{\partial t} = \nu \Delta u, \quad x \in R^2,$$

$$u|_{t=0} = \phi h_0.$$

Lemma If $h_0 \in L^2(\Omega)$, $h_0 \geq 0$, then there exists $T^* > 0$ such that $h \geq 0$, $\frac{\partial h}{\partial n} \leq 0$ on $\partial\Omega \times (0,T^*]$.

The solution can be expressed explicitly:

$$\begin{aligned} h(x,t) &= \int \frac{1}{4\pi t\nu} e^{-\frac{|x-\xi|^2}{4t\nu}} \phi h_0(\xi)\, d\xi \\ &= \int_{\Omega\setminus S} \frac{1}{4\pi t\nu} e^{-\frac{|x-\xi|^2}{4t\nu}} h_0(\xi)\, d\xi \\ &+ \int_{\Omega\cap S} \frac{1}{4\pi t\nu} \left(e^{-\frac{|x-\xi|^2}{4t\nu}} \right. \\ &\left. - e^{-\frac{|x-\Phi(\xi)|^2}{4t\nu}} \right) h_0(\xi)\, d\xi. \end{aligned}$$

$$|x - \Phi(\xi)| \geq |x - \xi|, \quad x \in \partial\Omega, \xi \in \Omega,$$

so $h|_{\partial\Omega} \geq 0$. The estimate for $\frac{\partial h}{\partial n}$ is technical.

lemma If $h_0 \in L^2(\Omega)$ then

$$\|h(t)\|_{0,\Omega}^2 + \nu \int_0^t |u(\tau)|_{1,\Omega}^2\, d\tau \leq \|h(0)\|_{0,\Omega}^2$$

for $t \in (0,T^*]$.

Applying the above results to the case of half plane it can be seen that convergence results are valid for these cases:

(1) Benfatto-Pulvirenti

$$(u_k^*)_1(x,ik) = \begin{cases} (\tilde{u}_k)_1(x_1,x_2,ik-0), & x \in \Omega, \\ -(\tilde{u}_k)_1(x_1,-x_2,ik-0), & x \notin \Omega. \end{cases}$$

$$(u_k^*)_2(x,ik) = \begin{cases} (\tilde{u}_k)_2(x_1,x_2,ik-0), & x \in \Omega, \\ (\tilde{u}_k)_2(x_1,-x_2,ik-0), & x \notin \Omega. \end{cases}$$

(2) Chorin-Marsden

$$(u_k^*)_1(x,ik) = \begin{cases} (\tilde{u}_k)_1(x_1,x_2,ik-0), & x \in \Omega, \\ -(\tilde{u}_k)_1(x_1,-x_2,ik-0), & x \notin \Omega. \end{cases}$$

$$(u_k^*)_2(x,ik) = \begin{cases} (\tilde{u}_k)_2(x_1,x_2,ik-0), & x \in \Omega, \\ -(\tilde{u}_k)_2(x_1,-x_2,ik-0), & x \notin \Omega. \end{cases}$$

(3) Euler equation for the convection step

$$\frac{\partial \tilde{u}_k}{\partial t} + (\tilde{u}_k \cdot \nabla)\tilde{u}_k + \frac{1}{\rho}\nabla\tilde{p}_k = f,$$

$$\nabla \cdot \tilde{u}_k = 0.$$

The result has been extended to 3-D cases (Lin).

3 VORTEX METHOD FOR THE EULER EQUATIONS

We consider the initial value problem

$$\frac{\partial u}{\partial t} + (u \cdot \nabla)u + \frac{1}{\rho}\nabla p = f,$$

$$\nabla \cdot u = 0,$$

$$u|_{t=0} = u_0, \quad u|_{|x|=\infty} = u_\infty.$$

For two dimensional problems let $\omega = -\nabla \wedge u = -\frac{\partial u_1}{\partial x_2} + \frac{\partial u_2}{\partial x_1}$, then ω satisfies

$$\frac{\partial \omega}{\partial t} + u \cdot \nabla\omega = F = -\nabla \wedge f.$$

By introduction the stream function ψ, we have $u = \nabla \wedge \psi$ and

$$-\Delta\psi = \omega.$$

The particle trajectory $x = x(t)$ is defined by

$$\frac{dx}{dt} = u(x,t),$$

along which it holds that

$$\frac{d\omega}{dt} = F.$$

For the convenience of exposition, we assume that f is potential, then $F = 0$.

A vortex blob function ζ is defined to satisfy

$$\int_{R^2} \zeta(x)\, dx = 1.$$

The following moment conditions are always assumed:

$$\int_{R^2} x^\alpha \zeta(x)\, dx = 0, \quad 1 \le |\alpha| \le k-1,$$

$$\int_{R^2} |x|^k |\zeta(x)|\, dx < \infty.$$

Let

$$K = \frac{1}{2\pi |x|^2}(-x_2, x_1),$$

$$\zeta_\varepsilon(x) = \frac{1}{\varepsilon^2}\zeta\left(\frac{x}{\varepsilon}\right),$$

then the semi-discretization scheme reads

$$\omega^\varepsilon(x,t) = \sum_j \alpha_j \zeta_\varepsilon(x - X_j^\varepsilon(t)),$$

$$\frac{dX_j^\varepsilon}{dt} = u^\varepsilon(X_j^\varepsilon(t), t), \quad X_j^\varepsilon(0) = X_j,$$

$$u^\varepsilon(x,t) = \int K(x-\xi)\omega^\varepsilon(\xi,t)\, d\xi + u_\infty,$$

where $\sum_j \alpha_j \zeta_\varepsilon(x - X_j)$ is the approximation of $\omega_0 = -\nabla \wedge u_0$.

For three dimensional problems the equations are different.

$$\omega = \mathrm{curl} u, \quad \omega_0 = \mathrm{curl} u_0,$$

$$\frac{\partial \omega}{\partial t} + (u\cdot\nabla)\omega - (\omega\cdot\nabla)u = 0,$$

$$K(x) = -\frac{1}{4\pi}\frac{x}{|x|^3},$$

$$u(x,t) = \int_{R^3} K(x-\xi)\times\omega(\xi,t)\, d\xi + u_\infty.$$

It holds that

$$\frac{d\omega}{dt} = (\omega\cdot\nabla)u,$$

or

$$\omega(\xi(t;x,0),t) = \frac{\partial \xi(t;x,0)}{\partial x}\omega_0(x),$$

where $\xi = \xi(\tau; x, t)$ satisfies

$$\frac{d\xi}{dt} = u(\xi(\tau;x,t),\tau).$$

If ω_0 is compactly supported, let $A = \mathrm{supp}\omega_0$. We set $\Phi_t(\alpha) = \xi(t;\alpha,0)$, and $\Phi_{-t}(\alpha) = \xi(0;\alpha,t)$, then we define a smooth homeomorphism $T : B\times[0,2\pi] \to A$ such that $T(b,0) = T(b,2\pi), b \in B$ and

$$\begin{aligned} \omega_0(T(\tilde\alpha)) &= c(T(\tilde\alpha))\cdot\frac{\partial}{\partial\varphi}T(\tilde\alpha), \\ \tilde\alpha \in B\times[0,2\pi], &\qquad \varphi\in[0,2\pi], \end{aligned}$$

where c is a scalar function and $\frac{\partial T}{\partial\varphi}$ indicates the direction of ω_0. It is deducted that

$$\frac{\partial\omega(\alpha,t)}{\partial t} = c(\alpha)\frac{\partial}{\partial\varphi}\left(\frac{\partial\Phi}{\partial t}\circ T\right)T^{-1}(\alpha).$$

These different formulas for ω along particle trajectories lead different schemes.

Convergence problems have been studied by many authors: Hald, DelPrete(10)(11)(12), Beale, Majda(13)(14), Raviart(15), Anderson, Greengard(16)(17), Goodman(18), Hou, Lowengrub(19)(20)(21), Cottet(22)(23), Xin, Liu. The following is a typical result for two dimensional problems.

Theorem Assume the following hypotheses:

(a) ζ satisfies the moment conditions with $k \ge 2$,

(b) $\zeta \in W^{m-1,\infty}(R^2), m \ge 2$, and ζ is compactly supported,

(c) $C^{-1}\varepsilon^\alpha \le h \le C\varepsilon^\beta$, $\alpha \ge \beta > 1$,

where h is the mesh size. Then

$$\begin{aligned} &\|u(\cdot,t) - u^\varepsilon(\cdot,t)\|_{0,\infty,R^2} + \|e(t)\|_{0,\infty,h} \\ \le\ &\frac{C}{\varepsilon^s}\left(\varepsilon^k + \frac{h^m}{\varepsilon^{m-1}}\right), \quad s > 0, \end{aligned}$$

where

$$\|e(t)\|_{0,\infty,h} = \max_j |X_j(t) - X_j^\varepsilon(t)|.$$

The error estimates for full discretization have also been obtained. We will discuss it in details for initial-boundary value problems.

Next let us consider two dimensional initial-boundary value problems,

$$\frac{\partial u}{\partial t} + (u\cdot\nabla)u + \frac{1}{\rho}\nabla p = f,$$

$$\nabla\cdot u = 0,$$

$$u\cdot n|_{x\in\partial\Omega} = 0,$$

$$u|_{t=0} = u_0, \quad u|_{|x|=\infty} = u_\infty.$$

We assume that the initial vorticity is approximated by

$$\omega_0 = \sum_{j\in J}\alpha_j\zeta_\varepsilon(x - X_j), \quad X_j \in \Omega.$$

If the explicit Euler scheme with step length Δt is applied to the time discretization, then in a straight forward way the scheme is

$$\omega^i(x) = \sum_{j\in J^i}\alpha_j^i\zeta_\varepsilon(x - X_j^i),$$

$$J^i = \{j \in J^{i-1}; X_j^i \in \Omega\}, \quad j = 1, 2, \cdots, J^0 = J,$$

$$u^i(x) = \frac{1}{2\pi} \int_\Omega \frac{(-x_2 + \xi_2, x_1 - \xi_1)}{|x - \xi|^2} \omega^i(\xi)\, d\xi + \nabla \phi^i,$$

$$\Delta \phi^i = 0,$$

$$\left.\frac{\partial \phi^i}{\partial n}\right|_{x \in \partial\Omega}$$

$$= -\frac{1}{2\pi} \int_\Omega \frac{(-x_2 + \xi_2, x_1 - \xi_1)}{|x - \xi|^2} \omega^i(\xi)\, d\xi \cdot n \Big|_{x \in \partial\Omega},$$

$$\alpha_j^{i+1} = \alpha_j^i + \Delta t h^2 F(X_j^i, i\Delta t), \quad j \in J^i, \alpha_j^0 = \alpha_j,$$

$$X_j^{i+1} = X_j^i + \Delta t u^i(X_j^i), \quad j \in J^i, X_j^0 = X_j,$$

where the formula from the vorticity to the velocity is the well known Biot-Savart law.

It is still unknown whether this scheme converges or not. However if all particles $j \in J$ are taken into account, the convergence result is nice. To obtain X_j^{i+1} an extrapolation of the velocity field is needed. It is defined that

$$g^i(x) = \sum_{m=1}^{M} a_m u^i(x^{(m)}), \quad x \in R^2 \setminus \Omega,$$

where $x^{(m)} = (m+1)Y - mx$, Y is the point on $\partial\Omega$ nearest to x, and a_m satisfy the following algebraic system:

$$\sum_{m=1}^{M} (-m)^l a_m = 1, \quad l = 0, 1, \cdots, M-1.$$

The coordinates of particles satisfy

$$X_j^{i+1} = X_j^i + \Delta t g^i(X_j^i), \quad j \in J.$$

To solve the potential function ϕ we assume that a second order finite element scheme is applied with mesh size δ.

Theorem Assume the following hypotheses:

(a) ζ satisfies the moment conditions with $k = M \geq 3$,

(b) $\zeta \in W^{m+1,\infty}(R^2), m \geq 1$, and ζ is compactly supported,

(c) $C^{-1}\varepsilon^a \leq h \leq C\varepsilon^{1+\frac{k-1}{m}}$, $a \geq 1 + \frac{k-1}{m}$,

(d) $\delta \leq C\varepsilon$, $C^{-1}\delta^b \leq h \leq C\delta$, $b > 0$,

(e) $\Delta t \leq C\delta^2$, $\varepsilon^{k-1} \leq C\delta$,

then for $p \in [1, \infty)$

$$\|e^i\|_{0,p,h} + \|u^i(\cdot) - u(\cdot, i\Delta t)\|_{0,p} \leq C(\varepsilon^k + \delta^2 + \Delta t),$$

where

$$\|e^i\|_{0,p,h} = \left(\sum_j |X_j^i - X_j|^p h^2 \right)^{\frac{1}{p}}.$$

The extrapolation approach requires smoothness of the velocity. There are boundary layers for viscous flows. The gradient of the velocity field near the boundary is large, so it causes a large error. We propose another boundary correction technique here, which gets rid of extrapolation and yields a second order accuracy.

For each particle near the boundary we fix one reference point, $X_{j_0}^\varepsilon$, on the boundary at $t = 0$. For definiteness we may take $X_{j_0}^\varepsilon$ to be the nearest point to X_j. Since $u \cdot n|_{x \in \partial\Omega} = 0$, $X_{j_0}^\varepsilon \in \partial\Omega$ for all t. We define

$$X_{j'}^\varepsilon(t) = 2X_{j_0}^\varepsilon(t) - X_j^\varepsilon(t), \quad \alpha_{j'} = \alpha_j,$$

The vorticity is defined as

$$\omega^\varepsilon(x, t) = \sum_{j \in J \cup J(t)} \alpha_j \zeta_\varepsilon(x - X_j^\varepsilon(t)),$$

where $J = \{j; X_j \in \bar{\Omega}\}$, $J(t) = \{j'\}$.

For semi-discretization scheme it is proved that

Theorem Under the same hypotheses it holds that

$$\|u - u^\varepsilon\|_{0,p} + \|e(t)\|_{0,p,h} \leq C\varepsilon^2.$$

We turn now to study the initial-boundary value problems of three dimensional Euler equations. There is an essential difference between the problems with or without boundaries. If there is no boundary the Biot-Savart law,

$$u = -\frac{1}{4\pi} \int_{R^3} \frac{x - \xi}{|x - \xi|^3} \times \omega(\xi, t)\, d\xi + u_\infty$$

can be applied to recover velocity from vorticity. While it does not work for a domain Ω with boundaries. Since generally speaking the function

$$\tilde{\omega}(x, t) = \begin{cases} \omega, & x \in \Omega, \\ 0, & x \notin \Omega \end{cases}$$

does not satisfy $\nabla \cdot \tilde{\omega} = 0$, which is not the curl of any velocity field.

The following approach can be used for the recovery, which is a nonstandard boundary value problem of the Stokes equation,

$$-\Delta \psi + \nabla z = \omega,$$

$$\nabla \cdot \psi = 0,$$

$$z|_{\partial\Omega} = 0, \quad \psi \times n|_{\partial\Omega} = 0.$$

It is proved that for all vector field ω, not necessarily divergence free, the above problem is well-posed, and if $\omega = \text{curl} u$, then it gives u as the solution.

To study convergence continuous norms are more convenient than discrete norms. Therefore a continuous set of characteristics $\xi^\varepsilon(t; \eta, \tau)$ is defined as

$$\frac{d}{dt}\xi^\varepsilon(t; \eta, \tau) = u^\varepsilon(\xi^\varepsilon(t; \eta, \tau), t),$$

$$\xi^\varepsilon(\tau;\eta,\tau)=\eta,$$

Set $x(\eta,t) = \xi(t;\eta,0)$, $x^\varepsilon(\eta,t) = \xi^\varepsilon(t;\eta,0)$, and $e(t)=x(\cdot,t)-x^\varepsilon(\cdot,t)$, In the Lagrangian coordinates the vorticities α, α^ε satisfy

$$\frac{d\alpha}{dt}=(\alpha\cdot\nabla)u(x(\eta,t),t)+h^3F(x(\eta,t),t),$$

$$\alpha(\eta,0)=\omega_0(\eta)h^3,$$

$$\frac{d\alpha^\varepsilon}{dt}=(\alpha^\varepsilon\cdot\nabla)u^\varepsilon(x^\varepsilon(\eta,t),t)+h^3F(x^\varepsilon(\eta,t),t),$$

$$\alpha^\varepsilon(\eta,0)=\omega_0(\eta)h^3.$$

We define $\bar{\omega}(t)=(\alpha(\cdot,t)-\alpha^\varepsilon(\cdot,t))/h^3$.

The convergence result is the following:

Theorem If $p\in[1,\infty)$, the moment condition holds for $k\geq 3$, ζ is compactly supported and $\zeta\in W^{m+1,\infty}(R^3)$, $m\geq 3$, and $h\leq C\varepsilon^2$, then

$$\|u-u^\varepsilon\|_{1,p}+\|e(t)\|_{1,p}+\|\bar{\omega}(t)\|_{0,p}\leq C\left(\varepsilon^k+\frac{h^m}{\varepsilon^m}\right),$$

and

$$\|u-u^\varepsilon\|_{2,p}+\|e(t)\|_{2,p}+\|\bar{\omega}(t)\|_{1,p}\leq C\left(\varepsilon^{k-1}+\frac{h^m}{\varepsilon^{m+1}}\right).$$

References

[1] Beale,J.T. and Majda,A.(1981),"Rate of convergence for viscous splitting of the Navier-Stokes equations",Math. Comp., Vol.37,p243-259.

[2] Alessandrini,G., Douglis,A. and Fabes,E.(1983), "An approximate layering method for the Navier-Stokes equations in bounded cylinders", Ann. Mat. pura Appl.,Vol.135,p329-347.

[3] Douglis,A. and Fabes,E.(1984), "A layering method for viscous incompressible L_p flows occupying R^n", *Research Notes in Mathematics*,Vol.108,Pitman.

[4] Zheng,Q. and Huang,M.(1992), "A simplified viscosity splitting method for solving the initial boundary value problems of Navier-Stokes equation", J. Comp. Math,Vol.10,p39-56.

[5] Zhang,P.-w.(1993), "A sharp estimate of simplified viscosity splitting scheme", J. Comp. Math.,Vol.11,p205-210.

[6] Ying,L.-a.(1991), "Optimal error estimates for a viscosity splitting formula", in *Proceedings of the Second Conference on Numerical Methods for PDE*, World Scientific,p139-147.

[7] Chorin,A.J., Hughes,T.J.R., McCracken,M.F. and Marsden,J.E.(1978), "Product formulas and numerical algorithms",Comm. Pure Appl. Math.,Vol.31,p205-256.

[8] Benfatto,G. and Pulvirenti,M.(1984), "Generation of vorticity near the boundary in planar Navier-Stokes flows", Comm. Math. Phys.,Vol.96,p59.

[9] Benfatto,G. and Pulvirenti,M.(1986), "Convergence of Chorin-Marsden product formula in the half plane",Comm. Math. Phys.,Vol.106,p427-458.

[10] Hald,O.H. and DelPrete,V.M.(1978), "Convergence of vortex methods for Euler's equations", Math. Comp.,Vol.32,p781-809.

[11] Hald,O.H.(1979), "The convergence of vortex methods II", SIAM J. Numer. Anal.,Vol.16,p726-755.

[12] Hald,O.H.(1987), "Copnvergence of vortex methods for Euler's equations III", SIAM J. Numer. Anal.,Vol.24,p538-582.

[13] Beale,J.T. and Majda,A.(1982),"Vortex methods I: Convergence in three dimensions", Math. Comp.,Vol.39,p1-27.

[14] Beale,J.T. and Majda,A.(1982),"Vortex methods II: Higher order accuracy in two and three dimensions" Math. Comp.,Vol.39,p29-52.

[15] Raviart,P.A.(1985), "An analysis of particle methods", *Lecture Notes in Mathematics*,Vol.1127,p243-324.

[16] Anderson,C.R. and Greengard,C.(1985), "On vortex methods", SIAM J. Numer. Anal.,vol.22,p413-440.

[17] Greengard,C.(1986), "Convergence of the vortex-filament method", Math. Comp.,Vol.47,p387-396.

[18] Goodman,J.(1987),"Convergence of the random vortex method", Comm. Pure Appl. Math.,Vol.40,p189-220.

[19] Goodman,J. and Hou,T.Y.(1991), "New stability estimates for the 2-D vortex method", Comm. Pure Appl. Math.,Vol.44,p1015-1031.

[20] Goodman,J. Hou,T.Y. and Lowengrub,J.(1990), "Convergence of the point vortex method for the 2-D Euler equation", Comm. Pure Appl. Math.,Vol.43,p415-430.

[21] Hou,T.Y. and Lowengrub,J.(1990), "Convergence of the point vortex method for the 3-D Euler equations", Comm. Pure Appl. Math.,Vol.43,p965-981.

[22] Cottet,G.H.(1987), "Convergence of a vortex in cell method for the two dimensional Euler equations", Math. Comp.,Vol.49,p407-425.

[23] Cottet,G.H., Goodman,J. and Hou,T.Y.(1991), "Convergence of the grid free point vortex method for the 3-D Euler equations", SIAM J. Numer. Anal.,Vol.28,p291-307.

[24] Ying,L.-a.(1995), "Convergence of vortex methods for three dimensional Euler equations in bounded domains", SIAM J. Numer. Anal.,Vol.32,No.2,p542-559.

[25] Ying,L.-a. and Zhang,P.-w.(1997), Vortex Methods, Science Press and Kluwer.

[26] Ying,L.-a.(1998),"Vortex method for two dimensional Euler equations in bounded domains with boundary correction", Math. Comp.,Vol.67,No.224,p1383-1400.

[27] Ying,L.-a.(1999), "Convergence of Chorin-Marsden formula for the Navier-Stokes equations on convex domains", J. Comp. Math.,Vol.21,No.1,p73-88.

3D Vortex methods: achievements and challenges

G-H. Cottet

LMC-IMAG, Université Joseph Fourier
Grenoble, France

1 Introduction

The question we wish to address in this talk is the following: is there a chance that vortex methods become in a near future a reliable tool for CFD. By reliable we mean simple and accurate. We are in particular in mind algorithms very much in the spirit of finite-differences: easy to implement, with no or little parameter tuning, and based on consistent and simple tools.

We have no definite answer yet for this question but we believe that some results already obtained give some insight and indicate several directions that should be followed.

Some of the difficulties that vortex methods have had in the past for 3D calculations are the need of fast solvers, the problem of the vorticity divergence and the treatment of boundaries. The need of fast solvers is certainly the most severe difficulty that vortex methods have and will continue to have in the years to come. We wish to stress here that, although vortex methods are often seen as models of turbulent flows which, with a limited numbers of particles can give a qualitative taste of the vorticity dynamics, we are here interested in methods more in the spirit of Direct or Large Eddy Simulations. If the relevant viscous scales are to be captured in a typical moderate to high Reynolds number flow, one must be ready to handle numbers of particles ranging from several hundred thousand to millions. In two dimensions the development of fast solvers, since the pioneering work of Greengard and Rockhlin, has led to efficient algorithms which routinely compute velocities for sets of millions of particles. In 3D, although several algorithms are now available (see [12, 16] and the references therein), they have not reached the same efficiency. All figures reported in the literature indicate that the best implementations of multipole methods are still several orders of magnitude more expensive than grid-based Poisson solvers. Of course the particle solvers have several built-in advantages: they incorporate far-field boundary conditions, they adapt "easily" to complex geometries and they are used on sets of points which occupy only portions of the flow. Despite these nice features, completely grid-free vortex methods seem far to be able to compete with conventional grid-based methods for DNS or LES of bluff-body flows.

Our approach has thus been to use Vortex-In-Cell methods. The idea in these method is to retain the Lagrangian character of vortex methods for the solution of the vorticity convection-stretching-diffusion equation while relying on finite-difference type calculations of the velocity field. Most of the computational work is then devoted to the interpolation formulas that are used to transfer informations between particles and the grid. The resulting codes are slightly more expensive than fully Eulerian codes but definitely less than grid-free methods. While Particle In Cell methods are very common in the field of plasma physics [5], for fluid applications the common belief is that they do not allow to retain the information that particles carry at the sub-grid level, except if local corrections are explicitly computed [1]. This is certainly true if the ratio of the grid-size versus the inter particle spacing is large. Our experience however is that for a ratio close to 1 and when accurate interpolation schemes are employed, VIC methods can match, with the same grid-size, the results obtained by centered finite-difference schemes.

The outline of the talk is as follows. We will first review the basic tools that we use in our VIC code: diffusion solver, vorticity creation algorithm, regridding scheme and discretization of the stretching term. We will show numerical results for 2 and 3D flows which aim at quantifying the sub-grid behavior of the method. We will also address the issue of the vorticity divergence in 3d simulations. We will show that if the appropriate tools are used at each stage of the code the problem if vorticity divergence essentially disappears. We will finally try to list the challenges that vortex methods still have to face to become a flexible tool for CFD.

2 The VIC tool-box

Our VIC codes for the simulation of 3D unsteady viscous flows involves the following ingredients: non-dissipative regridding, Particle Strength Exchange (PSE) scheme for the diffusion, conservative treatment of the stretching and vorticity-flux boundary conditions.

Regridding. It is a classical observation that any turbulent, 2D or 3D, flow produces high strain which distorts the particle distribution. To maintain the particle overlapping, which is a prerequisite for an accurate velocity evaluation, it is necessary to re-grid frequently the particles on regular locations. Such regridding must be done through interpolation formulas that are smooth but not dissipative. A good compromise is given by the following cubic-spline [10]

$$\phi(x) = \begin{cases} 0 & \text{if } |x| > 2 \\ \frac{1}{2}(2-|x|)^2(1-|x|) & \text{if } 1 \leq |x| \leq 2 \\ 1 - \frac{5x^2}{2} + \frac{3|x|^3}{2} & \text{if } |x| \leq 1 \end{cases}$$

This is a C^1 function which spreads the vorticity to the 64 (in 3D) nearest particles. Its non-dissipative nature is attested by the fact that it conserves the angular impulse. The truncation error is thus of order h^3 where h is the distance between particles. In a vortex code using at least a second order time advancing scheme, the time-step is typically scaled by $(\max|\nabla \mathbf{u}(\mathbf{x})|)^{-1}$. When regridding is performed at each time-step, which is a simple and efficient choice, the overall truncation effect is thus measured by $h^3 \max|\nabla \mathbf{u}(\mathbf{x})|$ (see [2]).

Diffusion solver. In our simulations, we make a constant use of the PSE scheme [4]. This is a straightforward method, easy to implement, accurate and with a marginal cost, even in a VIC code and we see no reason to switch to other, more involved schemes. The PSE schemes are based on the following formula

$$\frac{d\omega_p}{dt} = \nu\varepsilon^{-3} \sum_q v_q[\omega_q - \omega_p]\eta(\frac{\mathbf{x}_p - \mathbf{x}_q}{\varepsilon}) \tag{1}$$

where ω_q and v_q respectively denote the vorticity and volumes of the particles. The kernel η has to satisfy second order momentum properties. The accuracy of the schemes is again subject to an overlapping condition. In practice, in our calculation regridding is done at each step before the diffusion. This of course guarantees the desired overlapping. Our diffusion kernel extends to the 27 nearest points and is normalized in such a way that the moment properties are satisfied at the discrete level. The diffusion thus reduces to a generalized finite-difference formula.

Particle-grid operators and treatment of the stretching term. In a 3D VIC code velocities and their derivatives are computed on a fixed grid through a finite-difference or FFT-based solver which evaluates the 3 components of the stream function with the required boundary conditions. Vorticity values have first to be assigned to the grid points, by interpolation from the particle values. Conversely, once grid velocities and their derivatives are known, they can be interpolated to the particle locations. For these interpolation steps we use the same third order kernel ϕ as for regridding.

The treatment of the stretching term deserves some particular attention. If ω_p are the particle vorticities, the natural discretization of the stretching term inherited from the weak formulation of the vorticity equation would lead to

$$\frac{d\omega_p}{dt} = [\nabla u(\mathbf{x}_p, t)]\omega_p.$$

We have implemented a different scheme based on the conservative form $\text{div}(\boldsymbol{\omega} : \mathbf{u})$ of the stretching term: values of the vorticity and velocity are first multiplied on the grid, the divergence is then computed by finite-differences on the grid and finally interpolated on the particle locations. This scheme has the advantage of being conservative on the grid and, unlike other conservative schemes, does not amplify the vorticity divergence [2]. Clearly its accuracy is again conditioned by that of the interpolation scheme and it is here important to use at least fourth order differencing formulas.

Vorticity boundary conditions. Our approach here follows what has successfully been done in 2D [7]. It consists in rewriting no-slip boundary conditions as vorticity flux conditions. This method originally conceived for 2D flows easily extends to 3D flows.

In 3D wall-bounded flows one needs 3 vorticity conditions. One is clear: the normal vorticity component has to vanish, since it involves tangential derivatives of the velocity. As for the 2 normal components, it is natural to provide them with Neumann boundary conditions, each of them canceling the slip in the orthogonal direction. These boundary conditions have been proposed in [2]. An important feature of this method, which is proved in this reference, is that it does not produce any vorticity divergence at the wall.

Computational cost. A typical profile of the code for a 3D homogeneous turbulent flow (see below for some results on this test case) using a unit ratio grid-particle spacing indicates that about 50% of the CPU time is consumed in the various interpolation formulas. The Poisson solver - borrowed from a standard Fishpack library - requires about 15% and another 15% is spent for the diffusion and the regridding. For a 128^3 resolution (which means about 2 millions particles in this flow) this results in a CPU time of 55 seconds per time-step on a single processor DEC workstation, each time-step consisting of a second order Runge-Kutta time advancing scheme for the particle locations and vorticity, a first order Euler scheme for the diffusion, and regridding at every time-step.

Note that in most cases, particles would occupy only a portion of the computational box, which reduces the cost of all operations except that of the Poisson solver. For instance, in a run of the Crow instability (see below) where the number of particle would range between 300,000 and 800,00 the interpolation formulas amount to the same cost as the Poisson solver.

3 Results

Sub-grid effects of the interpolations. Our first concern is to quantify the sub-grid effects of the various interpolation formulas used in the VIC code. For 2D laminar flows, a first answer can be found in the high resolution ellipse calculations of Koumoutsakos [6]. These grid-free experiments show that frequent regridding with the third order kernel allows long time accurate calculations. The enstrophy decay is marginal and sharp gradients which appear during the filamentation are well captured. Other test cases have confirmed these results [8, 3]. In a VIC code using the same interpolation formula to transfer quantities between grid and particles, careful comparisons with centered finite-differences confirm that the particle-grid operators do no add significantly to the numerical errors that are inherent to vortex methods [9, 8].

For 3D flows the appearance of small scales resulting from the vorticity stretching poses another challenge to vortex methods. The sub-grid behavior of the involved interpolations is thus even more critical than in 2D. Figure 1 shows the evolution of the vorticity in a Crow instability at a Reynolds number of 2,500. The initial conditions are that of [11], and we used a 128^3 grid and mesh-size ratio grid/particle of 1. The results are in good agreement with those of [11]. These authors use a much finer grid, refined in the region a closest approaches of the tubes, but the only noticeable difference with our runs is a higher maximum vorticity value in their case. In that test case we have monitored the energy dissipation rate which must give the viscosity through the formula

$$\frac{dE}{dt} = 2\nu\mathcal{E}$$

where E and $\mathcal{E}$ respectively stand for the energy and enstrophy. Figure 2 reveals that except for a particular time, which, interestingly enough, roughly corresponds to the reconnection time, the discrepancy between the actual and desired dissipation never exceeds 10%.

A more challenging test is the case of an homogeneous turbulent flow. In Figure 3 we show the energy spectra obtained at successive times by the VIC code and a spectral non-dealiased code. The grid resolution is again 128^3 for both codes, with a unit ratio particle/grid ratio for the VIC code. The Reynolds number, based on the Taylor micro-scale is 95 at the beginning of the simulation. The initial condition consists of a field with a peak at low wave number and random phases. The comparisons of the spectra show that the large and intermediate scales are well-captured by the VIC code. The better behavior of the spectral code at high wave numbers results in a more pronounced enstrophy peak (Figure 4). Note however that, if the enstrophy is calculated on the basis of the 2/3 lower modes, the enstrophy-peak deficit of the VIC code is reduced from 10% to about 5%. All other statistics which are classically computed on this type of flow are very similar see [13] for more extensive comparisons). As an example the left picture of Figure 4 shows the evolution of the skewness factor for the velocity derivatives. Note that the vortex code gives for large times a value which is in good agreement with the theoretically predicted value of 0.5. More experiments are under way in particular to compare the results of under-resolved vortex simulations with spectral simulations using classical LES models.

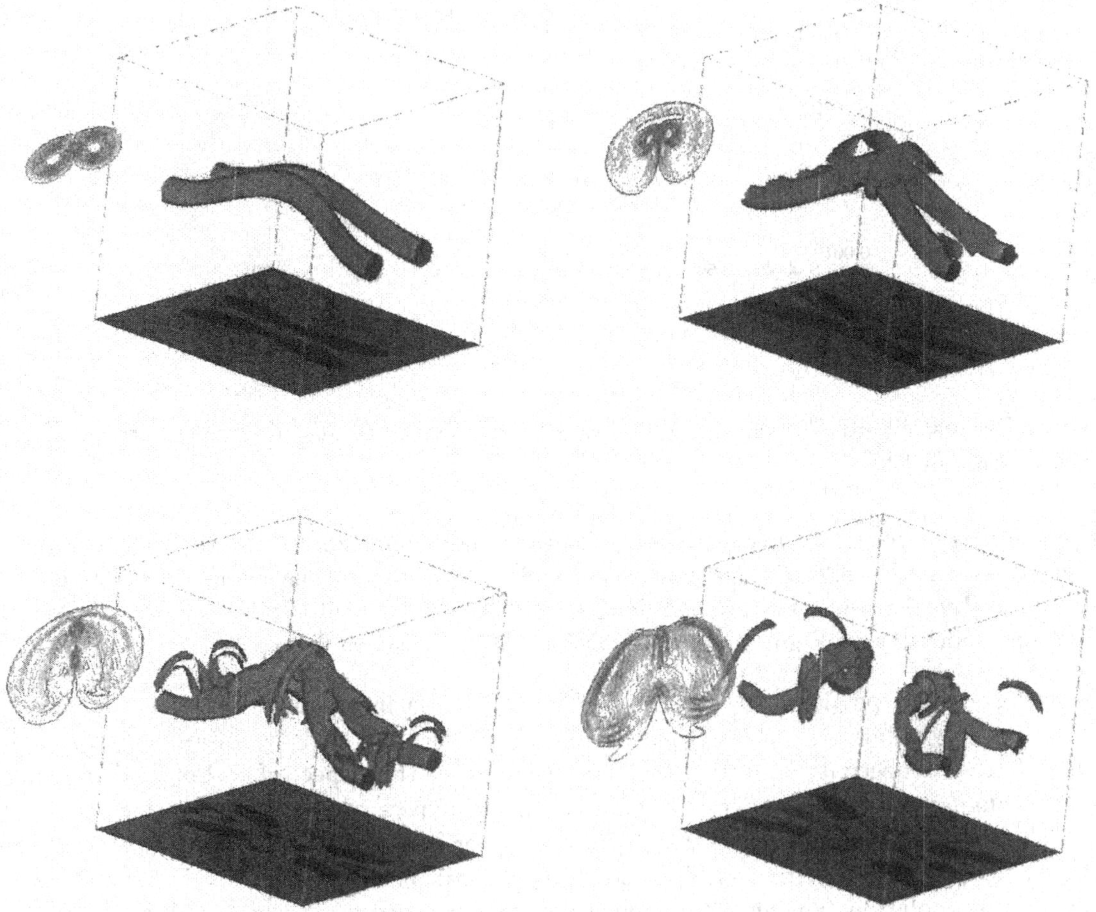

Figure 1: Crow instability at a Reynolds number of 2500. Vorticity isosurfaces and cross section in the plane of closest approach times 0, 0.75, 1.5 and 2.5 (form left to right, top to bottom).

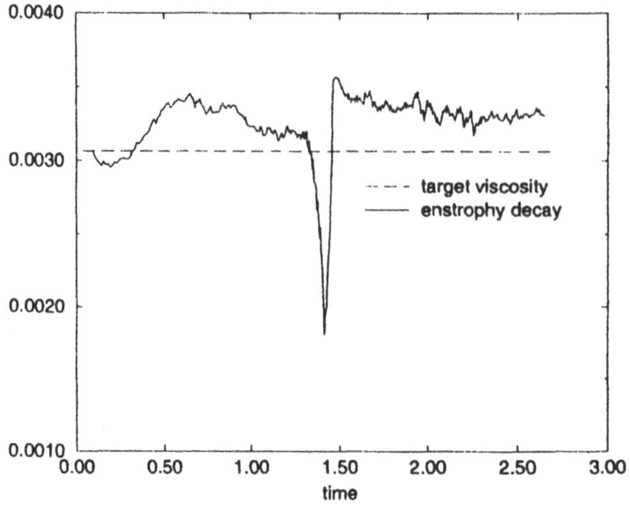

Figure 2: Dissipation rate for the simulation of a Crow instability at a Reynolds number of 2,500 for a resolution of 128^3

The issue of the vorticity divergence. The solenoidal condition for the vorticity is in general considered as a major difficulty for vortex methods which do not use filaments. Filaments are indeed natural in the context of inviscid flows but, although filament surgery is often invoked as a dissipation model, they are not adapted to DNS or LES of viscous flows.

When using vorticity particles, the possibility of having increasing vorticity divergence is related to the fact that only pointwise values of the vorticity are tracked and that the diffusion does not allow to conserve the connectivity of nearby elements. To improve this aspect of vortex particle methods several techniques have been proposed. One class of techniques consists of processing the particle vorticities to make them divergence free. This is more or less equivalent to using a pressure term in the vorticity equation. However, due to the regularization involved in the Poisson solver this involves a deconvolution-type algorithm which becomes ill-posed when the vorticity develops large gradients [15, 2]. This greatly reduces the practical efficiency of this technique. Another approach is to explicitly add a term in the vorticity equation that dissipates the vorticity divergence [2]. This is however at the expense of dissipating the vorticity itself.

One may view the divergence of the vorticity appearing in the course of the calculation as a mere result of truncation effects related in particular to the production of spurious small scales. This is of course the case only if the vortex scheme is based on a formulation that conserves the vorticity divergence. Note that this rules out schemes that use the transpose form of the velocity gradients tensor. For this truncation error to be minimal, the best is to use smooth high order schemes at every level of the algorithm. Our experience is actually that when the tools described in the previous section are used in a clean way, the problem of the vorticity divergence essentially disappears by itself. To substantiate this claim we have chosen the example of a vortex-wall interaction. This type of flow is challenging since small scales are created at the wall and stretched

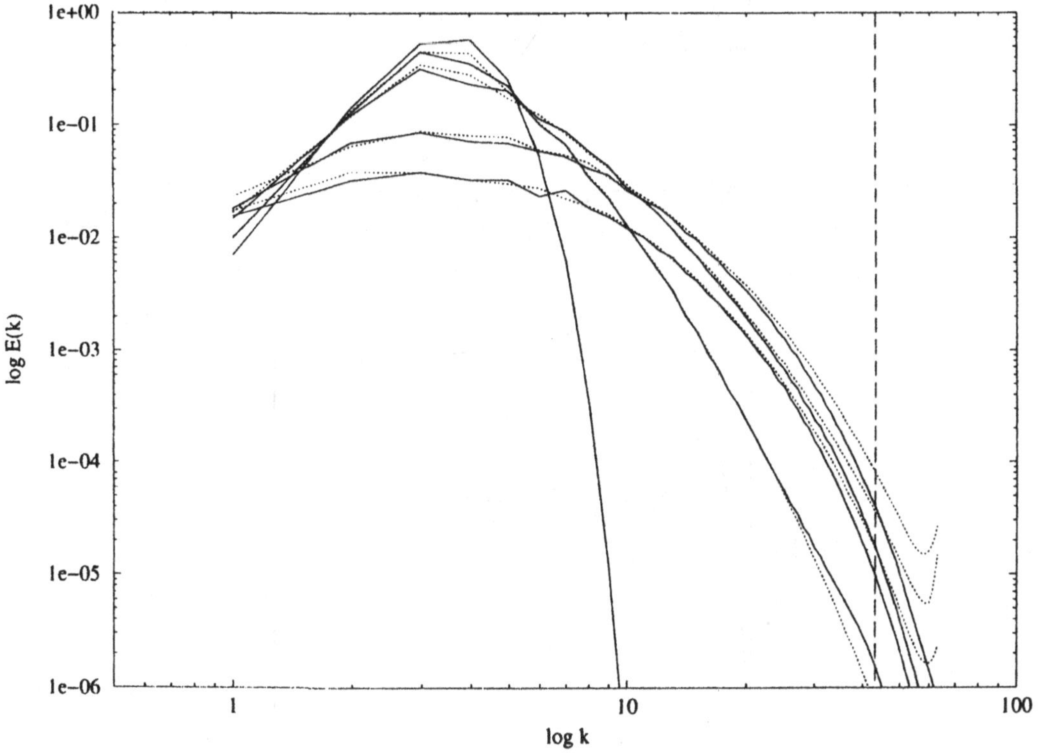

Figure 3: Energy spectra obtained by a spectral (dashed lines) and a VIC (solid lines) code in an homogeneous turbulence experiment in a 128^3 periodic box. Curves correspond to times equal to 1, 2 6, and 10 turnover times. $Re_\lambda = 95$.

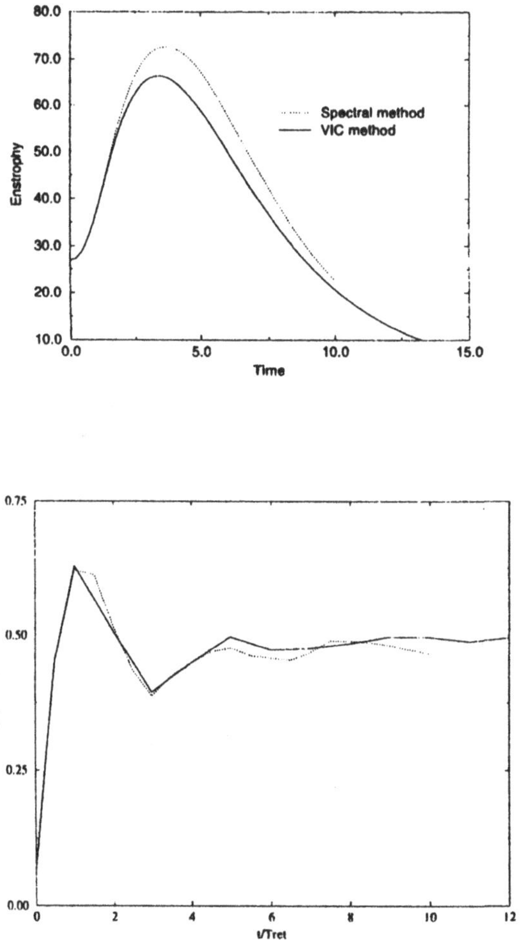

Figure 4: Homogeneous turbulent flow. Enstrophy and skewness factor in a spectral (dashed lines) and a VIC (solid lines) code. Times are normalized with respect to turnover time.

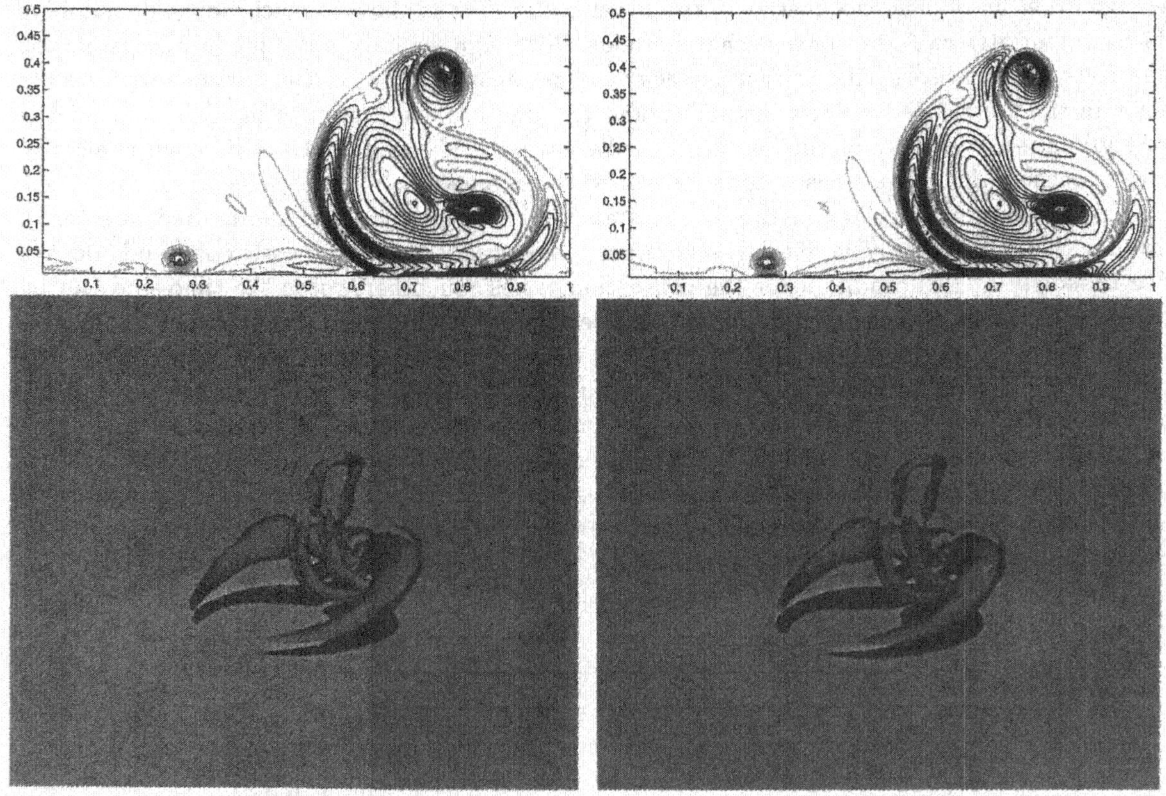

Figure 5: Rebound of a vortex ring impinging on a wall at an angle of 38.5 degrees. $Re = 1400$. Vorticity strength isosurface and contours in the symmetry plane. Left pictures: direct calculation; right pictures: rotational of the velocity.

as time goes on. It is also of crucial importance that the vorticity boundary conditions does not violate the divergence free constraint. Figure 5 shows the vorticity isosurface and contours in the symmetry plane, just after the rebound of a ring impinging a wall at an angle. The left pictures correspond to the brute vorticity field as it assigned to the grid points during the calculations, whereas the right pictures correspond to the - divergence free - vorticity field obtained by taking the curl of the velocity. The comparison does not show any significant difference despite the large vorticity values (the initial maximum vorticity value is roughly multiplied by a factor 4 at this stage of the computation) and the stretching that occurs in this flow. The monitoring of the divergence of the vorticity indeed reveals that it remains at a level of a few percents of the maximum vorticity derivatives. The same conclusions hold for all 3D flows that we have computed.

4 Perspectives

Although the results just shown do bring some partial answers to some important questions regarding vortex methods, there are still a number of issues which are unsettled. The main question is the following: to which extent vortex methods would be able to provide results comparable to, say high order finite-difference methods, at a lower cost and thus allow to reach Reynolds numbers which are not attainable by conventional methods. Although we have not done in 3D systematic quantitative comparisons with finite-difference type methods, qualitative comparisons with results shown in the literature (for Crow instability and the ring problem) seem to indicate that for the same grid-size, results are of similar quality. We have already said that a VIC code is slightly more expensive than a grid-based code. To provide an economic class of solvers, on must thus rely on two features of vortex methods: the stability of the method and the limited number of particles that are typically in the far field zones. It is well-known that vortex methods do not have to satisfy a CFL condition for the convection terms. In other words the time-step can be chosen independent of the spatial resolution. If a second or higher order particle advancing scheme is used, accuracy considerations lead to choose a value of the time-step that is a fraction of the inverse of the maximum vorticity. As for the diffusion, the stability condition is the same as for conventional explicit finite-difference solvers and allows large values for the time-step as soon as the Reynolds number is large enough. In 2D high resolution calculations, this can lead to CPU cost which outperform finite-difference codes. Table 1 is borrowed from [9] and shows a comparison of the CPU time used for a VIC code and a second-order compact finite-difference scheme for driven cavity experiments. For 3D flows, the stretching term imposes an additional stability condition

Reynolds number	100	2000	10000
N_{fd}	64	128	256
N_{vic}	64	128	256
δt_{fd}	0.01	0.008	0.004
δt_{vic}	0.01	0.02	0.04
CPU time for finite-difference scheme	3	24	225
CPU time for VIC scheme	5	16	32

Table 1: 2D driven cavity flow. Run parameters and CPU times with VIC and finite-difference methods for various Reynolds numbers and resolutions.

which is again independent of the spatial resolution. This condition relates the time-step to the maximum eigenvalue of the strain tensor. It is of the form $\Delta t \leq C/\rho(\nabla u)$. In practice $\rho(\nabla u)$ can easily be evaluated. It is always of the same order as - indeed less than, in all the cases we have dealt with - the maximum vorticity.

As in 2D, vortex methods thus offers a potentially economic alternative to finite-difference based methods. However in 3D high resolution calculations are in general not affordable and for the kind of spatial resolution which is typically used the gain is not so clear and might very much depend on the type of flow. For the homogeneous turbulence experiments, the time-step used in the VIC code is at time zero about ten times bigger than the time-step used in the spectral calculations. As the maximum vorticity increases, it then reaches about the same value, and eventually increases again. Overall, the cost of both methods is about the same. For the ring problem, we have not done direct comparisons ourselves but from the results and code features described in [14] on may anticipate that the time-steps are, for the 128^3 spatial resolution that we have used, of the same order in both methods. Consequently the finite-difference method is probably cheaper for this flow.

For unbounded flows, vortex methods enjoy two additional features: the fact that particles only occupy the support of the vorticity reduces the cost of the interpolation steps. Furthermore from 2D calculations, one may expect that the natural treatment of the far-field boundary condition would allow to use a very low resolution there, at the expense of course of some inaccuracy in the wake behavior but without feedback effect on the flow near the obstacles. This would be a definite advantage over finite-difference based schemes which require careful choices of the artifial downstream boundary conditions.

To finish with, another direction which we are following is the extension to 3D of the domain-decomposition strategies that have been successfully developed for 2D flows [8, 9]. These methods consist of using a finite-difference solver near the obstacles and a vortex methods in the wake. They offer the flexibility of dealing either with the vorticity/velocity or velocity/pressure formulation in the finite-difference zone. For 3D flow this flexibility is very much desirable since it allows to greatly simplify the treatment of boundary conditions.

References

[1] Anderson C.R. 1986, *A method of local correction for computing the velocity field due to a distribution of vortex blobs*, J. Comp. Phys., 62, 111-123.

[2] Cottet G.-H and Koumoutsakos P., *Vortex methods*, Cambridge University Press, to appear.

[3] Cottet G.-H., Ould Salihi M.-L. and El-Hamraoui M. 1999, *Multi-purpose regridding in vortex methods*, Vortex Fl ows and Related Numerical Methods, ESAIM Proceedings, 7, 94-103.

[4] Degond P. and S. Mas-Gallic S. 1989, *The weighted particle method for convection-diffusion equations*, Math. Comp., 53, 485-526.

[5] Hockney R.W. and Eastwood J.W. 1981, Computer Simulation Using Particles, McGraw-Hill Inc.

[6] Koumoutsakos P. 1997, *Inviscid Axisymmetrization of an Elliptical Vortex*, J. Comp. Phys., 138, 821-857.

[7] Koumoutsakos P., Leonard A. and Pepin F. 1994, *Viscous boundary conditions for vortex methods*, J. Comp. Phys., 52, 113

[8] Ould Salihi M.-L. 1998, *Couplage de méthodes numériques en simulation directe d'écoulements incompressibles*, Thèse de doctorat, Université Jospeh Fourier.

[9] Ould Salihi M.-L., Cottet G.-H., El Hamraoui M., *Blending finite-difference and vortex methods for incompressible flow computations*, submitted.

[10] Monaghan J.J. 1985, *Extrapolationg B-splines for interpolation*, J. Comp. Phys., 60, 253-262.

[11] Shelley, M.J., Meiron, D.I. and Orszag, S.A. 1993, *Dynamical aspects of vortex reconnection of perturbed anti-parallel vortex tubes*, 246, 613-652.

[12] Strickland J.H. and Baty R.S. 1996, *A pragmatic overview of fast multipole methods*, Lectures in Applied Mathematics, 32, 807-830.

[13] G. Van der Linden 1999, *Simulation de la turbulence par la méthode particulaire*, Technical Report, LMC-IMAG.

[14] Verzicco R. and Orlandi P. 1994, *Normal and oblique collision of a vortex ring with a wall*, Meccanica, 29, 383-391.

[15] Winckelmans G. and Leonard A. 1993, *Contributions to vortex particle methods for the computation of three dimensional incompressible unste ady flows*, J. Comp. Phys., 109, 247-273.

[16] Winckelmans, G. S., Salmon, J. K., Warren, M. S., Leonard, A. and Jodoin, B. 1995, *Application of fast parallel and sequential tree codes to comput ing three-dimensional flows with the vortex element and boundary elements method*, Proc. of 2nd International Workshop on Vortex Flows and Related Numerical Methods, Montreal, August 20-24.

Development of a Vortex and Heat Elements Method and its Application to Analysis of Unsteady Heat Transfer around a Circular Cylinder in a Uniform Flow

Kyoji KAMEMOTO* and Toji MIYASAKA**

*Department of Mechanical Engineering and Materials Science, Yokohama National University

79-5 Tokiwadai Hodogaya-ku Yokohama, 240-8501 Japan / Email: kame@post.me.ynu.ac.jp

**Mitsubishi Heavy Industry Co Ltd.

ABSTRACT

In this study, a vortex element method based on Biot-Savalt law and a core spreading scheme is extended to analysis of unsteady and forced-convective heat transfer around a circular cylinder in a uniform flow. In the preset method, considering that the formulation of energy equation is very similar to the vorticity transport equation for a two-dimensional and incompressible flow, discrete heat elements are introduced into the flow field close to a wall surface in addition to nascent vortex elements with a viscous core. Each heat element has a thermal core which spreads with time according to thermal diffusion, and it drifts about at the mercy of the flow as same as the convection of each vortex element.

Using the present method, characteristics of forced-convective heat transfer around a two dimensional circular cylinder in a uniform flow were investigated for the Reynolds numbers of Re = 10^3, 10^4 and 10^5, and Prandtl number Pr = 0.71. As a result of the calculation, it was found that the local and time-averaged Nusselt numbers are in good agreement with those of experimental ones obtained in the past researches. Furthermore, interesting characteristics of unsteady heat transfer and instantaneous temperature distributions in the vortical flow field behind the cylinder were known.

1. INTRODUCTION

In relation with unsteady separated flow problems and flow-induced vibration problems being experienced in various engineering fields, the vortex methods based on the Biot-Savalt law method or the vortex in cell method, have been developed owing to the pioneering mathematical models like the random walk method by Chorin[1][2], the core spreading method by Leonard [3] and the surface element method by Lewis [4] for reasonable expression of viscous effects in a flow. Since the vortex methods based on the Biot-Savalt law consist of simple algorithm based on physics of flow and they do not need the complex work of grid generation for numerical treatment, they have been applied to analysis of practical problems in a wide range of engineering fields. Recently an advanced vortex method was proposed by Kamemoto and Matsumoto[5], in which characterized nascent vortex elements were introduced in the flow close to a body surface to increase computational stability and to reduce computational time. And the usability of the method has been confirmed from simulations of two-dimensional flows through centrifugal impellers by Zhu et al.[6][7] and Zhu and Kamemoto[8], and three-dimensional flows around a sphere and a spheroid by Ojima and Kamemoto[9].

On the other hand, a surface element algorithm for

simulation of vorticity and heat transport was firstly considered by Smith and Stansby [10]. Using the vortex in cell method incorporated with the random walk method, they introduced both vortex and temperature particles according to the similarity of equations of vorticity transport and forced-convective heat transfer in a two-dimensional flow. Results for the flow around a circular cylinder at constant temperature have shown reasonable agreement with experiment for moderate Reynolds numbers from $Re = 23$ to 289.

However, from the view point of engineering application, numerical investigations of unsteady heat transfer for much larger Reynolds numbers are usually required. Therefore, in the present research, extension of the advanced vortex method has been studied in consideration of introduction of nascent heat elements in addition to vortex elements. In this paper, the algorithm of the introduction of heat elements is explained, and application of the present method to unsteady forced-convective heat transfer around a circular cylinder in a uniform flow for the Reynolds numbers of $Re = 10^3$, 10^4 and 10^5, and Prandtl number $Pr = 0.71$ is described showing calculation results in comparison with experimental ones

2. ALGORITHMS OF VORTEX AND HEAT ELEMENTS METHOD

2.1 Mathematical basis

The governing equations of viscous and incompressible flow are described by the vorticity transport equation and pressure Poisson equation which are derived from the rotation and divergence of Navier-Stokes equations, respectively

$$\frac{\partial \omega}{\partial t} + (\boldsymbol{u} \cdot grad)\,\omega = (\omega \cdot grad)\,\boldsymbol{u} + \nu \nabla^2 \omega \qquad (1)$$

$$\nabla^2 p = -\rho\, div(\mathbf{u} \cdot grad\mathbf{u})$$

(2)

where $\boldsymbol{u}$ is a velocity vector. The vorticity ω is defined as

$$\omega = rot\ \boldsymbol{u} \qquad (3)$$

The energy equation of the forced convective heat transfer is expressed as

$$\frac{\partial T}{\partial t} + (\boldsymbol{u} \cdot grad\ )T = \alpha \nabla^2 T \qquad (4)$$

where T is temperature and α is the thermal diffusivity. Lagrangian expression for both the vorticity transport equation (1) and the energy equation (4) are given by

$$\frac{d\omega}{dt} = (\omega \cdot grad\ )\,\boldsymbol{u} + \nu \nabla^2 \omega \qquad (5)$$

$$\frac{dT}{dt} = \alpha \nabla^2 T \qquad (6)$$

It is clear that the energy equation (6) is of the similar form to the vorticity transport equation Eq.(5). When a two-dimensional flow is dealt with, the first term of the right hand side in Eq.(5) disappears and so the vorticity transport equation is simply expressed as

$$\frac{d\omega}{dt} = \nu \nabla^2 \omega \qquad (7)$$

Then, the form of Eq.(6) becomes completely the same as Eq.(7). This fact seems to suggest that the energy equation (6) can be solved in an analogous way, with nascent temperature elements, in place of vortex elements using a time splitting scheme.

In the vortex element method developed by the group of the present authors, the viscous diffusion expressed by Eq.(7) is approximately taken into account by the core spreading method. Therefore, in the present study, the thermal diffusion expressed by Eq.(6) is similarly taken into account by introducing a thermal core to a discrete heat element which spreads with the increase of time, and as same as that of a vortex element, the trajectory of each heat element in flow is represented by

$$\frac{d\boldsymbol{r}}{dt} = \boldsymbol{u} \qquad (8)$$

where $\boldsymbol{r}$ denotes location of a heat element. It should be noted here that assuming a forced heat convection in a flow of a high Reynolds number and a not-small Prandtl number in this study, we ignored the effects of natural heat convection and radiation, where Reynolds number and Prandtl number are respectively defined as $Re = Ud\,/$

ν and $Pr = \nu/\alpha$.

2.2 Generalized Biot-Savalt law

As explained by Wu and Thompson [11], the Biot-Savart law can be derived from the definition equation of vorticity (3) as

$$u=\int_V \omega_0 \times \nabla_0 G dv - \int_S [(\mathbf{n}_0 \cdot u_0)\cdot \nabla_0 G + (\mathbf{n}_0 \times u_0)\times \nabla_0 G] ds \tag{9}$$

Here, subscript "$_0$" denotes variable, differentiation and integration at a location $\mathbf{r}_0$, and $\mathbf{n}_0$ denotes the normal unit vector at a point on a boundary surface S. And G is the fundamental solution of the scalar Laplace equation with the delta function $\delta(\mathbf{r}-\mathbf{r}_0)$ in the right hand side, which is written as

$$G = -\frac{1}{2\pi}\log R \quad \text{(2-D)} \tag{10}$$

or

$$G = \frac{1}{4\pi R} \quad \text{(3-D)} \tag{11}$$

here, $\boldsymbol{R}=\boldsymbol{r}-\boldsymbol{r}_0$, $R=|\boldsymbol{R}|=|\boldsymbol{r}-\boldsymbol{r}_0|$.

In Eq. (9), the inner product, $\mathbf{n}_0 \cdot u_0$ and the outer product $\mathbf{n}_0 \times u_0$ stand for respectively normal and tangential velocity components on the boundary surface, and they respectively correspond to source and vortex distributions on the surface. Therefore, it is mathematically understood that a velocity field of viscous and incompressible flow is arrived at the field integration concerning vorticity distributions in the flow field and the surface integration concerning source and vortex distributions around the boundary surface.

2.3 Calculation of pressure

The pressure in the field is obtained from the integration equation formulated by Uhlman[12], instead of the finite difference calculation of the Eq.(2) as follows.

$$\beta H + \int_S H \frac{\partial G}{\partial n} ds = -\int_V \nabla G (u \times \omega) dv$$
$$-\nu \int_S \left\{ G \cdot \mathbf{n} \cdot \frac{\partial u}{\partial t} + \mathbf{n} \cdot (\nabla G \times \omega) \right\} ds \tag{12}$$

Here, β is $\beta = 1$ inside the flow and $\beta = 1/2$ on the boundary S. G is the fundamental solution given by Eq. (10) or (11), and H is the Bernoulli function defined as

$$H = \frac{p}{\rho} + \frac{u^2}{2} \tag{13}$$

here, $u = |\,u\,|$.

2.4 Introduction of nascent vortex elements

The vorticity field near the solid surface must be represented by proper distributions of vorticity layers and discrete vortex elements so as to satisfy the non-slip condition on the surface. In the present method, a thin vorticity layer with thickness h_i is considered along the surface and the surface of outer boundary of the thin vorticity layer is dicretized by a number of vortex sheet and/or source panels as shown in Fig.1.

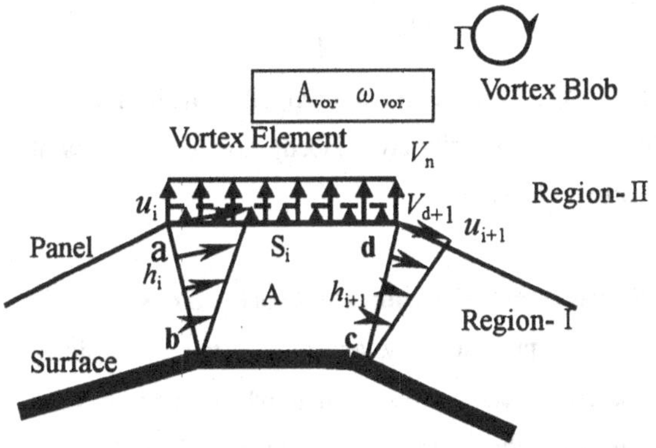

Fig.1 Thin vorticity layer and nascent vortex element.

If the flow is considered to be two-dimensional for convenience, and a linear distribution of velocity in the thin vorticity layer is assumed, the normal velocity υ_n on a panel can be expressed using the relation of continuity of flow and the non-slip condition on the solid surface for the element of the vorticity layer [abcd].

$$\upsilon_n = \frac{1}{s_i}\left(\frac{h_i u_i}{2} - \frac{h_{i+1} u_{i+1}}{2}\right) \tag{14}$$

here, s_i, h_i and u_i respectively denote the panel length, vorticity layer thickness and tangential velocity at a panel edge. Using the relation between the normal and tangential velocities for each panel expressed by Eq.(14), the strength of the vortex sheet and/or source of the panel for the following step can be calculated numerically from Eq.(9). On the other hand, the vorticity of the thin layer diffuses through the panel into the flow field with the diffusion velocity calculated from the suddenly accelerated plane wall problem, which is expressed as

$$\upsilon_d = \frac{1.136^2 \nu}{h_i + h_{i+1}} \quad (15)$$

here, ν is kinematic viscosity of the fluid. If $\upsilon_n + \upsilon_d$ becomes positive, a nascent vortex element is introduced in the flow field, where the thickness and vorticity of the element are given as follows.

$$h_{vor} = (\upsilon_n + \upsilon_d) \cdot dt \quad (16)$$

$$\omega_{vor} = \frac{\Gamma}{A} \quad (17)$$

Here, Γ is the circulation originally involved in the element of the vorticity layer [abcd], and A is the area of the vorticity layer element.

2.5 Replacement with equivalent vortex blobs

In the present study, assuming a two-dimensional flow, every nascent vortex element which moves beyond a boundary at the distance of four times h_i from the solid surface, is replaced with an equivalent two-dimensional vortex blob of the core spreading model. The total circulation and the sectional area of the blob core are determined to be the same as those of the rectangular nascent vortex element. As explained by Leonard [3], if a vortex blob has a core of radius ε_i and total circulation $\Delta\Gamma_i$, a Gaussian distribution of vorticity around the center of the blob is given as

$$\omega(r) = \frac{\Delta\Gamma_i}{\pi\varepsilon_i^2} \exp\left\{ -\left(\frac{r - r_i}{\varepsilon_i} \right)^2 \right\} \quad (18)$$

here r_i denotes a position of center of the blob. As explained by Kamemoto [13], the spreading of the core radius ε_i according to the viscous diffusion expressed by Eq.(7) is represented as

$$\frac{d\varepsilon_i}{dt} = \frac{2.2418^2 \nu}{2\varepsilon_i} \quad (19)$$

2.6 Introduction of nascent heat blobs

The transfer of heat between a solid body and a fluid must be represented by proper distributions of thermal layers surrounding the body and discrete heat elements so as to satisfy the energy balance for a fluid element in motion. In the same manner as the introduction of nascent vortex elements described above, a thin thermal layer with thickness h_{ti} is considered along the body surface and the outer boundary of the thin thermal layer is discretized as shown in Fig.2.

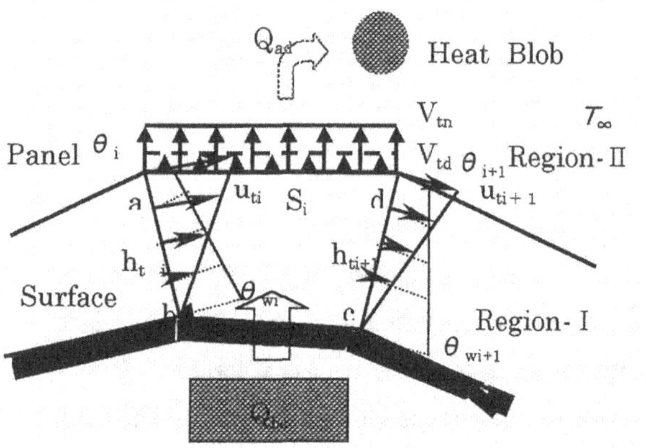

Fig.2 Thin thermal layer and nascent heat element

If we neglect the effects of natural convection and radiation on the heat transfer in a flow at a high Reynolds number, the energy balance in an element of the layer [abcd] is determined by the conduction of heat and the forced convection of heat with the stream. When linear distributions are assumed for both velocity and temperature in the thin thermal layer, quantities of heat $Q_{a\text{-}b}$ and $Q_{c\text{-}d}$ brought into or out from the element through [a-b] and [c-d] respectively during small time dt are approximately expressed

$$Q_{a-b} = \int_0^{h_{ti}} c \cdot \theta \cdot \rho u dt dy$$

$$= \frac{1}{2} c\rho\theta_{wi} u_{ti} h_{ti} dt \left(1 - \frac{2}{3} \frac{\theta_{wi} - \theta_i}{\theta_{wi}} \right) \quad (20)$$

$$Q_{c-d} = \int_0^{h_{ti+1}} c \cdot \theta \cdot \rho u dt dy$$
$$= \frac{1}{2} c \rho \theta_{wi+1} u_{ti+1} h_{ti+1} dt \left(1 - \frac{2}{3} \frac{\theta_{wi+1} - \theta_{i+1}}{\theta_{wi+1}} \right) \quad (21)$$

Here c and ρ denote the specific heats at constant pressure and density of fluid respectively. θ_{wi} and θ_i are the relative temperatures at a location on the wall ($\theta_{wi} = T_{wi} - T_o$) and at a location on the outer boundary of the thermal layer element ($\theta_i = T_i - T_o$), where T_{wi} and T_o denote the temperature on the wall and the temperature of the approaching flow, respectively. If we neglect the change of temperature distribution in the fluid element [abcd] during the short time interval, and if the quantities of heat transferred through the solid surface [b-c] and the outer boundary [d-a] during dt are written as Q_{b-c} and Q_{d-a} respectively, the relation of energy balance for the fluid element in the thin thermal layer is given by

$$Q_{a-b} + Q_{b-c} - Q_{c-d} - Q_{d-a} = 0 \quad (22)$$

where positive sign show the heat transfer into the element and vice versa. Once the temperature T_i is known and T_w or Q_{b-c} is given as the boundary condition, the quantity of heat transferred into the stream through the boundary [d-a] can be evaluated, and then a corresponding nascent heat blob with heat $\Delta Q_i = Q_{d-a}$ is introduced for the next step of calculation.

In the case of the constant heat flux q around the body, the value of Q_{b-c} is directly calculated as $Q_{b-c} = q s_i dt$, where s_i denotes the area (3-D) or length (2-D) of the surface [b-c], and the wall temperature T_{wi} can be calculated from the relation of heat flux expressed as

$$q = \frac{1}{2} \lambda \left(\frac{T_{wi} - T_i}{h_{ti}} + \frac{T_{wi+1} - T_{i+1}}{h_{ti+1}} \right) \quad (23)$$

Here λ denotes the conductivity ($\lambda = c \rho \alpha$). Then Q_{d-a} is evaluated by substituting Eqs.(20) and (21) into Eq.(22). On the other hand, using the local coefficient of heat transfer H_i, the relation of heat flux is expressed as

$$q = H_i \left(\frac{T_{wi} + T_{wi+1}}{2} - T_0 \right) \quad (24)$$

Therefore, the local Nusselt number Nu_i is evaluated from

$$Nu_i = \frac{H_i d}{\lambda} \quad (25)$$

Here d denotes the representative length of the body.

In the case of the constant temperature Tw around the body surface, once the temperature T_i is known at a calculation step, the local heat flux q_i can be evaluated as

$$q_i = \frac{1}{2} \lambda \left(\frac{T_w - T_i}{h_{ti}} + \frac{T_w - T_{i+1}}{h_{ti+1}} \right) \quad (26)$$

and so the value of Q_{b-c} is directly calculated as $Q_{b-c} = q_i s_i dt$, Then Q_{d-a} is evaluated by substituting Eqs.(20) and (21) into Eq.(22) and also the local Nusselt number Nu_i is evaluated from Eq.(25).

On the other hand, the transfer of heat through [a-d] is consisting of forced convection and temperature diffusion which are respectively based on the convection velocity υ_{tn} evaluated from the condition of continuity in the similar form as Eq.(14) and the temperature diffusion velocity υ_{td} derived from the solution of the suddenly heated plane wall problem in the similar form as Eq.(15).

$$\upsilon_{tn} = \frac{1}{s_{ti}} \left(\frac{h_{ti} u_{ti}}{2} - \frac{h_{ti+1} u_{ti+1}}{2} \right) \quad (27)$$

$$\upsilon_{td} = \frac{1.136^2 \alpha}{h_{it} + h_{ti+1}} \quad (28)$$

here, α is the thermal diffusivity of the fluid. When the value of $\upsilon_{tn} + \upsilon_{td}$ becomes positive, a nascent heat element is considered in the flow field over the thin thermal layer element [abcd] as shown in Fig.2. The value of the total heat for the nascent heat element ΔQ_i is equal to Q_{d-a} which is the heat brought out through the section [a-d] during small time dt, and the thickness of the element is given as

$$h_{heat} = (\upsilon_{tn} + \upsilon_{td}) \cdot dt \quad (29)$$

In order to simplify the numerical procedure in the heat calculation, the nascent rectangular heat element is replaced with an equivalent heat blob which has a thermal core, and the blob is introduced in the flow field in stead of the rectangular element. In the replacement, the total heat and the sectional area of the thermal core are determined so as to be the same as those of the corresponding nascent heat element. If a heat blob has a core of radius ε_{ti} and total heat ΔQ_i, a Gaussian distribution of the relative temperature around the center of the blob is given as

$$\theta(r) = \frac{\Delta Q_i}{\pi \varepsilon_{ti}^2} \exp\left\{-\left(\frac{r - r_i}{\varepsilon_{ti}}\right)^2\right\} \tag{30}$$

here r_i denotes a position of center of the blob and θ is temperature difference defined as $\theta = T - T_\infty$. From the similarity of the equation form of the thermal diffusion Eq.(6) to that of viscous diffusion Eq(7), the spreading of the thermal core radius ε_{ti} is expressed as

$$\frac{d\varepsilon_{ti}}{dt} = \frac{2.2418^2 \alpha}{2\varepsilon_{ti}} \tag{31}$$

2.7 Approximation in the constant temperature case

If the temperature of the body surface is kept always constant and the thickness of the thermal layer is taken very thin, further approximation can be considered in the introduction process of nascent heat blobs.

In Eq.(20), when both the temperature of body surface T_w and the temperature of approaching flow T_0 are constant and the thickness h_{ti} is thin, the relative temperature θ_{wi} is also constant $\theta_{wi} = \theta_w = T_w - T_0$ and $(\theta_{wi} - \theta_i)/\theta_{wi}$ becomes very small. Therefore, Eq.(20) can be approximated by

$$Q_{a-b} \approx \frac{1}{2} c \rho \theta_w u_{ti} h_{ti} dt \tag{32}$$

In the same manner, Eq.(21) is approximated by

$$Q_{c-d} \approx \frac{1}{2} c \rho \theta_w u_{ti+1} h_{ti+1} dt \tag{33}$$

It is an interesting point in the above equations that either the heat through the section [a-b] or [c-d] during a time interval dt can be approximately evaluated by using the constant relative temperature θ_w on the body surface and it is independent from the relative temperature at the outer boundary of the thin thermal layer θ_i. Therefore, if the evaluation of θ_i at the beginning of every time step is not needed in the procedure of introducing nascent heat blobs, the computing time will be reduced very much. Considering that the transfer of the heat $Q_{d\text{-}a}$ into the flow field through the section [d-a] is consisting of the convection of heat and the thermal diffusion, we assume an approximate expression of $Q_{d\text{-}a}$ using the velocities of convection and diffusion defined by Eqs.(27) and (28) as follows.

$$Q_{a-d} = c_o \theta_w c \rho s_i (\upsilon_{tn} + \upsilon_{td}) dt \tag{34}$$

where c_o is the corrective coefficient to be chosen as $0 < c_o < 1$.

3. CALCULATION RESULTS

3.1 Calculation conditions

As an example of application of the present method, a two-dimensional unsteady flow around a heated circular cylinder was calculated, where the effects of natural convection and radiation on the heat transfer were neglected.

The conditions of flow and calculations are as follows: Reynolds number $Re = Ud/\nu = 10^3$, 10^4 and 10^5, Prandtl number $Pr = \nu/\alpha = 0.71$, the temperature of body wall $T_w = 363$ K (constant), the temperature of approaching flow $T_0 = 293$ K (constant), the number of vortex sheet panels distributed around the cylinder $N = 100$, the non-dimensional time step $\Delta t = Udt/d = 0.05$. Referring the development of a laminar and thermal boundary layer along a flat plate, the thickness of vorticity layer was taken as $h_i = 5\sqrt{1/R_e}$, and the thickness of thermal layer as $h_{ti} = h_i \times \sqrt[3]{13/14} \cdot P_r^{-1/3}$. In this study, the approximation treatment was used assuming the corrective coefficient in Eq.(34) as $c_o = 0.2$.

3.2 Results and discussions

Figure 3 shows an instantaneous temperature distribution in the flow field around a circular cylinder and the corresponding flow pattern at a non-dimensional time T= 25.0 after the impulsive start of flow, where $Re = 10^4$ and $Pr = 0.71$. It is clarified that thermal crowds are formed behind the cylinder and periodically shed in the wake, corresponding to the periodical shedding of vortices. On the other hand, a time-averaged temperature distribution in the same flow field is shown in Fig.4. It is clear that the high temperature in the wake just behind the cylinder rapidly diffuses in the time-averaged flow field.

Figure 5 shows the comparison of time histories of the local velocity over the surface, the local pressure coefficient and the local Nusselt number at the rear stagnation ($\phi = 180^0$) during the non-dimensional time period from T=32 to 42 in the same flow as shown in Figs.(3) and (4) ($Re = 10^4$ and $Pr = 0.71$), where the scale of the ordinate has no meaning.

Figure 6 shows instantaneous distributions of the local coefficient of heat transfer (local Nusselt number) around the cylinder at T= 38.5 and 40.5 for $Re = 10^4$ and $Pr = 0.71$, together with the corresponding flow patterns, where $\phi = 0$ and 180 (degree) correspond to the front and rear stagnation points of the cylinder. It is seen that the two distributions of local Nusselt number are anti-symmetrical each other because the two flow patterns show a opposite phase relation in vortex shedding. The time-averaged local Nusselt number distribution in the same flow is shown in Fig.7, compared with experimental results by Igarashi(14) and Schmidt & Wenner(15). It is known that the calculated result reasonably coincides with the experiments.

Figure 8 shows the time-averaged and surface-averaged Nusselt numbers calculated for $Re = 10^3$, 10^4 and 10^5, in comparison with empirical curves proposed by Hilpert(16) and Douglas and Churchill(17). The agreement between the calculated values and the empirical curves is very well.

Figure 9 shows the time-averaged local Nusselt numbers at $\phi = 0^0$ and 180^0 calculated for for $Re = 10^3$, 10^4 and 10^5. It is confirmed that the value of $N_{u\phi}$ at $\phi = 0^0$ (front stagnation) increases with Reynolds number in proportion to $Re^{1/2}$ and the value of $N_{u\phi}$ at $\phi = 180^0$ (rear stagnation) increases in proportion to $Re^{2/3}$.

4. CONCLUSIONS

In this study, a vortex and heat elements method was proposed for the analysis of unsteady heat transfer in a flow around a body. Applying the method to two-dimensional unsteady and forced-convective heat transfer around a heated circular cylinder in uniform flows of $Re = 10^3$, 10^4 and 10^5, and $Pr = 0.71$, the following conclusions were obtained.

(1) It was clarified that thermal crowds are formed behind the cylinder and periodically shed in the wake, corresponding to the periodical shedding of vortices.

(2) Instantaneous distributions of the local Nusselt number were reasonably calculated.

(3) The time-averaged distribution of the local Nusselt number around the cylinder was almost coincident with experimental results.

(4) The time-averaged and surface-averaged Nusselt numbers calculated for $Re = 10^3$, 10^4 and 10^5 were in good agreement with empirical curves.

(5) It was shown that the relations between the local Nusselt number and Reynolds number are $N_{u\phi} \propto Re^{1/2}$ for $\phi = 0^0$ (front stagnation) and $N_{u\phi} \propto Re^{2/3}$ for $\phi = 180^0$ (rear stagnation).

Finally, it is concluded that the vortex and heat elements method proposed in this study will be very efficient and convenient for investigation of unsteady and forced-convective heat transfer in a flow around a heated body of arbitrary shape.

5. ACKNOWLEDGEMENT

The authors wish to thank Prof. T. Igarashi, the National Defense Academy, Japan, for his significant discussions and comments.

6. REFERENCES

(1) Chorin, A.J., (1973) Numerical study of slightly viscous flow. J. Fluid Mech. 57, 785-796.

(2) Chorin, A.J., (1978) Vortex sheet approximation of boundary layers. J. Comp. Phys. 74, 283-317.
(3) Leonard, A., (1980) Vortex methods for flow simulations. J. Comp. Phys. 37, 289-335.
(4) Lewis, R.I., (1981) Surface vorticity modelling of separated flows from two-dimensional bluff bodies of arbitrary shape. J. Mech. Engng. Sci. 23-1, 1-12.
(5) Kamemoto, K. and Matsumoto, H., (1995) On the attractive features of the advanced vortex methods as a Lagrangian large eddy simulation. Proc. of 4th KSME-JSME Fluid Engng. Conf. Pusan Oct.18-21, 1995, 293-296.
(6) Zhu, B., Kamemoto, K. and Matsumoto, H., (1998) Direct simulation of unsteady flow through a centrifugal pump impeller using fast vortex method. Comp. Fluid Dynamics J. 7-1, 15-25.
(7) Zhu, B., Kamemoto, K. and Matsumoto, H., (1998) Computation of unsteady viscous flow through centrifugal impeller rotating in volute casing by direct vortex method. Comp. Fluid Dynamics J. 7-3, 313-323.
(8) Zhu, B. and Kamemoto, K., (1999) Simulation of the unsteady interaction of a centrifugal impeller with its diffuser by an advanced vortex method. Proc. of the 3rd ASME/JSME Joint Fluids Engng. Conf. San Francisco, July 18-22, FEDSM99-6821.
(9) Ojima, A. and Kamemoto, K., (1999) Numerical analysis of the unsteady flow around three-dimensional bluff bodies by use of an advanced vortex method. Proc. of the 3rd ASME/JSME Joint Fluids Engng. Conf. San Francisco, July 18-22, FEDSM99-6822.
(10) Smith, P.A. and Stansby, P.K., (1989) An efficient surface algorithm for random particle simulation of vorticity and heat transport. J. Comp. Phys. 81, 349-371.
(11) Wu, J.C. and Thompson, J.E., (1973) Numerical solutions of time-dependent incompressible Navier-Stokes Equations using an integro-differential formulation. Computers & Fluids 1, 197-215.
(12) Uhlman, J.S., (1992) An integral equation formulation of the equation of motion of an incompressible fluid. Naval Undersea Warfare Center T.R. 10-086.
(13) Kamemoto, K., (1995) On attractive features of the vortex methods. Computational Fluid Dynamics Review 1995. Edt. By M.Fafez and K. Oshima JOHN WILEY & SONS, 334-353.
(14) Igarashi, T., (1984) Flow and heat transfer in the separated region around a circular cylinder. Trans. JSME. B 50-460, 3008-3014. (in Japanese)
(15) Schmidt, E. and Wenner, K., (1941) Warmeabgabe uber den Unfang eines angeblasenen geheizten Zylinders. Forshg. Ing.-Wes. 12, 65-73.
(16) Hilpert, R., (1933) Warmeabgabe von geheizten Drahten und Rohren in Luftstrom. Forschg. Ing.-Wes. 4, 215-224.
(17) Douglas, W.Y.M. and Churchill, S.W., (1955) Recorrelation of data for convective heat transfer between gases and single cylinders. AICE Heat Tran. Symp. National Meeting Louisville, Ky., Preprint 16.

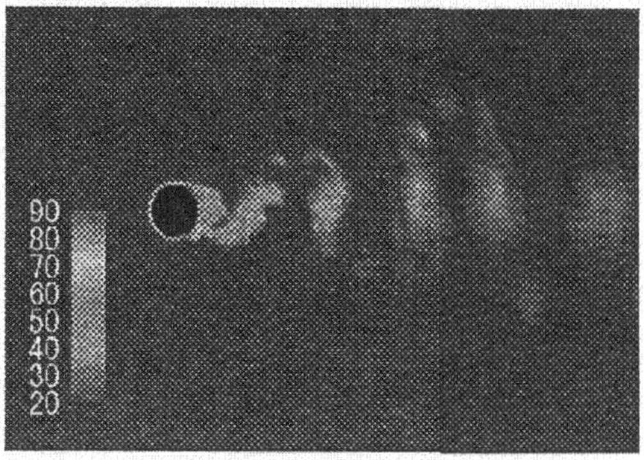

(a) Temperature distribution

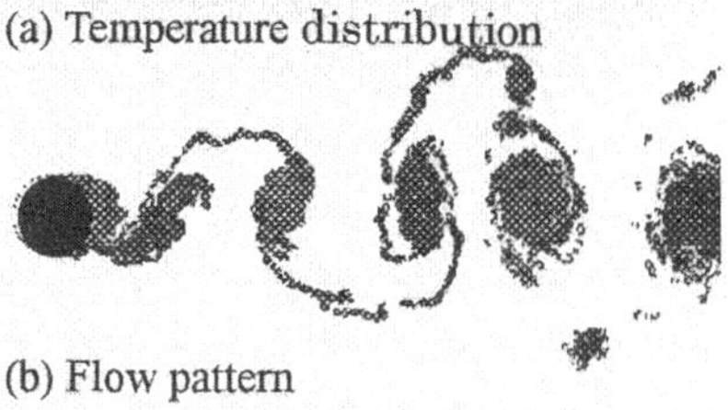

(b) Flow pattern

Fig.3 Instantaneous temperature distribution and flow pattern.(Re=10^4, Pr=0.71, Time=25.0)

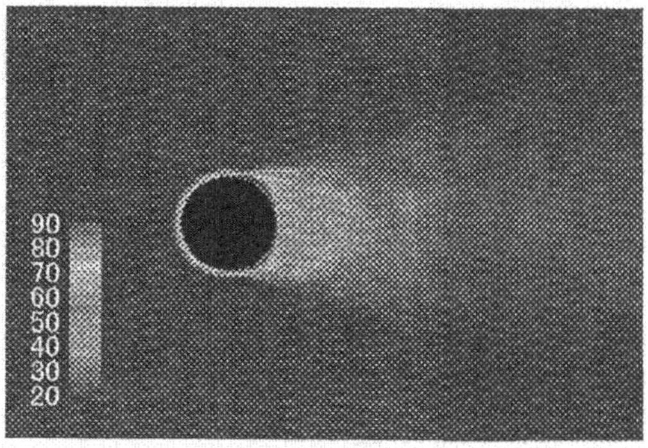

Fig.4 Time-averaged temperature distribution. (Re=10^4, Pr=0.71)

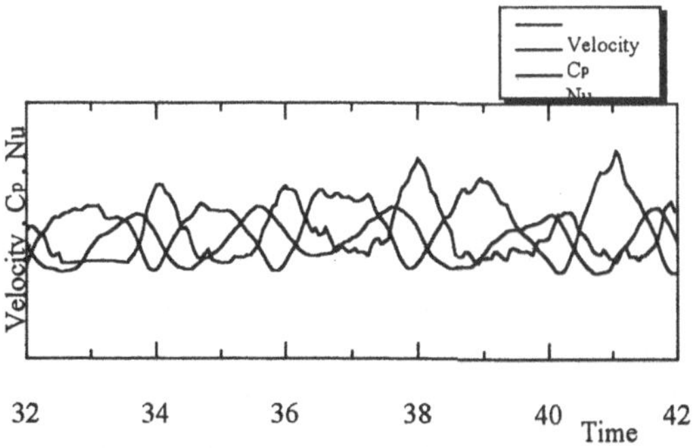

Fig.5 Comparison of time histories of the velocity, pressure coefficient and Nusselt number at the rear stagnation point. (Re=10^4, Pr=0.71)

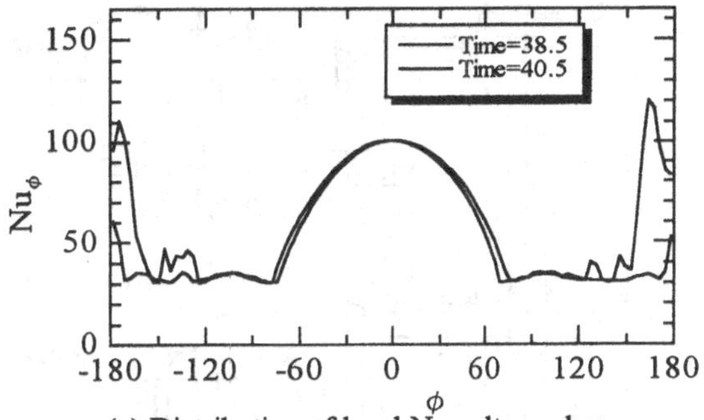

(a) Distribution of local Nusselt number

(b) T=38.5

(c) T=40.5

Fig.6 Instantaneous distributions local Nusselt number. (Re=10^4, Pr=0.71)

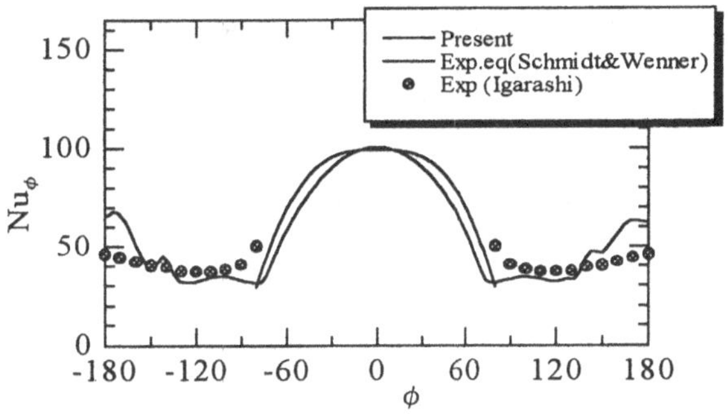

Fig.7 Time-averaged local Nusselt number distribution. (Re=10^4, Pr=0.71)

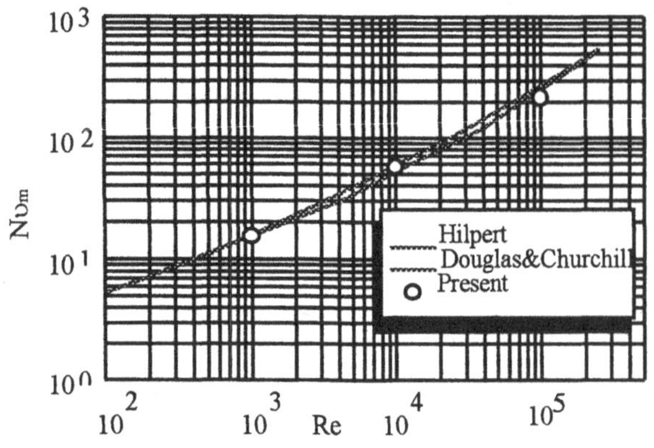

Fig.8 Time-averaged and surface-averaged Nusselt numbers. (Pr=0.71)

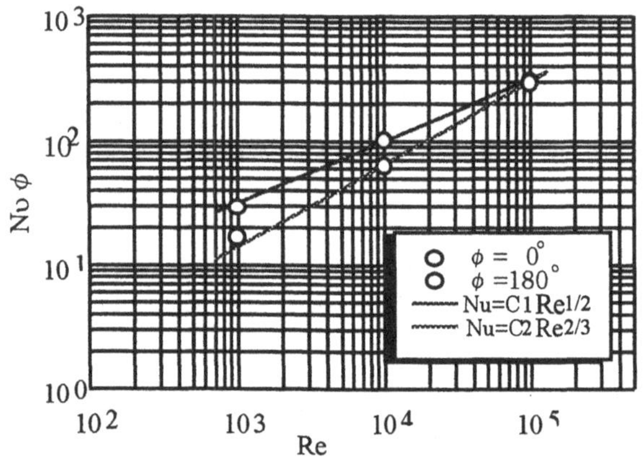

Fig.9 Time-averaged local Nusselt numbers at the front stagnation point and the rear stagnation point. (Pr=0.71)

A VORTEX METHOD FOR HEAT-VORTEX INTERACTION AND FAST SUMMATION TECHNIQUE

Yoshifumi Ogami
Department of Mechanical Engineering, Ritsumeikan University
1-1-1 Noji-higashi, Kusatsu 525-8577, Japan / Email: ogami@cfd.ritsumei.ac.jp

ABSTRACT

The solutions to the two problems on the diffusion velocity method are presented. It is shown that the remeshing technique is useful for portioning a vortex in regions where vortices are sparse and more vortices are required to make them overlapped and continue to diffuse, and that treating separately the negative particles and the positive particles avoids producing an unreasonably large velocity and gives very moderate solutions. Also very simple and effective fast summation methods which treat the Lagrangian particle motions due to convection and diffusion are presented.

1. INTRODUCTION

The vortex methods have been applied to a variety of physical flows such as vortex sheets, shear layers, external and internal flows, or reactive flows (1,2). However, not so many researchers have worked on heat-transfer problems by the vortex methods.

The extension of the vortex methods to the heat transfer problem requires models for

1. the representation of temperature/heat with particles,
2. the creation of vorticity from heat, and
3. the diffusion of heat and vorticity.

Ghoniem and Sherman (3) investigated one-dimensional and quasi-one-dimensional diffusion using temperature elements for representing temperature and random walks for diffusion. The creation of vortex from the temperature is taken into account. Also Ghoniem et al (4,5) studied shear layer and plume rise using the core spreading method for diffusion. The vortex strength is updated by the transport element method in which scalar gradients are used in the transport process. Smith and Stansby (6) treated one– and two–dimensional flows using temperature particles and the random walks. The creation of vorticity from heat is not considered.

Ogami (7,8) presented two models for creating vortices from temperature particles. The first model is based on the direct interpretation of the vorticity equation and is considered to be an extension of the scheme adopted by Ghoniem and Sherman (3) to two-dimension. The second model is a new idea that one temperature particle creates one vortex pair (two vortices with the strength Γ and $-\Gamma$). He treated a boundary condition problem in one–dimension where the temperature of a wall is kept constant, and a two–dimensional problem with no boundary. For treating the diffusion, a deterministic Lagrangian technique called the diffusion velocity method(9,10) is employed.

Regarding the diffusion velocity method, one serious problem was pointed out by Clarke and Tutty (12) that the diffusion is limited to regions where the vortices are overlapped. The *core* of this problem is considered that the strength/circulation of each vortex is too large to represent the regions of small vorticity. Therefore, a solution to this problem would be to divide a vortex into multi vortices. For this purpose, we use the remeshing (re–initializing) technique, which usually is used for fixing the distortion of the Lagrangian gird (13). However, this technique also works for portioning a vortex in regions where vortices are sparse and more vortices are required to make them overlapped and continue to diffuse.

Another problem of the diffusion velocity method is that an unreasonably large velocity is produced in the regions where plus vortices and minus vortices are mixed. This is because the vorticity at some point can be extremely small (the diffusion velocity is inversely proportional to the vorticity). A solution to this problem is to treat plus vortices and minus vortices separately.

This paper shows that the two solutions mentioned above compensate for the defect of the diffusion velocity method. Also two fast summation techniques, one is for the diffusion velocity and the other for the convection velocity, are presented.

2. GOVERNING EQUATIONS AND DIFFUSION VELOCITY

We consider the vorticity equation, the energy equation and the continuity equation in two dimension to briefly explain our method. The motion law of our Lagrangian scheme is found if these equations are put into a conservation form as follows.

$$\frac{\partial \omega}{\partial t} + \mathrm{div}(\omega \boldsymbol{u}_c + \omega \boldsymbol{u}_\omega) = -g\beta \frac{\partial T}{\partial x} \qquad (1)$$

$$\frac{\partial T}{\partial t} + \mathrm{div}(T \boldsymbol{u}_c + T \boldsymbol{u}_T) = 0 \qquad (2)$$

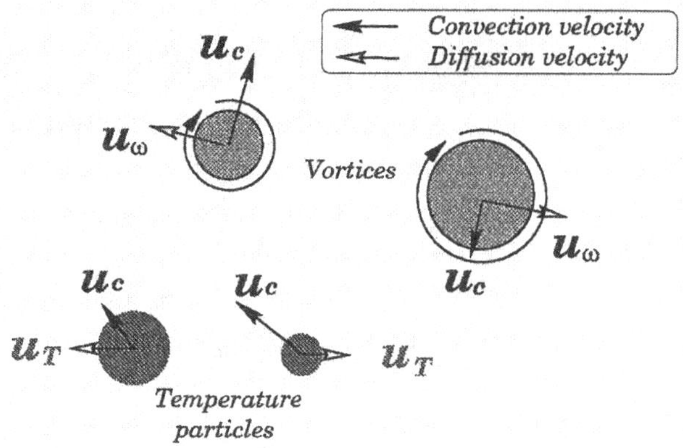

Figure 1. The convection velocity and the diffusion velocity

where

$$\boldsymbol{u}_c = (u, v), \quad \boldsymbol{u}_\omega = -\frac{\nu}{\omega}\nabla\omega, \quad \boldsymbol{u}_T = -\frac{\alpha}{T}\nabla T \qquad (3)$$

ω, T, ν, g, β and α are the vorticity, the temperature, the kinematic viscosity, the gravitational acceleration, the modulus of compressibility and the thermal diffusivity. u and v are the velocity components in x and y directions.

Equation (1) states that the vorticity, ω, moves both with the convection velocity, $\boldsymbol{u}_c$, induced by the vorticity (the Biot-Savart law), and with the diffusion velocity, $\boldsymbol{u}_\omega$ (9,10) (Fig.1). It also indicates that the strength of vorticity varies according to the right hand side of this equation. Similarly, Eq.(2) gives the law stating that the temperature distribution, T, moves both with the convection velocity, $\boldsymbol{u}_c$, and with the diffusion velocity, $\boldsymbol{u}_T$ (Fig.1).

The vorticity $\omega(\boldsymbol{x})$ and the temperature $T(\boldsymbol{x})$ are expressed by a summation of Gaussian-cored particles located respectively at $\boldsymbol{x}_{\omega j}$ and $\boldsymbol{x}_{Tj}$, with the strength Γ_j and Θ_j, and the core radius σ_ω and σ_T as

$$\omega(\boldsymbol{x}) = \sum_j \frac{\Gamma_j}{\pi {\sigma_\omega}^2} \exp\left(-\frac{|\boldsymbol{x} - \boldsymbol{x}_{\omega j}|^2}{{\sigma_\omega}^2}\right) \qquad (4)$$

$$T(\boldsymbol{x}) = \sum_j \frac{\Theta_j}{\pi {\sigma_T}^2} \exp\left(-\frac{|\boldsymbol{x} - \boldsymbol{x}_{Tj}|^2}{{\sigma_T}^2}\right) \qquad (5)$$

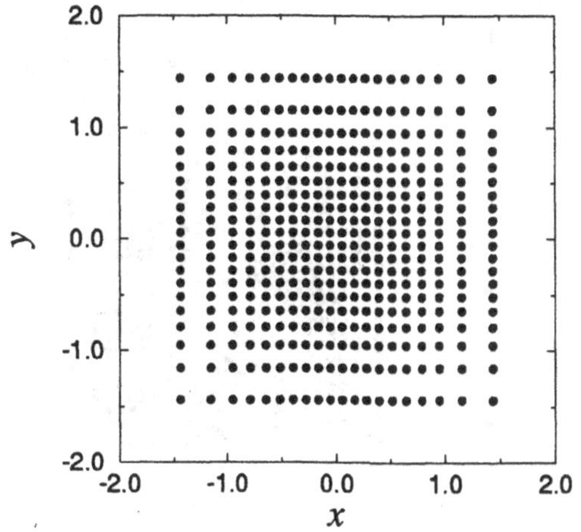

(a) Without remeshing

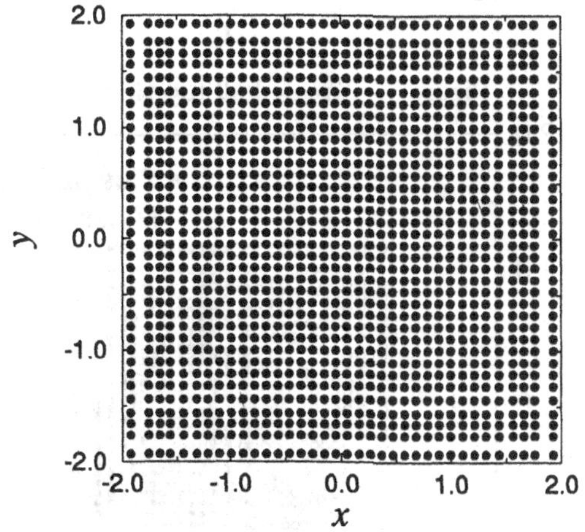

(b) With remeshing

Figure 2. Comparison of particles with and without remeshing

Finally, the Lagrangian scheme for the heat–vortex interaction is given by the ordinary differential equations which determine the center point of ith vortex, $\boldsymbol{x}_{\omega i}$ and that of ith temperature particle, $\boldsymbol{x}_{Ti}$,

$$\frac{d\boldsymbol{x}_{\omega i}(t)}{dt} = \boldsymbol{u}_c(\boldsymbol{x}_{\omega i}, t) + \boldsymbol{u}_\omega(\boldsymbol{x}_{\omega i}, t) \tag{6}$$

$$\frac{d\boldsymbol{x}_{Ti}(t)}{dt} = \boldsymbol{u}_c(\boldsymbol{x}_{Ti}, t) + \boldsymbol{u}_T(\boldsymbol{x}_{Ti}, t) \tag{7}$$

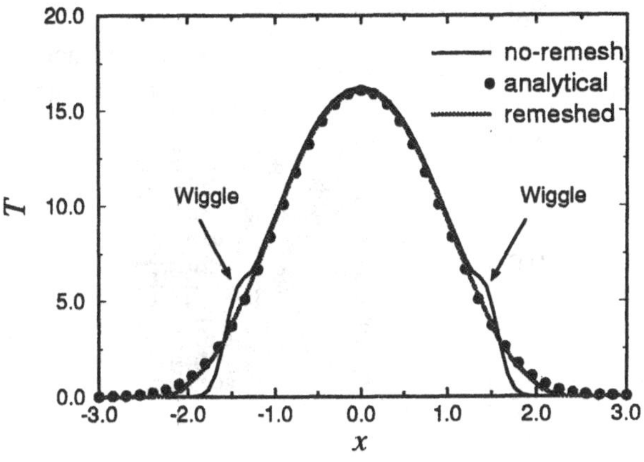

Figure 3. Comparison of temperature

3. TWO PROBLEMS OF THE DIFFUSION VELOCITY METHOD AND SOLUTIONS

In this section, we will see how the two problems of the diffusion velocity method stated in **INTRODUCTION** are resolved in the following simulations.

3.1 Portioning a Particle of Large Strength by Remeshing Method

In this subsection, simulations are conducted to show how the solution in the low-density regions is ruined by a large-strength particle and how this is resolved.

Initially, 20×20 temperature particles are placed each with the same distance apart in a square region of 1×1. Figure 2(a) shows the positions of the particles, at 1000 time steps, moving only with the diffusion velocity (with no convection velocity). The time step is $\Delta t = 0.0025$, the thermal diffusivity is $\alpha = 0.072$, the initial temperature is 20 and the ratio of the initial distance $\Delta \boldsymbol{x}$ to the core radius σ_T is 0.5. The remeshing technique is not used. Figure 2(b) shows the positions of the particles at the same time calculated with the remeshing technique used at every 200 steps.

Figure 3 compares the temperature distributions

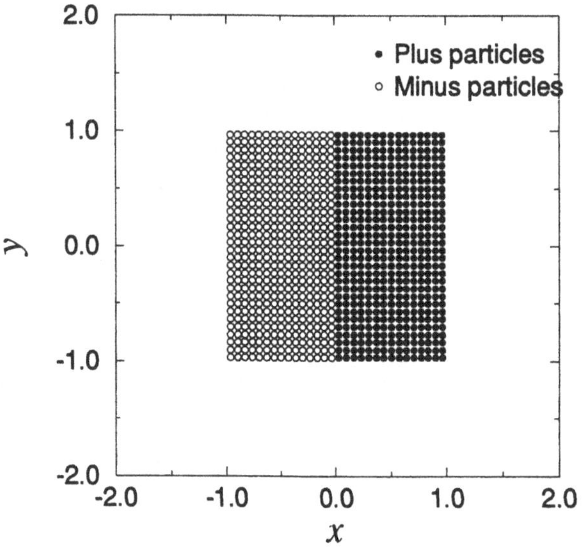

Figure 4. Initial position

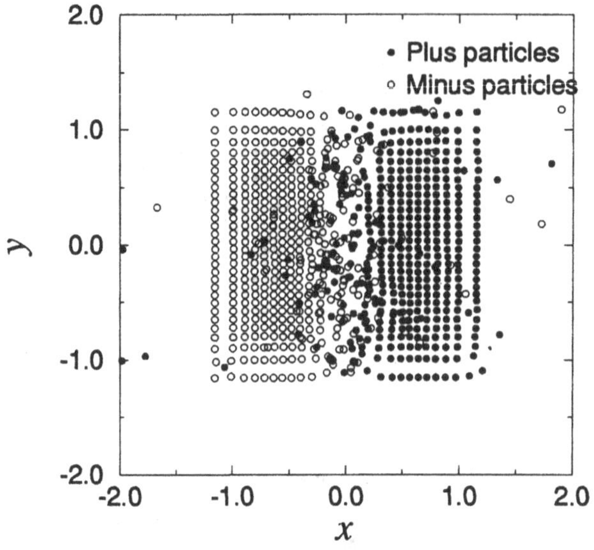

Figure 5. Treated all together

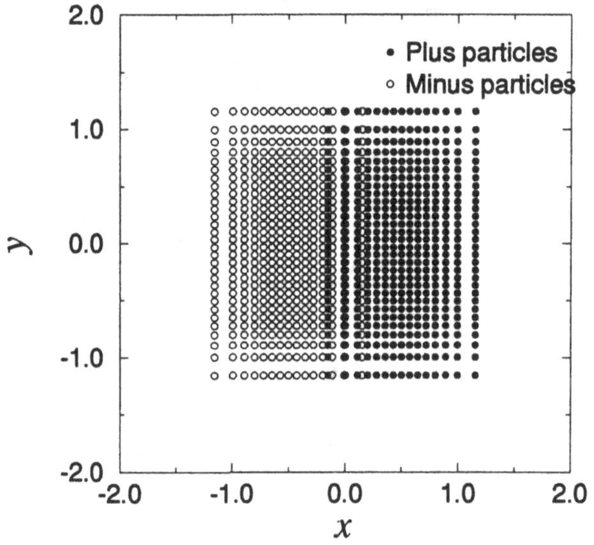

Figure 6. Separately treated

on the line $y = 0$. We see that the distribution obtained without the remeshing contains "wiggles," which means diffusion stops around here, and differs from the analytical solution. However, we also see that the distribution with the remeshing technique agrees well with the analytical solution and thus the problem of the diffusion velocity method is resolved.

3.2 Plus and Minus Particles

When particles of positive strength and negative strength are treated at the same time, the diffusion velocity at the position where the density is extremely small becomes unreasonably large and the solution is destroyed. One way to avoid this is to treat particles separately as two groups (positive and negative ones) when calculating the density and the diffusion velocity. The final density is obtained by adding the density of positive particles and that of negative ones. This may be allowed since the diffusion process is linear. To demonstrate these the following simulations are conducted.

Initially, 15×30 temperature particles of the same positive strength and of the same negative strength are arranged with the same interval in the square region as shown in Fig.4. The parameters used here are as same as the previous subsection.

Figure 5 shows the positions of the particles, at 200 time steps, moving only with the diffusion velocity. The positive particles and the negative particles are treated at the same time. It is clearly shown that the

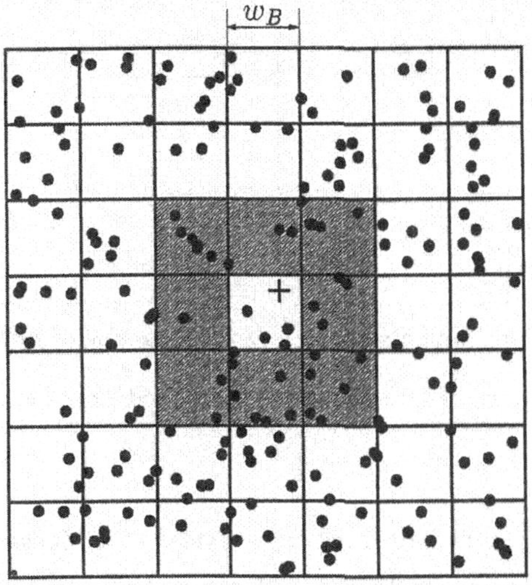

Figure 7. Blocks for diffusion velocity

closer the positions of the particles are to the center line ($x = 0$, where the initial density is zero), the more chaotic the solutions become. However, when the negative and the positive particles are treated separately, we can obtain very moderate solutions as shown in Fig.6.

4. FAST SUMMATION METHODS

4.1 Fast Summation Method for the Diffusion Velocity

The diffusion velocity is calculated with the density and its derivatives created by the particles. Since the density of a particle with a Gaussian core decays exponentially, the particles which exist with certain distance apart from the calculation point do not have to be taken into account.

As shown in Fig.7, the simulation region is divided into small blocks of the same width w_B, and only the particles in the block where the calculation point (+) exists and in the surrounding eight blocks (shaded blocks) are considered. This reduces the particle number for calculating the diffusion velocity.

Table 1. Ratio w_B/σ_T and relative error for particle number 2000

w_B/σ_T	Relative error%	CPU s
2.0	7.8×10^{-2}	0.28
3.0	1.6×10^{-4}	0.38
4.0	5.6×10^{-8}	0.54
5.0	4.6×10^{-12}	0.75

Table 1 shows how the relative error between the diffusion velocities calculated by the block method and by the traditional method reduces with the ratio w_B/σ_T. The particle number is 2000 and the particles are placed at random. The ratio of the average distance between the particles to the particle core radius is 0.5. The total CPU time (the time for creating blocks and for calculating the velocities at all the particle points) increases with the ratio and the relative error decreases. The CPU is much less than that by the traditional method, 1.8s. For this calculation, an HP Exemplar (equipped PA8200 processors) was used without paralellization.

Table 2 compares the CPU time by the block method with $w_B/\sigma_T = 3$ and by the traditional method. Roughly the former CPU time is proportional to N and the latter N^2, and the efficiency of this block method is remarkable. For this calculation, the code was paralellized and twelve processors were used.

4.2 Fast Summation Method for the Convection Velocity

Since the velocity induced by a vortex does not decay fast unlike the Gaussian density distribution, all the vortices have to be taken into account when calculating the convection velocity. One of the techniques to reduce the computational time is the mono-pole

Table 2. Particle number N, CPU time and relative error for $w_B/\sigma_T = 3$

N	CPU s (bl.)	CPU s (trad.)	Relative error %
200	0.03125	0.01563	1.80×10^{-4}
500	0.08594	0.11719	1.24×10^{-4}
4000	0.96094	11.17188	1.82×10^{-4}
32000	8.46875	644.09375	1.96×10^{-4}

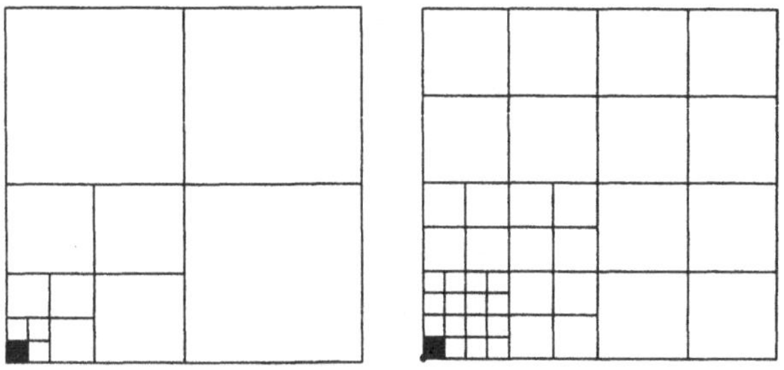

Figure 9. Block structure of different precision level

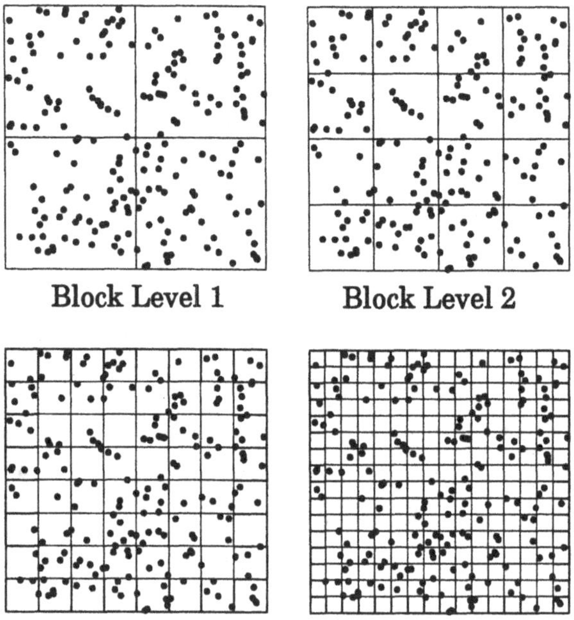

Figure 8. Level of blocks with different width

method (12) (13), where the computational region is divided into blocks and the vortices in a block are merged into one *representative* vortex, resulting in decrease of the vortex number and computational time. The representative vortex has the total circulation of the vortices in each block and placed at the center of gravity.

Here, a very simple and effective method using the mono-pole technique is introduced. First, the computational region is divided into blocks with the same width w_L, and the calculation regarding the representative vortex in each block is conducted. This procedure is done with the block widths $w_L/2$, $w_L/2^2$, $w_L/2^3 \cdots$ (Fig.8).

Second, as shown in Fig.9, the smaller blocks are placed closer to the block where the velocity is calculated (black block) without any overlap nor gap. The velocity is calculated using all the vortices in the black block and in the blocks surrounding this block, and using the representative vortices in the rest of the blocks. This procedure is done about all the positions of the vortices.

The precision of the calculated velocity depends on the structure of the blocks. For example, the block structure shown on the right in Fig.9 produces more accurate velocity than the left one because more small blocks are arranged. Figure 10 shows examples of the block structure at precision level 2, where the calculated points are at the center of the region (left) and at the up-right corner (right).

Tables 3 and 4 show CPU time and relative error respectively at precision level 2 and 3. We see that the error becomes smaller at higher precision level and that the block method reduces CPU time effectively. The computational time of the present method is proportional to $N \log N$.

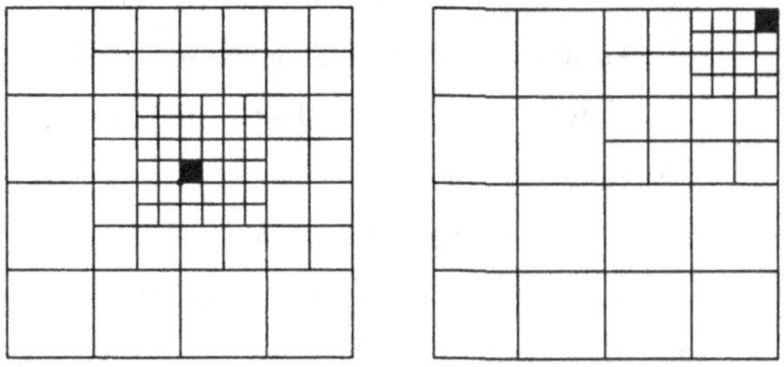

Figure 10. Examples of block structure at precision level 2

Table 3. Precision level 2

N	CPU s (bl.)	CPU s (trad.)	Relative error %
200	0.03125	0.02344	2.72×10^{-1}
500	0.07813	0.13281	3.14×10^{-1}
4000	0.60156	8.71094	1.62×10^{-1}
32000	6.58984	558.17969	4.85×10^{-1}

Table 4. Precision level 3

N	CPU (bl.)	CPU (trad.)	Relative error %
200	0.06250	0.02344	4.91×10^{-2}
500	0.11719	0.13281	5.52×10^{-2}
4000	1.21875	8.72656	5.29×10^{-2}
32000	9.65625	553.42969	4.11×10^{-2}

5. EXAMPLE

To test our temperature–vorticity algorithm with fast summation technique, the following initial value problem is treated.

$$\begin{cases} T = 20°\text{C} \quad \text{and} \quad \omega = 0 \quad \text{everywhere} \\ \text{except} \quad T = 40°\text{C} \quad (|\boldsymbol{x}| \leq 1\text{cm}, \quad |\boldsymbol{y}| \leq 1\text{cm}) \end{cases}$$

Equations (6) and (7) including both the convection velocity and the diffusion velocity are solved so that the interaction of the temperature and the vorticity is simulated. The parameters of water are used and dependency on the temperature is considered. The time step is 0.0025s. The high temperature region ($|\boldsymbol{x}| \leq 1$cm, $|\boldsymbol{y}| \leq 1$cm) is expressed by 30×30 temperature particles. The remeshing is done at every 100 steps.

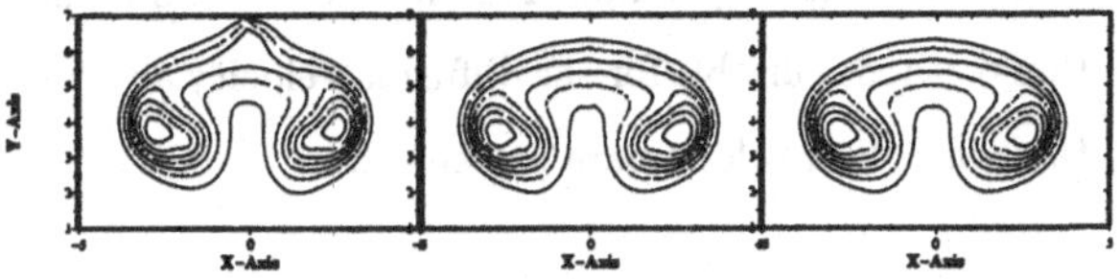

(a) Precision level 2 (19min) (b) Precision level 3 (22min) (c) Traditioal method (112min)

Figure 11. Temperature at time 5 (2000 steps)

Figure 11 shows temperature distributions at time 5 calculated with the fast algorithm (a, b) and without it (c). The parameter w_B/σ_T, which determines the precision of the diffusion velocity, is 4 and precision levels for the convection velocity are chosen to be 2 (Fig.11-a) and 3 (Fig.11-b) for comparison. By this time, the temperature particle number has increased to 4583 from the initial number 900 and the vortex number to 9743 from 1800. We can see that the fast summation method, which takes 19min and 22min, runs several times faster than the traditional method (112min), and that the precision level 3, of which order of relative error is 10^{-2}%, gives solutions comparable to the traditional one.

6. CONCLUSION

The solutions to the two problems on the diffusion velocity method are presented. It is shown that the remeshing technique is useful for portioning a vortex in regions where vortices are sparse and more vortices are required to make them overlapped and continue to diffuse, and that treating separately the negative particles and the positive particles avoids producing an unreasonably large velocity and gives very moderate solutions. Also very simple and effective fast summation methods suitable for the diffusion velocity and for the convection velocity are presented.

References

(1) Sarpkaya, T., (1989), "Computational Methods with Vortices — the 1988 Freeman Scholar J. Fluids Engineering, Vol. 111, p.5–52.

(2) Sethian, J.I., (1991), "A Brief Overview of Vortex Methods," Vortex Methods and Vortex Motion, SIAM. Philadelphia. p.1–32.

(3) Ghoniem, A. F., and Sherman F. S., (1985), "Grid-free Simulation of Diffusion Using Random Walk Methods," J. Comp. Phys., Vol.61, p.1–37.

(4) Ghoniem, A. F., Heidarinejad, G., and Krishan A., (1988), "Numerical Simulation of a Thermally Stratified Shear Layer Using the Vortex Element Method," J. Comp. Phys., Vol. 79, p.135–166.

(5) Zhang X., and Ghoniem, A. F., (1993), "A Computational Model for the Rise and Dispersion of Wind-Blown, Buoyancy-Driven Plumes - I. Neutrally Stratified Atmosphere," Atmospheric Environ., Vol.27A, p.2295–2311.

(6) Smith P. A., and Stansby P. K., (1989), "An Efficient Surface Algorithm for Random-Particle Simulation of Vorticity and Heat Transport," J. Comp. Phys., Vol.81, p.349–371.

(7) Ogami, Y., (1998), "Heat-Vortex Interaction by the Vortex Method with the Diffusion Velocity," 3rd International Workshop on Vortex Flows and Related Numerical Methods, Toulouse, France, p.81–82.

(8) Ogami, Y., (1998), "Heat-Vortex Interaction by the Vortex Method with the Diffusion Velocity," 3rd International Workshop on Vortex Flows and Related Numerical Methods, ESAIM Proceedings, to appear.

(9) Ogami, Y., and Akamatsu, T., (1987), "Numerical Study of Viscous Flow around a Circular Cylinder by Vortex Method with Diffusion–Velocity (in Japanese)," 1st Symposium of Computational Fluid Dynamics, Tokyo, p.279–282.

(10) Ogami, Y., and Akamatsu, T., (1991), "Viscous Flow Simulation using the Discrete Vortex Method — the Diffusion Velocity Method," Computers & Fluids, Vol.19, p.433–441.

(11) Kempka, S.N., and Strickland, J.H., (1993), "A Method to Simulate Viscous Diffusion of Vorticity by Convective Transport of Vortices at a Non-Solenoidal Velocity," SAND93-1763. UC-700.

(12) Clarke N. R., and Tutty O. R., (1994), "Construction and Validation of a Discrete Vortex Method for the Two-Dimensional Incompressible Navier–Stokes Equations," Computers & Fluids, Vol. 23, pp.751–783.

(13) Koumoutsakos P., and Leonard A., (1995), "High–Resolution Simulations of the Flow around an Impulsively Started Cylinder using Vortex Methods," J. Fluid Mech., Vol.296, p.1–38.

Three-dimensional Vortex Method Using the Ferguson Spline

Michihisa Tsutahara, Akira Miura, Kazuhiko Ogawa, and Katsuhiko Akita
Department of Mechanical Engineering, Kobe University
Rokko, Nada, Kobe 657-8501 JAPAN

ABSTRACT

A new technique for three dimensional vortex methods has been presented. The vortex tubes are represented by the Ferguson spline and the induced velocity is calculated analytically by expanding the integrand of the Biot-Savart law into the Maclaurin expansion. The vortex tube is assumed to be a vortex filament without thickness. In order to avoid the singularity of self-induced velocity, two technique are employed. One is to neglect the vicinity of the control point in calculating the induced velocity. The other is to consider a core at the control point. The effectiveness of this method is shown in calculating the translating velocity of a vortex ring. A more complex example, a soliton, propagating a strong vortex tube, is also successfully calculated.

1. INTRODUCTION

In three-dimensional vortex methods, the vortex filaments are divided into discrete vortex elements, such as the vortex sticks or the vortex blobs. However, as well known, the vortex filaments must be continuous and should not be cut inside the flow field. On the other hand, calculating the induced velocity due to vortex segments by the Biot-Savart law is the most time consuming burden for CPU. Moreover the vortex lines are usually divided into vortex sticks or blobs and the discretization error is not negligible especially when the vortex lines stretch. A method describing continuous vortex lines as they are and calculating the induced velocity efficiently has been desired.

In this paper, a new method is presented, in which a vortex line is divided into several segments and each segment is represented by the Ferguson function (one of spline functions), which will be explained later. The induced velocity can be obtained analytically by using this function.

Real vortex tube may have a core, in which the vorticity is constant or appropriately distributed. But in this method, the vortex tube is represented, for simplicity, by a vortex filament without core. In three-dimensional vortex filament, the induced velocity becomes logarithmically infinite if the vortex line has a finite curvature. But this logarithmic singularity is eliminated by considering a finite core locally, or by neglecting the contribution from the vortex vicinity of the control point. Moreover the both technique give the similar form of the leading term. Therefore we employ the both techniques.

2. THE FERGUSON FUNCTION

The Ferguson function is one of the parametric spline functions[1], and each vortex filament is divided into finite number of the segments as shown in Fig.1(a), one of which is expressed by

$$\vec{p}_i(t) = \vec{a}_i + \vec{b}_i t + \vec{c}_i t^2 + \vec{d}_i t^3 \quad (-0.5 \le t \le 0.5) \tag{1}$$

$$\begin{cases} \vec{a}_i = 0.5(\vec{Q}_i + \vec{Q}_{i+1}) + 0.125(\vec{Q}'_i - \vec{Q}'_{i+1}) \\ \vec{b}_i = 1.5(\vec{Q}_{i+1} - \vec{Q}_i) - 0.25(\vec{Q}'_i + \vec{Q}'_{i+1}) \\ \vec{c}_i = 0.5(\vec{Q}'_{i+1} - \vec{Q}'_i) \\ \vec{d}_i = (\vec{Q}'_i + \vec{Q}'_{i+1}) - 2(\vec{Q}_{i+1} - \vec{Q}_i) \end{cases} \tag{2}$$

where t is the parameter along the vortex line, $\vec{Q}_i$ is an control point, and primes represent the differentiation with respect to t. The Ferguson function is described by the coordinates for two end points of the curve and the differentials at the same points. Each segment is represented by the parameter t from -0.5 to 0.5 as shown in Fig.1 (b). It should be noted that the parameter t does not refer to the length of the vortex filament.

Time evolution of the vortex filaments is obtained as follows. The end points are considered to be the control points, so that each vortex filament has finite number of the control points. The induced velocities at these points are calculated by the Biot-Savart law, and the control points move with the induced velocities. After moving the all control points, new positions of the vortex filaments are determined by reconstructing a new Ferguson functions.

The induced velocity by the vortex filament for an unbounded domain can be obtained by the Biot-Savart law as mentioned above[2-4],

$$\mathbf{u}(\mathbf{x}) = -\frac{\Gamma}{4\pi}\int_C \frac{(\mathbf{x} - \mathbf{r}(s)) \times \dfrac{\partial \mathbf{r}}{\partial s'}\left|\dfrac{ds'}{dt}\right|}{|\mathbf{x} - \mathbf{r}|^3} dt \tag{3}$$

However, direct integration of the above integral by, for example, the Simpson's method will take long time like other discrete vortex methods, even the integrand is expressed in the form of a rational function of the parameter t. In this paper, integration of the Biot-Svart law is performed segment by segment.

The control point is taken at each end point of the segment. In this case, the integrand of the Biot-Savart law is very smooth function except the segment including the control point,

$$\mathbf{u}_i = \frac{\Gamma}{4\pi}\int_{-0.5}^{+0.5}(\mathbf{v}_0 + \mathbf{v}_1 t + \mathbf{v}_2 t^2 + \mathbf{v}_3 t^3 + \mathbf{v}_4 t^4) /(\alpha_0 + \alpha_1 t + \alpha_2 t^2 + \alpha_3 t^3 + \alpha_4 t^4 + \alpha_5 t^5 + \alpha_6 t^6)^{3/2} dt \qquad (4)$$

and the integrand can be Maclaurin expanded about $t = 0$. Therefore, the induced velocity at the control point due to this segment can be obtained analytically by integrating the polinomial function of t.

$$\mathbf{u}_i = \frac{\Gamma}{4\pi}\times \left[\beta_0 t + \beta_1 t^2 + \beta_2 t^3 + \beta_3 t^4 + \beta_4 t^5 + \beta_5 t^6 + \beta_6 t^7\right]_{-0.5}^{+0.5}$$

Definition of t as $-0.5 \leq t \leq 0.5$ leads the integration of the even terms to zero, and then we have to consider the odd terms only. The terms are taken up to 7th order in this paper as,

$$\mathbf{u}_i = \frac{\Gamma}{4\pi}\left[\beta_0 t + \beta_2 t^3 + \beta_4 t^5 + \beta_6 t^7\right]_{-0.5}^{+0.5} \qquad (5)$$

The accuracy is sufficient for calculating the induced velocity from the segments not so close to the considered control points. For example, induced velocity from the next segment to the segment including the control point is shown in Fig.2. The difference between the exact velocity and the approximated velocity is almost unrecognized.

However, the induced velocity from the segment including the control point is not approximated appropriately in the Maclaurin series, because there is a logarithmic singularity at the control point.

When, in general the vortex line has a curvature the induced velocity due to very close vortex element becomes infinite as the distance approaches to zero. Two techniques have been used to avoid the difficulty.

One is to set a kind of core and to limit the self induced velocity finite as

$$r \rightarrow r + \varepsilon \quad \text{when} \quad r \rightarrow 0 \qquad (6)$$

This technique is similar to the so-called softening parameter used for ordinary discrete vortex method.

The other one is exclude the vicinity of the control point from the integrand of the Biot-Savart law. Then we have[2]

$$u = \frac{\Gamma}{4\pi}\frac{\ln a}{R} + \text{bounded terms} \qquad (7)$$

where Γ is the circulation of the vortex ring, a is the omitted length of the vortex element when the integration is performed, and R is the radius of curvature.

Extending this idea, the integral is evaluated from $-L$ to L along the vortex filament excluding the vicinity of the control point. This technique is called the Local Induced Approximation (LIA)[3],[4]

$$\mathbf{u} = \frac{\Gamma}{4\pi}\left[\log\left(\frac{L}{a}\right)\right]\kappa\mathbf{b}\ , \qquad (8)$$

where κ is the local curvature, $\mathbf{b}$ the unit binormal vector, and L the segment length considered in calculating the induced velocity.

In our calculations, Model 1 in which the vicinity of the control point and Model 2 in which the core is considered are employed. In order to verify the accuracy of these methods, the translation velocity of a circular ring and the propagation of the twisted-shaped soliton on the vortex tube are simulated.

The following are the results obtained by these method.

3. CIRCULAR VORTEX RING

Kinematics of circular vortex ring has been well established. Circular vortex ring translates in the direction of its axis with a constant velocity. If we do not consider the core of the vortex tube, the velocity becomes infinity, and the velocity strongly depends on the core size. The translation velocity U can be given for rigidly rotating core by[3]

$$U = \frac{\Gamma}{4\pi R}\left(\log\frac{8R}{d} - \frac{1}{4}\right) \qquad (9)$$

where Γ is the circulation of the vortex tube, R the ring radius, and d the core radius. We can see, comparing (7) and (9), the omitted length and the core radius contribute to the leading term in the same manner.

We divide the circular ring into eight segments as shown in Fig.3. The calculated results are compared with this analytical result in Fig.4. The result by Model 2 which considers the core, which is not exactly the same core considered in the theory, agrees very well to the analytical result with an error of about 1%. On the other hand, Model 1 with excluding distance a gives a result of about 5% smaller induced velocity if a and d have the same value. This is obviously because Model 1 neglects the induced velocity due to the vorticity inside the core.

4. SOLITON ON VORTEX TUBE

Hashimoto[5],[6] obtained a loop soliton solution, which is a deformation of the vortex tube and propagates along the vortex tube. He started the expression in (8), and let L/d be constant. He showed that the complex function given by

$$\psi = \kappa \exp\left[\int^{s} \tau(s',t)ds'\right] \qquad (10)$$

satisfies the nonlinear Schrodinger equation

$$\frac{1}{i}\frac{\partial \psi}{\partial t} = \frac{\partial^2 \psi}{\partial s^2} + \frac{1}{2}|\psi|^2 \psi . \qquad (11)$$

Then he showed the existence the soliton solution as

$$x + iy = 2\beta \operatorname{sech}\xi \exp\left[i\{(1-\beta)^{1/2}\xi + \theta_0\}\right]$$
$$z = \xi - 2\beta \tanh \beta\xi \qquad (12)$$

where

$$\kappa = 2\beta \operatorname{sech}(\beta\xi), \quad \xi = s - c_0 t . \qquad (13)$$

This single soliton propagates in z-direction, along which the vortex filament extends, with the velocity c_0.

This soliton is simulated by Model 1 with 142 segments. Initial shape is given by (12) and (13), and vortex tube including far distant point from the deformation is considered in calculation.

The birds eye view of the vortex tube soliton is shown in Fig.5, in which $\beta = 1$ and $t = \theta_0 = 0$.

The calculated results are given in Fig. 6. The twisted deformation rotates about the axis of the vortex tube and propagates along the tube. The results show that the shape of the deformation does not change

Collision of two same size solitons, but propagating in the opposite directions, is also simulated. In this case, the calculation is performed by the LIA, that is, the induced velocity at a control point is obtained only from the segments including the point. Otherwise, the shapes of the solitons are broken, because other parts of the vortex filament approach the control point and induce strong velocities.

As shown in Fig.7, two solitons collide and are united. Then the two appear again and are separated without changing their shapes. The typical characteristics of the solitons are realized.

5. CONCLUDING REMARKS

A new technique of three-dimensional vortex method is presented. Vortex tubes are represented by the Ferguson spline function and the calculation of the induced velocity is analytically obtained. In this sense this method is not of the discrete vortex method but a direct vortex method.

The accuracy is very good. For the induced velocity of the circular vortex ring, the results agree well with the analytical result. The soliton traveling through a vortex tube, in which a very delicate balance among the induced velocities is established, is also well described.

REFERENCES

[1] Miura, A. et al. (1994) "Non-Uniform Rational B-Spline for CAD/CG Engineers" (in Japanese) , Kogyo Chosakai.

[2] Batchelor,G.K.(1967) "An Introduction to Fluid Dynamics," Cambrdge UP, Chap.7.

[3] Saffman, P.G. (1992) "Vortex Dynamics," Cambridge UP, Chaps 2 and 10..

[4] Leonard, A.,(1985) "Computing Three-Dimensional Incompressible Flows with Vortex Elements" Ann. Rev. Fluid Mech., Vol.17, pp.523-559.

[5] Hasimoto, H. (1972) "A Soliton on a Vortex Filament," J. Fluid Mech., Vol.51, pp.477-485.

[6] Lamb, G.L. Jr. (1980) "Elements of Soliton Theory," John Wiley & Sons Inc. Chap. 6.

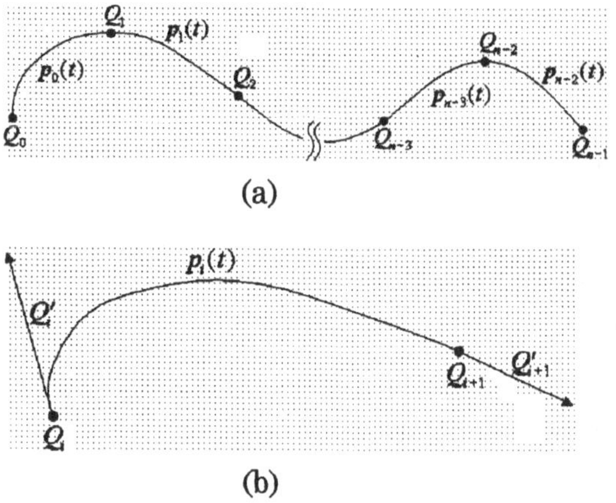

Fig.1 Ferguson spline. $\mathbf{p}(t)$ represents the curve of the vortex filament. $\mathbf{Q}_i$ are the endpoints of the segment.

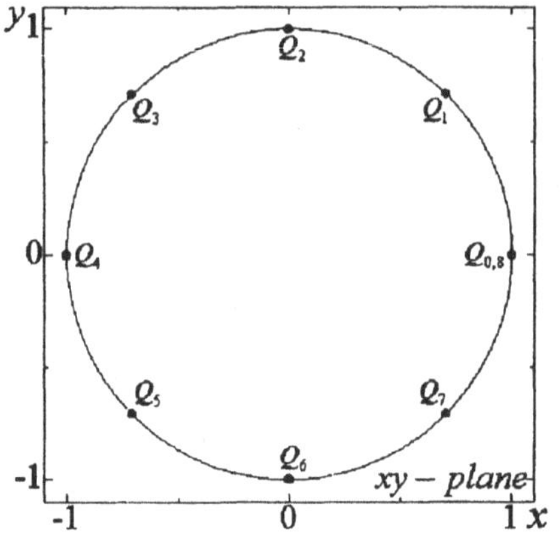

Fig. 3. Circular vortex ring represented by eight spline functions.

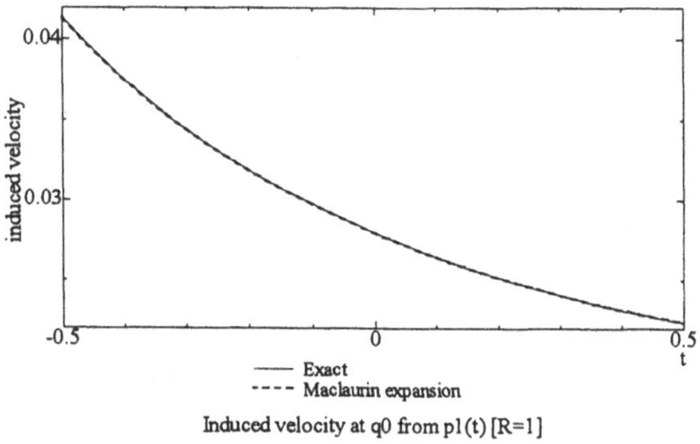

Fig.2 Accuracy of Maclaurin series for induced velocity from each vortex element.

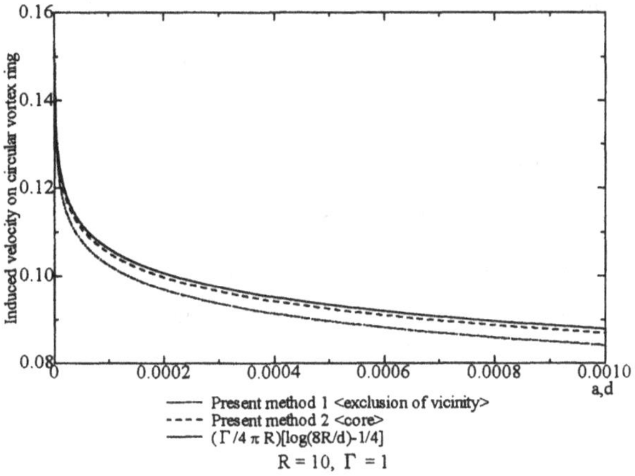

Fig.4 Comparison of the velocity of circular vortex ring. The solid line is the analytical solution by (6).

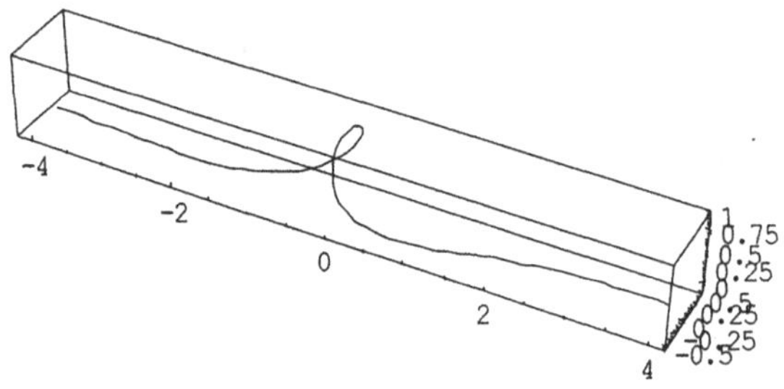

Fig. 5 The shape of a soliton on a vortex tube. The variables are non-dimensionalized and the amplitude is taken to be unity.

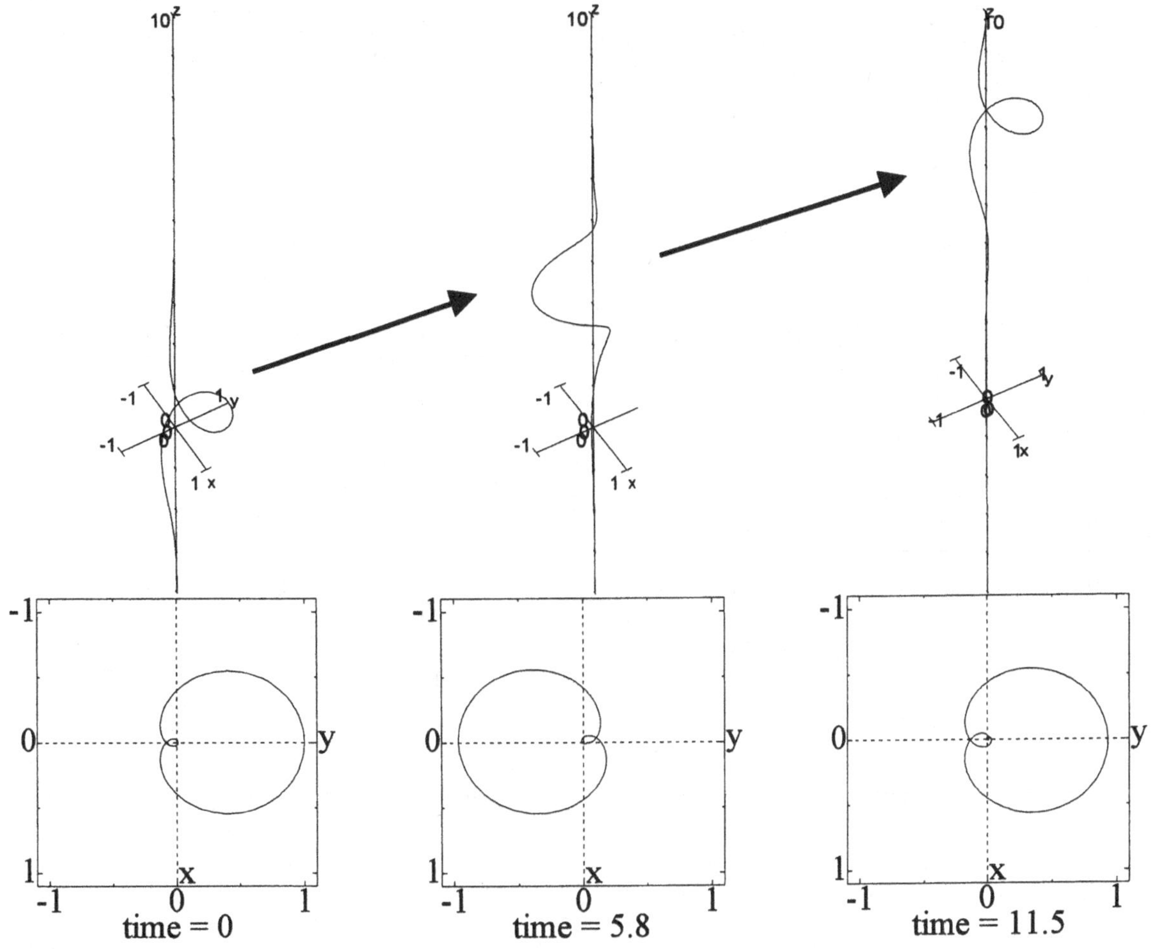

Fig.6 Simulation of a soliton propagating the vortex tube. Time increment Δt is taken as 0.001. The lines of the upper figures represent the shape of the vortex tube at t = 0, 5.8, 11.5. Lower figures show the projections of the deformation onto the x-y plane. At t=11.5 the soliton is in-phase with the initial phase, but at t=5.8 it is out-of-phase.

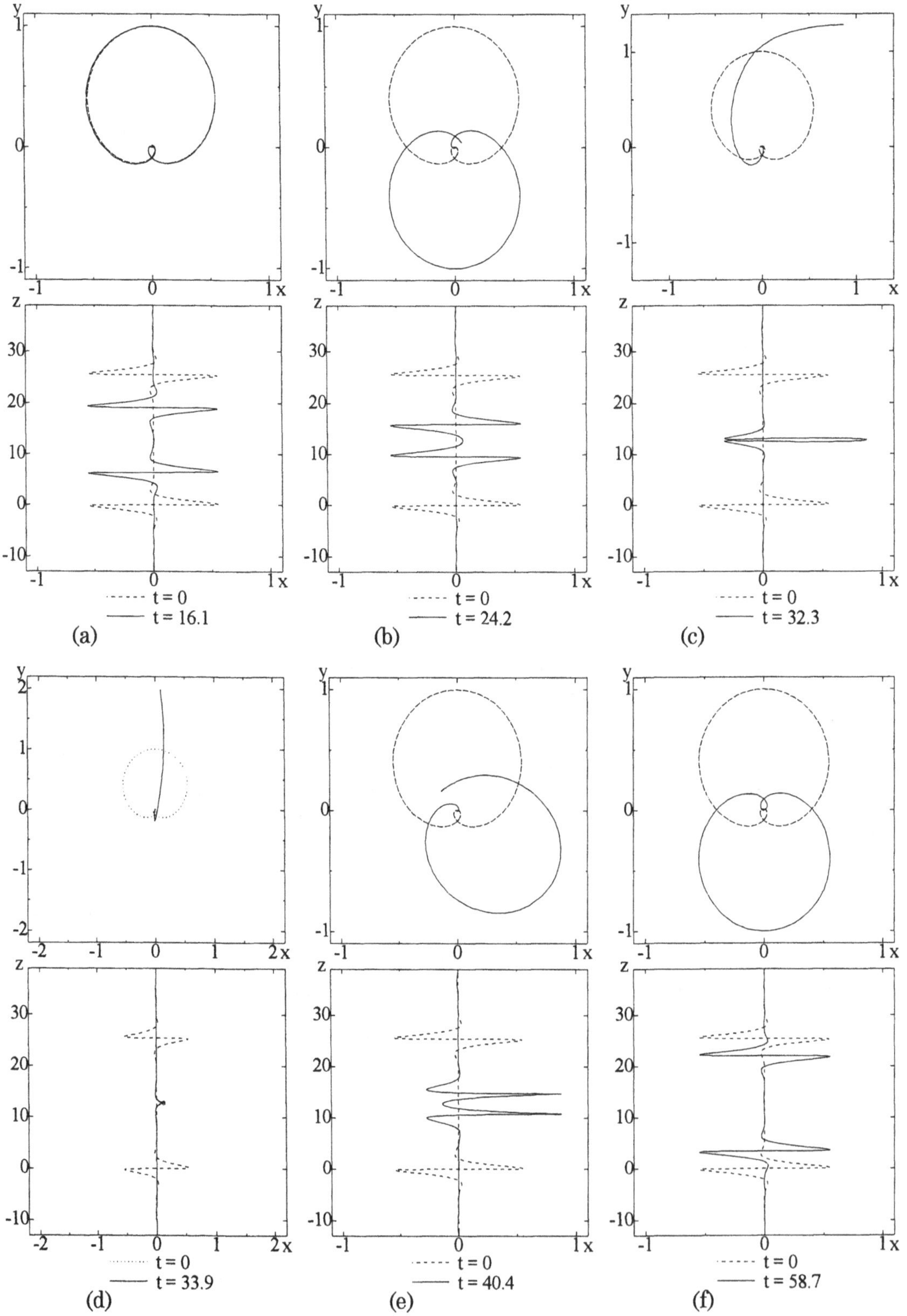

Fig.7 Collision of two solitons. Upeer figures show the projection of the vortex filament on the x-y plane, and lower figures show those on the x-z plane, where the solid lines represent the shape at the given time and the dashed lines represent the initial shape.

NUMERICAL SIMULATION OF GAS-SOLID TWO-PHASE FREE TURBULENT FLOW BY A VORTEX METHOD

Tomomi Uchiyama
Center for Information Media Studies, Nagoya University
Furo-cho, Chikusa-ku, Nagoya 464-8603, Japan / Email:uchiyama@info.human.nagoya-u.ac.jp

Kiyoshi Minemura, Masaaki Naruse and Hikaru Arai
Graduate School of Human Informatics, Nagoya University
Furo-cho, Chikusa-ku, Nagoya 464-8601, Japan

ABSTRACT

This paper proposes a two-dimensional numerical method for gas-solid two-phase free turbulent flow, which can take account of the interaction between the particle and the gas-phase. The computation of the gas flow by a vortex method and the Lagrangian calculation of the particle motion are simultaneously performed. The present method is applied to the analysis of a gas-solid two-phase mixing layer. The numerical results, such as the number density, velocity and turbulent intensity of the particle, are favorably compared with the experimental data, suggesting the validity of the method.

1. INTRODUCTION

Free turbulent flows containing small solid particles are frequently observed in various industrial equipment. Investigations have been made on the particle behavior and the change in the fluid flow due to the interaction between the particle and the fluid (1). Brown and Roshko (2) had clarified that the development and momentum transport of single-phase mixing layer are dominated by the large-scale organized eddies. Thus, the relation between such eddies and the particle behavior has received considerable attention. Crowe *et al.* (3) made a hypothesis that the particle motion could be classified by the responsiveness of the particle to the large-scale eddies, and proposed the Stokes number, which is the ratio of the particle response time to the characteristic time of the fluid flow, as a parameter estimating the particle motion. The validity of the classification method was confirmed by the measurements of the number density and velocities of particle in various free turbulent flows such as plane mixing layers (4,5) and jets (6). A Large Eddy Simulation and a Direct Numerical Simulation were also performed by Yuu (7) for gas-solid two-phase jets to elucidate the influence of the particle on the velocity fluctuation and development of the gas-phase flow.

Recently, vortex methods have been successfully applied to the analysis of single-phase flows relating closely to various engineering problems (8,9). The methods can directly calculate the development of vortex structure without employing any turbulent models. Therefore, the methods promise to be usefully applicable to simulate gas-solid two-phase free turbulent flows, in which the eddies of the gas-phase play a dominant role. A few methods have been thus far proposed for dispersed two-phase flows, and every method calculates the fluid flow by vortex methods and estimates the particle motion by Lagrangian approach. Chen and Chung (10) analyzed the behavior of the organized eddies in a plane mixing layer and simulated the particle motion in the layer by assuming that the fluid flow is not affected by the particle. The analysis using such assumption, which is called one-way method, can yield a reasonable result only when the particle-loading ratio is sufficiently low. To analyze the flow with higher loading ratio, some methods considering the interaction between the particle and the fluid, which are called two-way method, have been presented. Joia *et al.* (11) proposed a two-dimensional vortex method, which can take account of the change in the vorticity due to the particle, and calculated a particle-laden jet by the method. Yokoi *et al.* (12) presented a method based on an assumption that the rotation of the particle induces fluid velocity, and applied the method to simulate two-dimensional liquid-solid two-phase flows around a circular cylinder. Since the abovementioned numerical results were not compared with the experimental data, the validity of the numerical methods was not satisfactorily clarified. Furthermore, the solution procedure was not fully described (11).

This paper proposes a two-dimensional vortex method for gas-solid two-phase free turbulent flows. The method is also applied to calculate a gas-solid two-phase mixing layer, which has been experimentally investigated by Hishida *et al.* (4). The numerical results are compared with the experiment to verify the applicability of the method.

2. BASIC EQUATIONS

2.1 Assumptions

The following assumptions are used for the formulation.

(1) The gas-phase is incompressible.

(2) The density of the particle is much larger than that of the gas-phase.

(3) The particle has a spherical shape with uniform diameter and density.

(4) The particle concentration is so low that the collision between the particles is negligible.

2.2 Governing equations

The conservation equations for the gas-phase are expressed as follows under the assumption (1).

$$\nabla \cdot \boldsymbol{u}_g = 0 \tag{1}$$

$$\frac{\partial \boldsymbol{u}_g}{\partial t} + \boldsymbol{u}_g \cdot \nabla \boldsymbol{u}_g = -\frac{1}{\rho_g}\nabla p + \nu \nabla^2 \boldsymbol{u}_g - \frac{1}{\rho_g}\boldsymbol{F}_D \tag{2}$$

where $\boldsymbol{F}_D$ is the force exerted by the particle acting on the gas-phase per unit volume.

Using the assumption (2), the dominant forces on the particle are the drag and gravitational forces, while the virtual mass force, the lift force, the Basset force and the pressure gradient force are negligible. The equation of motion for a particle (mass m) is written as:

$$m\frac{d\boldsymbol{u}_p}{dt} = \boldsymbol{f}_D + m\boldsymbol{g} \tag{3}$$

where the drag force $\boldsymbol{f}_D$ is given by the following from the assumption (3).

$$\boldsymbol{f}_D = (\pi d^2 \rho_g/8) C_D \mid \boldsymbol{u}_g - \boldsymbol{u}_p \mid (\boldsymbol{u}_g - \boldsymbol{u}_p) \tag{4}$$

Here the drag coefficient C_D is estimated as (13):

$$C_D = (24/Re_p)(1 + 0.15 Re_p^{0.687}) \tag{5}$$

where $Re_p = d \mid \boldsymbol{u}_g - \boldsymbol{u}_p \mid / \nu$.

The simultaneous calculation of Eqs. (1) through (3) yields a gas-solid two-phase flow, in which the interaction between the particle and the gas is considered. In this paper, a vortex method is used to solve Eqs. (1) and (2), and Lagrangian approach is applied to Eq. (3).

3. NUMERICAL METHOD

3.1 Discretization of vorticity field

The calculations are made on a two-dimensional flow field. When taking the curl of Eq. (2) and substituting Eq. (1) into the resulting equation, the vorticity equation for the gas-phase is obtained.

$$\frac{D\omega}{Dt} = \nu \nabla^2 \omega - \frac{1}{\rho_g}\nabla \times \boldsymbol{F}_D \tag{6}$$

The gas velocity $\boldsymbol{u}_g$ at $\boldsymbol{x}$ is given by the Biot-Savart equation.

$$\boldsymbol{u}_g(\boldsymbol{x}) = -\frac{1}{2\pi}\int \frac{(\boldsymbol{x} - \boldsymbol{x}') \times \omega(\boldsymbol{x}')}{\mid \boldsymbol{x} - \boldsymbol{x}' \mid^2} d\boldsymbol{x}' + \boldsymbol{u}_{g0} \tag{7}$$

where $\boldsymbol{u}_{g0}$ stands for the velocity of uniform flow or potential flow.

In this paper, a vortex element having a viscous core proposed for single-phase flow (14) is used. It is postulated that a vortex element α at $\boldsymbol{x}^\alpha$ has a circulation Γ_α and a core radius σ_α. The vorticity at $\boldsymbol{x}$ induced by the vortex element α is expressed as:

$$\omega^\alpha(\boldsymbol{x}) = \frac{\Gamma_\alpha}{\sigma_\alpha^2} f\left(\frac{|\boldsymbol{x} - \boldsymbol{x}^\alpha|}{\sigma_\alpha}\right) \tag{8}$$

where the core function $f(\varepsilon)$ is determined as follows.

$$f(\varepsilon) = \begin{cases} 1/(2\pi\varepsilon) & \varepsilon \le 1 \\ 0 & \varepsilon > 1 \end{cases} \tag{9}$$

When the vorticity field is discretized into a set of N vortex elements, $\boldsymbol{u}_g$ at $\boldsymbol{x}$ is given by the following equation derived from Eqs. (7) and (8).

$$\boldsymbol{u}_g(\boldsymbol{x}) = -\frac{1}{2\pi}\sum_{\alpha=1}^{N} \frac{(\boldsymbol{x} - \boldsymbol{x}^\alpha) \times \boldsymbol{k}\Gamma_\alpha}{|\boldsymbol{x} - \boldsymbol{x}^\alpha|^2} g\left(\frac{|\boldsymbol{x} - \boldsymbol{x}^\alpha|}{\sigma_\alpha}\right) + \boldsymbol{u}_{g0} \tag{10}$$

where

$$g(\varepsilon) = \begin{cases} \varepsilon & \varepsilon \le 1 \\ 1 & \varepsilon > 1 \end{cases} \tag{11}$$

This study employs the boundary layer approximation proposed for single-phase free turbulent flows. The velocity shear layers due to boundary layer separation are represented by introducing vortex elements. The change in the vorticity caused by the viscous effect and the particle is separately simulated through the following method.

3.2 Change in vorticity due to viscosity

The vorticity is decreased due to the viscous effect. The decrement is simulated by applying the core spreading method.

$$\frac{d\sigma_\alpha}{dt} = \frac{\nu c^2}{2\sigma_\alpha} \tag{12}$$

where c=2.242.

3.3 Change in vorticity due to particle

To consider the change in the vorticity due to the interaction between the particle and the gas, one estimates the time rate of change in the circulation Γ along any closed curve (line element $d\boldsymbol{r}$, surface element $d\boldsymbol{S}$) fixed in the flow field.

$$\frac{d\Gamma}{dt} = \int \frac{D\boldsymbol{\omega}}{Dt} \cdot d\boldsymbol{S} \tag{13}$$

When substituting Eq. (6) into Eq. (13) with neglecting the viscous term, the following relation is obtained.

$$\frac{d\Gamma}{dt} = -\int \frac{1}{\rho_g} \nabla \times \boldsymbol{F}_D \cdot d\boldsymbol{S} = -\frac{1}{\rho_g}\int \boldsymbol{F}_D \cdot d\boldsymbol{r} \tag{14}$$

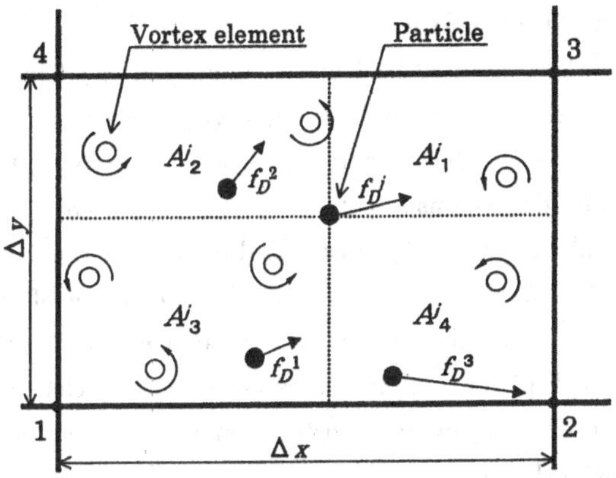

Fig. 1 Computational grid with vortex element and particle

In this paper, the calculating domain is discretized into rectangular grids. A grid is shown in Fig. 1. For the values of $\boldsymbol{F}_D$ on each grid point, $\boldsymbol{F}_D^\beta$ (β=1~4), the components (F_{Dx}^β, F_{Dy}^β) are postulated to be known. If $\boldsymbol{F}_D$ is supposed to vary linearly between the grid points, the time rate of change in Γ, $\Delta\Gamma/\Delta t$, is expressed by the following derived from Eq. (14).

$$\frac{\Delta\Gamma}{\Delta t} = -\frac{1}{\rho_g}\left[\frac{\Delta x}{2}(F_{Dx}^1 + F_{Dx}^2 - F_{Dx}^3 - F_{Dx}^4) + \frac{\Delta y}{2}(F_{Dy}^2 + F_{Dy}^3 - F_{Dy}^4 - F_{Dy}^1)\right] \tag{15}$$

In the case that the number of vortex elements in the grid is n_v, the change in the circulation for each vortex element during Δt is supposed to be $\Delta\Gamma/n_v$. In the case that there are no vortex elements in the grid, a vortex element with a circulation $\Delta\Gamma$ is generated at the center of the grid.

Joia *et al.* (11) also evaluated $\Delta\Gamma$ from Eq. (15). But they did not present the calculating method for $\boldsymbol{F}_D^\beta$, and it is not clear whether they correctly estimates the interaction between two phases. To calculate $\boldsymbol{F}_D^\beta$, this paper proposes the following area-weighting method.

When the number of particles in a grid is n_p and the drag force $\boldsymbol{f}_D^j$ acts on the j-th particle, as illustrated in Fig. 1, the value $\boldsymbol{F}_D^\beta$ on each grid point is determined as:

$$\boldsymbol{F}_D^\beta = \frac{1}{A}\sum_{j=1}^{n_p} \frac{A_\beta^j}{A} \boldsymbol{f}_D^j \qquad (\beta = 1, 2, 3, 4) \tag{16}$$

where $A=\Delta x\Delta y$ and $A_1^j + A_2^j + A_3^j + A_4^j = A$.

3.4 Numerical procedure

When the flow field at $t=t$ is known, the flow at $t=t+\Delta t$ is estimated by the following procedure.

(1) Calculate the particle motion by applying the 2nd-order Adams-Bashforth method to Eq. (3).

(2) Calculate F_D^β on each grid point from Eq. (16).

(3) Calculate the change in the circulation of the vortex elements $\Delta\Gamma$ due to the particle in each grid from Eq. (15). When there are no vortex elements in the grid, generate a vortex element at the center of the grid.

(4) Calculate the core radius of the vortex element by Eq. (12).

(5) Calculate the position of the vortex elements by integrating the gas velocity given by Eq. (10) with use of the 2nd-order Adams-Bashforth method.

(6) Calculate the velocity distribution of the gas-phase from Eq. (10).

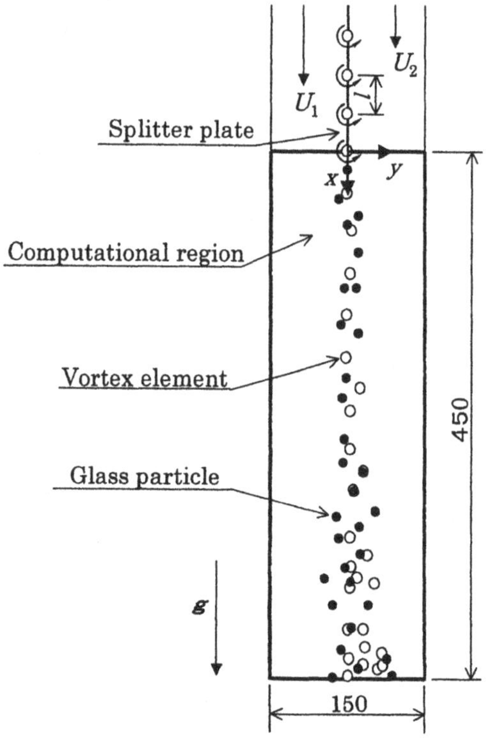

Fig. 2 Configuration of flow field

4. CALCULATION OF GAS-SOLID TWO-PHASE MIXING LAYER

4.1 Calculating condition

The present method is applied to calculate a two-dimensional gas-solid two-phase mixing layer. The flow properties, such as the number density and velocities of the particle, were measured by Hishida *et al.* (4). Figure 2 shows the flow configuration. The streamwise direction is vertical downward. The air streams having the velocities U_1=13 m/s and U_2=4 m/s at higher and lower sides of a splitter plate, respectively, are introduced to a mixing region (450 mm × 150 mm). Spherical glass particles are loaded from an opening at the tip of the splitter plate into the origin of the shear layer, where the particle velocity at the plate tip is 0.9 m/s. The particle diameter d is 72 μm, the density ρ_p is 2590 kg/m^3. The massflow rate of the particle is 20.9 g/s.

The calculating domain corresponds to the mixing region. It is discretized into 50×20 grids. The splitter plate is treated as a vortex sheet and expressed by arranging vortex elements with a circulation Γ at the space l in accordance with the Inoue's method (15).

$$\Gamma = (U_1 - U_2)l, \quad l = (U_1 + U_2)\Delta t_v/2 \tag{17}$$

The vortex elements are introduced from the plate tip into the mixing region at a time interval of Δt_v. To exclude the vortex element leaving the calculating domain from the calculation, a vortex sheet expressed by Eq. (17) is set downstream of the domain (16). The velocity u_{g0} in Eq. (10) is given by the superposition of the velocities induced by these vortex sheets and the mean velocity $(U_1 + U_2)/2$. A single-phase gas flow obtained by the present method is used as the initial condition for two-phase flow analyses.

The time interval Δt is set at 0.0001 s, and a relation of $\Delta t_v=\Delta t$ is used. The time interval to load the particle into the flow field is determined from the mass flow rate.

4.2 Numerical results

Figure 3 shows the distribution of the mean streamwise velocity of the gas-phase $\overline{u}_{gx}$ under single-phase flow condition. The results on four cross-stream sections are superimposed. The velocities at every section distribute on a hyperbolic tangent curve, showing a good agreement with measured results.

When one subtracts the background velocity $(U_1 + U_2)/2$ from the gas velocity under two-phase flow condition, the velocity distribution of the gas-phase can be easily recognized, as shown in Fig. 4. Large-scale eddies are calculated at $x \geq 125$ mm.

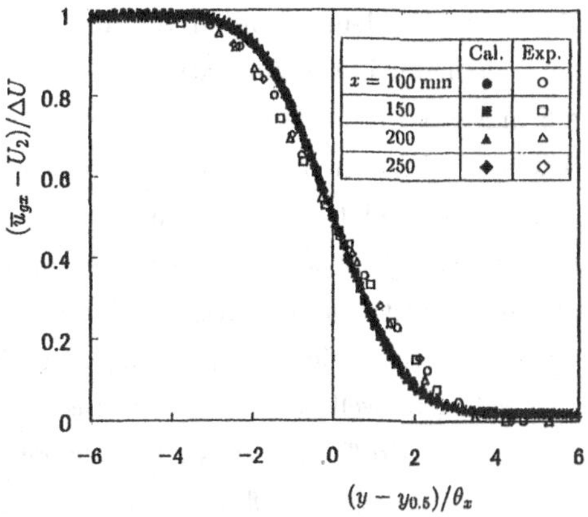

Fig. 3 Distribution of mean streamwise velocity of gas-phase under single-phase flow

Figure 5 indicates the particle distribution at the same instance as Fig. 4. The particles distribute on a vertical line (y=0) at x <125 mm, but they spread in the cross-stream (y) direction at $x \geq 125$ mm owing to the motion of the large-scale eddies of the gas-phase. Most particles concentrate on the periphery of the large-scale eddies, while there are few near the center of each eddy. Computation of an isotropic turbulence (17) has revealed that particles preferentially concentrate on regions of low vorticity and high strain rate. The distribution obtained here is accordant with this preferential concentration.

Figure 6 compares the mean streamwise velocities of the gas-phase $\overline{u}_{gx}$ at single-phase and two-phase flow conditions, where the velocities on four cross-stream sections are presented. Though the velocity is not affected by the particle on the section of x=100 mm, the velocity at two-phase flow is lower than that at single-phase flow downstream of the section. This is because the gas-phase loses its momentum due to the friction between the two phases, as discussed later.

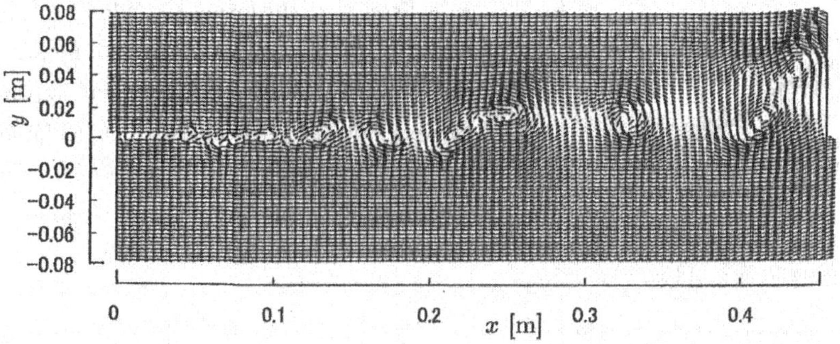

Fig. 4 Instantaneous velocity field of gas-phase

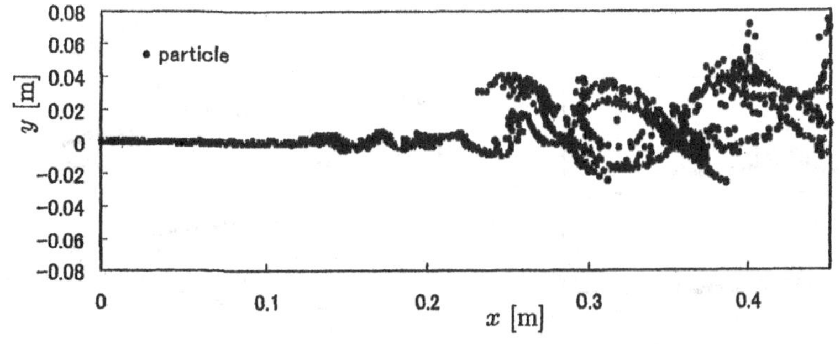

Fig. 5 Instantaneous distribution of particle

Figure 7 shows the distributions of the particle number density N_p on four cross-stream sections, where N_p is normalized by the maximum value N_{pmax} on the section of x=100 mm. With increasing streamwise direction, the particles disperse in the cross-stream (y) direction and the maximum value of N_p decreases. The position, at which the value N_p reaches its maximum on each section, locates on a vertical line (y=0). In the region of $x \geq 150$ mm, the value N_p on the low-speed side ($y > 0$) is larger than that on the high-speed side ($y < 0$). Since the mixing layer spreads more toward the low-speed side, the particles added from the origin of the mixing layer encounter the vortex structure of the gas-phase in the high-speed region with a higher probability.

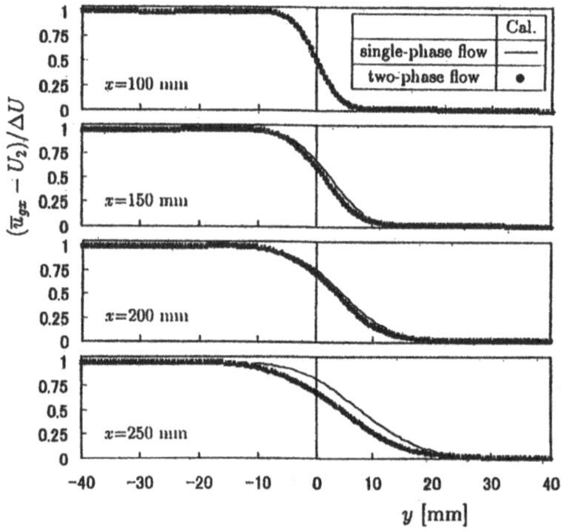

Fig. 6 Comparison of mean streamwise velocities of gas-phase under single-phase and two-phase flows

Consequently, the particles are affected by the gas motion from the high-speed side toward the low-speed side, and more disperse to the low-speed side. The measured results are also plotted in Fig. 7. The present numerical method favorably simulates the measurements, except the simulated values are slightly smaller in the region around y=0. This discrepancy may be due to the fact that the particles are supposed to have an identical initial velocity as well as a uniform diameter in this simulation.

Figure 8 presents the mean streamwise velocity of the particle $\overline{u}_{px}$, where the calculated velocity of the gas-phase $\overline{u}_{gx}$ is also plotted by solid lines. Though the particle velocity gradually increases with an increment in the downstream distance, it is lower than the gas velocity on the region around y=0, where N_p takes larger value, as well as on the high-speed side ($y < 0$). This velocity difference causes the decrement in the gas velocity at two-phase flow, as shown in Fig. 6. It should also be mentioned that the particle velocity exceeds the gas velocity on the low-speed side at $x \geq 150$ mm. Because the particles on the high-speed side disperse to the low-speed side with higher velocity owing to the large-scale eddies of the gas-phase. But the velocity difference does not yield an increment in the gas velocity at two-phase flow, as found from Fig. 6. This is because the particle number density is quite low there. The calculated particle velocity is favorably compared with the measurement.

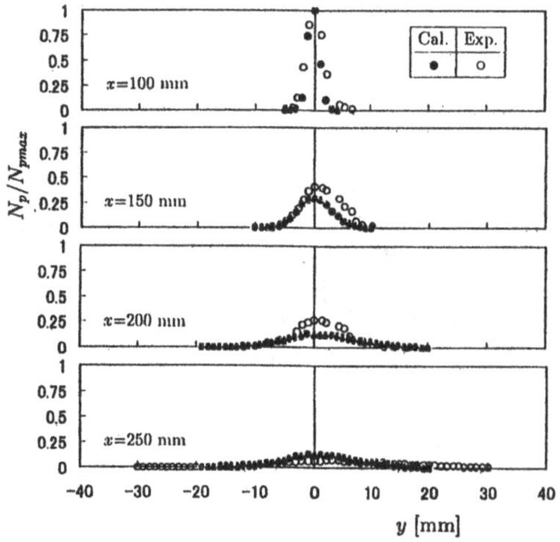

Fig. 7 Distribution of particle number density

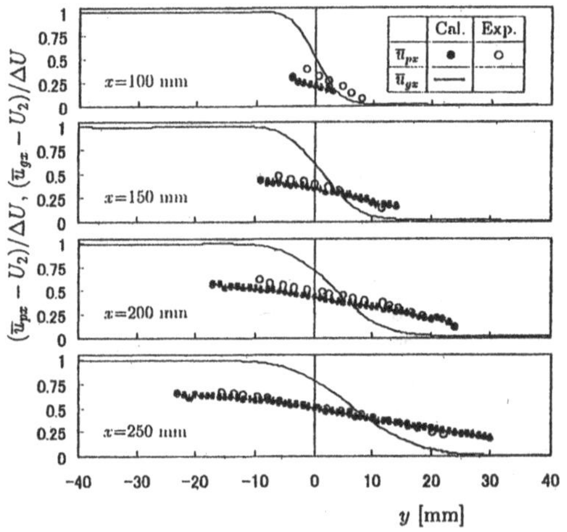

Fig. 8 Distribution of mean streamwise velocity of particle

The particle turbulent intensities in the streamwise and cross-stream directions, u'_{px} and u'_{py}, distribute as shown in Figs. 9 and 10, respectively. They are smaller than the turbulent intensities of the gas-phase indicated by solid lines, and $u'_{py} < u'_{px}$. These suggest that the particle can not follow the turbulent motion of the gas-phase, especially in the cross-stream (y) direction. The numerical result is somewhat larger than the measurement. This may be due to the aforementioned initial conditions of the particles.

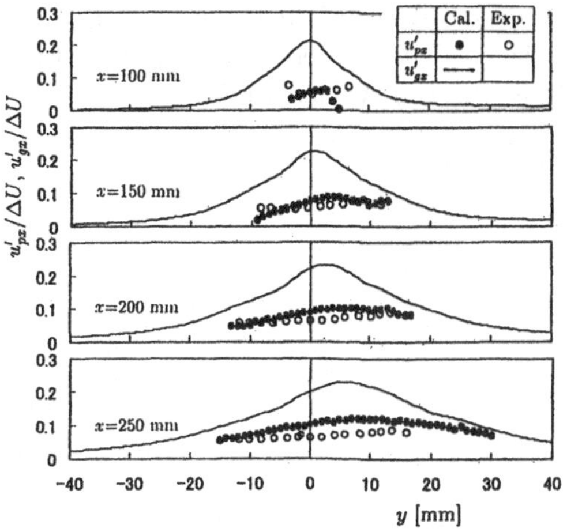

Fig. 9 Distribution of streamwise turbulent intensity of particle

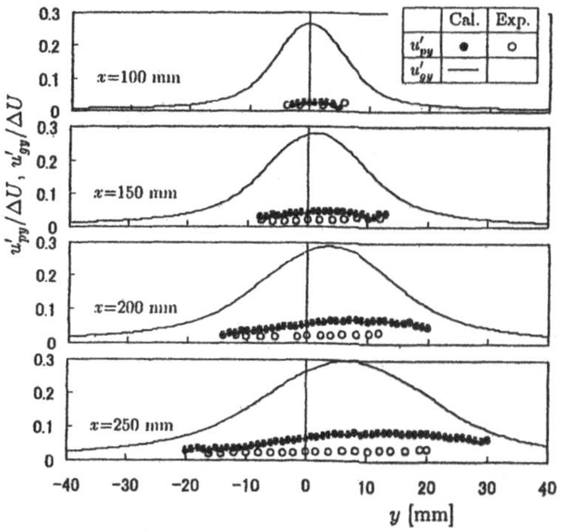

Fig. 10 Distribution of cross-stream turbulent intensity of particle

5. SUMMARY

A numerical method for gas-solid two-phase free turbulent flow is presented, in which the gas flow is computed by a vortex method and the particle motion is calculated by the Lagrangian approach. To take account of the interaction between the two phases, the change in the vorticity of the gas due to the particle is simulated.

The method is also applied to calculate a two-dimensional gas-solid two-phase mixing layer to demonstrate the applicability. The numerical results are favorably compared with the measurement.

NOMENCLATURE

A : area of computational grid
C_D : drag coefficient of particle
d : particle diameter
$\boldsymbol{F}_D$: force exerted by particle acting on gas-phase
$\boldsymbol{f}_D$: drag force acting on particle
$\boldsymbol{g}$: gravitational constant
p : pressure
t : time
$\boldsymbol{u}$: velocity
$\boldsymbol{u}'$: turbulent intensity
$\overline{\boldsymbol{u}}$: mean velocity
U_1 : gas velocity of higher velocity stream
U_2 : gas velocity of lower velocity stream
x, y : orthogonal coordinates
$y_{0.5}$: location at which $\overline{u}_{gx}=(U_1+U_2)/2$
ΔU : velocity difference $=U_1-U_2$
Γ : circulation
θ_x : momentum thickness of mixing layer

$$= \frac{1}{\Delta U^2}\int_{-\infty}^{\infty}(U_1-\overline{u}_{gx})(\overline{u}_{gx}-U_2)dy$$

ν : kinematic viscosity of gas-phase
ρ : density
σ : core radius
ω : vorticity of gas-phase $=\nabla \times \boldsymbol{u}_g$

Subscripts

g : gas
p : particle
x, y : component in direction of x or y

REFERENCES

(1) Tsuji, Y. (1991), "Turbulence modification in fluid-solid flows," *Proc. Turbulence Modification in Multiphase Flows*, ASME, pp.1-6.

(2) Brown, G. L. and Roshko, A. (1974), "On density effects and large structure in turbulent mixing layers," *J. Fluid Mech.*, Vol.64, pp.775-816.

(3) Crowe, C. T., *et al.* (1985), "Particle dispersion by coherent structures in free shear flows," *Particle Sci. Tech.*, Vol.3, pp.149-158.

(4) Hishida, K., *et al.* (1992), "Experiments on particle dispersion in a turbulent mixing layer," *Int. J. Multiphase Flow*, Vol.18, No.2, pp.181-194.

(5) Ishima, T., *et al.* (1993), "Effects of particle residence time on particle dispersion in a plane mixing layer," *Trans. ASME, J. Fluid Eng.*, Vol.115, No.4, pp.751-759.

(6) Longmire, E. and Eaton, J. K. (1992), "Structure of particle-laden round jet," *J. Fluid Mech.*, Vol.236, pp.217-257.

(7) Yuu, S. (1997), "Effects of particle existence on low and high Reynolds number gas-particle jets," *Proc. High Performance Computing on Multiphase Flows*, JSME, pp.67-72.

(8) Kimura, T., *et al.* (1992), "Wake of a rotating circular cylinder," *AIAA J.*, Vol.30, pp.555-556.

(9) Kamemoto, K., *et al.* (1997), "Numerical simulation of flow around an in-line forced oscillating circular cylinder by a vortex method combined with boundary layer calculation," *Proc. Int. Conf. Fluid Engng*, JSME, pp.303-308.

(10) Chein, R. and Chung, J. N. (1987), "Effects of vortex pairing on particle dispersion in turbulent shear flows," *Int. J. Multiphase Flow*, Vol.13, No.6, pp.785-802.

(11) Joia, I. A., *et al.* (1998), "A discrete vortex model of particle laden jets," *Proc. 3rd Int. Conf. Multiphase Flow*, P644.pdf, (on CD-ROM).

(12) Yokoi, Y., *et al.* (1996), "Numerical simulation of flow around a circular cylinder in a solid-liquid two-phase flow using a vortex method," *Trans. JSME, Ser.B*, Vol.62, No.603, pp.3824-3831, (in Japanese).

(13) Clift, R. *et al.* (1978), *Bubbles, Drops and Particles*, Academic Press, New York.

(14) Ghoniem, A. F. and Cagnon, Y. (1987), "Vortex simulation of laminar recirculating flow," *J. Comput. Phys.*, Vol.68, pp.346-377.

(15) Inoue, O. (1985), "Vortex simulation of a turbulent mixing layer," *AIAA J.*, Vol.23, No.3, pp.367-373.

(16) Leonard, A. (1980), "Vortex methods for flow simulation," *J. Comput. Phys,*, Vol.37, pp.289-335.

(17) Squires, K. D. and Eaton, J. K. (1991), "Preferential concentration of particles by turbulence," *Phys. Fluid, A*, Vol.3, pp.1169-1178.

SIMULATION OF PARTICULATE FLOWS USING VORTEX METHODS

Jens H. Walther, Julian T. Sagredo and Petros Koumoutsakos
Institute of Fluid Dynamics, ETH Zürich,
Sonneggstrasse 3, CH-8092 Zürich, Switzerland

ABSTRACT

We present a three-dimensional vortex-in-cell (VIC) method for the simulation of particulate flows with two-way coupling. The solid particles move subject to viscous drag and gravity creating vorticity which is discretised using vortex particles.

The method uses a high order moment conserving interpolation scheme for all particle-mesh interpolation and assignment, and conservative finite-differences for the vorticity stretching term. Diffusion is computed on the mesh using finite-differences or the method of particle strength exchange.

The two-way coupling is accomplished through a vorticity source term in the context of a fractional step algorithm. The vortex particles are remeshed to secure a regular flow map and to create new vortex particles according to the vorticity source term.

The method is applied to the study of a viscous drop falling due to gravity in a miscible, initially quiescent fluid. The expansion rate of the drop is found to be in good agreement with related experiments and a Rayleigh-Taylor instability is observed for a drop starting from rest.

1 INTRODUCTION

Particle (vortex) methods have previously been applied in a number of studies of two-phase flows, for two-dimensional [1, 2, 3, 4, 5, 6, 7], and three-dimensional problems [8]. The majority of the implementations are limited to inviscid flows, and confine the vortex particles to the interface between the two phases, or treat a limited physics [9].

The present work extends the VIC method by considering three-dimensional, two-phase particulate viscous flows with a two-way coupling of the phases.

The solid particles are advected using relatively well-established particle-particle and particle-fluid interactions cf. [10, 11, 12].

The method is applied to the study of a viscous drop falling due to gravity and subject to viscous drag in a miscible and initially quiescent fluid. The drop is modelled by a spherical 'blob' of small solid particles suspended in the fluid, a model approach suggested by the experimental work of Powell and Mason [13], and by Kojima, Hinch and Acrivos [14], and by recent calculations by Nitsche and Batchelor [15].

In the pioneering study, Thomson and Newall [16] observed that a drop of ink falling from some height into a bath of water deforms into a disk and later develops into a vortex ring. They proposed that the drop penetrates the bath while retaining its shape and that the dynamics of the 'vortex film' created around the drop is responsible for the formation of the ring. Furthermore, they found that for drops of higher density than the bath the vortex ring is unstable and forms bulges that eventually develop into new drops connected with 'threads' of ink. If the surface tension is not too large, these drops form new rings repeating the cycle. They also observed *'that alteration of the height from which a drop falls before reaching the surface of the column modifies the formation of the rings considerably'* and found this to be related to the oscillation of the drop at impact.

From photographs of a falling liquid drop into a bath of the same liquid Chapman and Critchlow [17] found that vortex rings are best formed when the drop is spherical as it falls into the bath and change from an oblate to prolate spheroid. The analysis indicated that the surface energy contributes significantly to the energy of the vortex ring.

The experiments by Kojima and co-workers [14] of a viscous drop falling at low Reynolds numbers in a miscible fluid provided further information of the stability of the drop. They observed that the drop shortly after entrance forms an elongated tail, and that it flattens and develops an intrusion at the rear stagnation point. The tail subsequently detaches and the drop ring is formed as the intrusion penetrates the drop. The ring finally becomes unstable forming bulges as described in [16].

Experiments of a vortex ring in an immiscible fluid by Baumann, Joseph, Mohr and Renardy [18] showed that the formation of the ring depends on the rupture of the membrane formed at the centre of the deformed drop. The ratio of the drop and bath viscosities (μ_d/μ) was found to be an important parameter for the formation of the ring (rings were not observed for $\mu_d/\mu < 5$). The vortex cascade was ascribed to Rayleigh-Taylor instability and the wavelength related to the Reynolds number and surface tension.

Using different miscible fluids, Mitts [19] studied the breakup of drops falling into a bath for different viscosity ratios and impact velocities. Four primary regimes were found for the breakup of the drop depending on the Reynolds number and viscosity ratio. The present work uses parameters in the 'multi-mode deformation and breakup regime' by Mitts, in which the vortex cascade is most prominent.

The remaining part of the paper is organised as follows: § 2 outlines the governing equations, and § 3 the numerical method. The results are presented in § 4, and summarised in § 5.

2 GOVERNING EQUATIONS

2.1 Fluid Motion

The governing equation for an incompressible fluid with constant kinematic viscosity (ν) is given in terms of the vorticity transport equation

$$\frac{D\vec{\omega}}{Dt} = (\vec{\omega}\cdot\vec{\nabla})\vec{v} + \nu\vec{\nabla}^2\vec{\omega} + \vec{\phi}, \tag{1}$$

where $\vec{v}$ is the velocity, $\vec{\omega} = \vec{\nabla}\times\vec{v}$ the fluid vorticity, and $\vec{\phi}$ a vorticity source term

$$\vec{\phi} = \vec{\nabla}\times\vec{f}, \tag{2}$$

and $\vec{f}$ is the body force per unit volume.

The fluid velocity and vorticity are related through the equations

$$\nabla\cdot\vec{v} = 0, \qquad \vec{\omega} = \vec{\nabla}\times\vec{v}, \tag{3}$$

or defining a solenoidal vector potential, $\vec{\Psi}$ through a Poisson equation

$$\vec{\nabla}^2\vec{\Psi} = -\vec{\omega}, \tag{4}$$

where $\vec{v} = \vec{\nabla}\times\vec{\Psi}$.

Periodic boundary conditions are applied in all spatial directions, e.g, $\vec{v}(x) = \vec{v}(x+L)$, where L is the length of the computational domain.

2.2 Particle Motion

The solid particles are described by their density (ρ_p), diameter (d_p), and by the instantaneous position and velocity ($\vec{x}_p, \vec{u}_p$).

The governing equation for a solid particles is Newton's law

$$\rho_p vol_p \frac{\mathrm{d}\vec{u}_p}{\mathrm{d}t} = \vec{f}_p, \tag{5}$$

where $vol_p = \frac{\pi}{6}d_p^3$ is the volume of the particle, and $\vec{f}_p$ is the total force acting on the particle.

In this study we consider viscous drag

$$\vec{f}_d = \frac{1}{2}\rho C_d \frac{\pi}{4} d_p^2(\vec{v}-\vec{u})|\vec{v}-\vec{u}|, \tag{6}$$

where C_d is the drag coefficient

$$C_d = \begin{cases} \frac{24}{\mathrm{Re}_p}(1+0.15\mathrm{Re}_p^{0.687}), & \mathrm{Re}_p < 1000, \\ \frac{24}{\mathrm{Re}_p}, & \mathrm{Re}_p > 1000, \end{cases} \tag{7}$$

and Re_p the particle Reynolds number

$$\mathrm{Re}_p = \frac{|\vec{u}_p - \vec{v}(\vec{x}_p)|d_p}{\nu}, \tag{8}$$

and $\vec{v}(\vec{x}_p)$ is the fluid velocity at the position of the particle.

The solid particle are furthermore subject to gravity

$$\vec{f}_g = vol_p(\rho_p - \rho)\vec{g}, \tag{9}$$

where $\vec{g}$ is the acceleration due to gravity, and ρ the fluid density.

3 PARTICLE METHOD

3.1 Fluid Motion

The fluid vorticity is discretised using Lagrangian (vortex) particles defined by their position ($\vec{x}_p$) and strength ($\vec{\alpha}_p = vol_p\vec{\omega}(\vec{x}_p)$), where vol_p is the volume occupied by the vortex particles.

The particles are advanced in time using a fractional step algorithm, employing Runge-Kutta timestepping for the inviscid problem of particle convection and stretching

$$\frac{\mathrm{d}\vec{x}_p}{\mathrm{d}t} = \vec{v}(\vec{x}_p), \tag{10}$$

$$\frac{\mathrm{d}\vec{\alpha}_p}{\mathrm{d}t} = (\vec{\alpha}_p\cdot\vec{\nabla})\vec{v}(\vec{x}_p). \tag{11}$$

The inviscid step is followed by a diffusion step including the vorticity source term, with

$$\frac{\mathrm{d}\vec{x}_p}{\mathrm{d}t} = 0, \tag{12}$$

$$\frac{\mathrm{d}\vec{\alpha}_p}{\mathrm{d}t} = \nu\vec{\nabla}^2\vec{\alpha}_p + vol_p\vec{\phi}_p. \tag{13}$$

3.2 Vortex-in-Cell

The local fluid velocity in Eq. (10) is computed using a vortex-in-cell (VIC) algorithm [20, 21]. The physical domain is discretised with a ($N_x \times N_y \times N_z$) mesh of equidistant spacing (h). The mesh vorticity is assigned from the particle strength and the Poisson Eq. (4) is solved for the unknown vector potential using efficient fast Fourier transforms. The fluid velocity and the vorticity stretching are computed on the mesh using 4th order finite-differences and interpolated back onto the vortex particles to update their position and strength.

The stretching term is discretised using conservative finite differences to minimise spurious production of $\nabla \cdot \vec{\omega}$ cf. [22] as

$$\left(\vec{\omega} \cdot \vec{\nabla}\right) \vec{v} = \vec{\nabla} \cdot (\vec{\omega} : \vec{v}) = \left(\vec{\omega} \cdot \vec{\nabla}\right) \vec{v} + \vec{v} (\nabla \cdot \vec{\omega}) . \quad (14)$$

3.3 Interpolation

The assignment of mesh vorticity from the particle strength and the interpolation of particle values from the mesh use the moment conserving M'4 scheme proposed by Monaghan [23]. The first three moments of the vorticity field are conserved, hence satisfying the Kelvin theorem (zero order moment), and avoids spurious fluid forcing, since the rate-of-change of the first and second moments are proportional to the fluid forces and moments cf. [24].

The M'4 scheme is further used to remesh the vortex particles onto a regular mesh in order to preserve the regularity of the flow map and to create vortex particles as required by the vorticity source term.

3.4 Particle Diffusion

Viscous diffusion is computed on the mesh using finite-differences or by the particle strength exchange scheme proposed by Degond and Mas-Gallic [25]

$$\frac{d\vec{\alpha}_p}{dt} = \frac{\nu}{\sigma^2} \sum_q \zeta_\sigma(\vec{x}_p - \vec{x}_q)(vol_q \vec{\alpha}_q - vol_q \vec{\alpha}_p), \quad (15)$$

where $\zeta_\sigma(\vec{x})$ is a smooth function approximating the heat kernel, and satisfying some moment properties, cf. [25], and σ is the smoothing length scale. The function proposed by Cottet [26] is used

$$\zeta_\sigma(\vec{x}) = \frac{15}{\pi^2} \left(\left(\frac{|\vec{x}|}{\sigma} \right)^{10} + 1 \right)^{-1} . \quad (16)$$

Thus, for inviscid flows and neglecting the source term, the remeshing is effectively an interpolation (using the M'4 scheme) of the particles strength onto a regular set of particles (mesh).

3.5 Particle Motion

The forces on the solid particles are computed from Eqs. (6) – (9), where the fluid velocity is interpolated using the M'4 scheme. The position and velocity is updated using the leap frog scheme

$$\vec{u}_p^{n+1/2} = \vec{u}_p^{n-1/2} + \delta t \vec{f}_p^n / (\rho_p vol_p), \quad (17)$$

$$\vec{x}_p^{n+1} = \vec{x}_p^n + \delta t \vec{u}_p^{n+1/2}. \quad (18)$$

Thus, the velocity and position are temporally staggered.

The implementation also allows tracking of fluid tracers, in which case Runge-Kutta methods are used.

4 RESULTS

We present preliminary results from the simulation of a viscous drop falling in an initially quiescent fluid. The physical parameters governing the flow are the fluid viscosity (μ) and density (ρ), the acceleration due to gravity ($\vec{g}$), the volume fraction of the solid particles (ϵ), the density of the solid particles (ρ_p), and their diameter (d_p). Assuming an uniform volume fraction the density of the drop is ($\rho_d = \epsilon\rho_p + (1 - \epsilon)\rho$), and the viscosity is given by the classical result of Einstein ($\mu_d = \mu(1 + 2.5\epsilon)$) [27], or by the expression due to Lundgren [28] valid for higher volume fraction

$$\mu_d \approx \frac{\mu}{1 - \frac{5}{2}\epsilon}. \quad (19)$$

The fluid and solid particles are initially at rest, hence no characteristic velocity is directly available. However, assuming Stokes flow the velocity of the drop is [27]

$$U = \frac{1}{12} \frac{D_d^2 g}{\mu} (\rho_d - \rho) \left(\frac{\mu + \mu_d}{\mu + \frac{3}{2}\mu_d} \right), \quad (20)$$

where D_d is the diameter of the drop. Baumann *et al.* [18] found large discrepancies between Eq. (20) and the measured velocity of the drop.

The non-dimensional parameters governing the flow are the viscosity ratio

$$\lambda = \mu_d / \mu, \quad (21)$$

the density ratio (ρ_d/ρ), or the Atwood number ,

$$\mathrm{A} = \frac{\rho_d - \rho}{\rho_d + \rho}, \quad (22)$$

the Reynolds number $\mathrm{Re}_d = \rho D_d U/\mu$, and the Froude number $\mathrm{Fr}_d = U^2/g D_d$ of the drop. Following the experiments of Mitts [19], an Atwood number of 0.24 and viscosity ratio of $\lambda = 35$ is used in the present simulation.

The results are given in non-dimensional form based on D_d, and time is non-dimensionalised as tU/D_d where U is obtained from Eq. (20).

The simulations are conducted using 10^4 to 10^5 solid particles placed inside a sphere of diameter D_d. The position of the particles is chosen at random using an uniform distribution and with an initial zero particle velocity unless otherwise specified.

The computational domain is cubic with a physical dimension of $10D_p$ and a mesh resolution of 64^3 unless otherwise specified. The fluid equations are advanced in time using a 3rd order Runge-Kutta integration and

Table 1: Physical and numerical parameters used in the simulations. λ = is the viscosity ratio, A the Atwood number, N_p the number of solid particles, $(v/U)_i$ the initial velocity of the drop, Re the Reynolds number, and Fr is the Froude number. †: Values based on initial velocity.

Case	λ	A	N_p	grid	$(v/U)_i$	Re	Fr
1	35	0.24	10^5	64^3	0.0	180	3.2
2	35	0.24	10^5	64^3	2.3	230†	5.3†
3	35	0.24	10^4	64^3	0.0	180	3.2
4	35	0.24	10^5	96^3	0.0	180	3.2

a non-dimensional time step of 0.0036. The parameters of the simulations are summarised in Table 1.

The non-dimensional velocity of the drop (centre of gravity) is shown in Figure 1 for the case of 10^4 and 10^5 solid particles for the 64^3 mesh, and for the case of 10^5 particles for the 96^3 mesh, respectively. The agreement between the two cases indicates a converged solution in terms of the number of solid particles and mesh resolution. However, the velocity of the drop is an order of magnitude smaller than the Stokes solution ($v/U \approx 0.1$), but it is found to be in good agreement with the experimental values of Baumann *et al.* [18]. The effective Reynolds and Froude numbers are 10 and 0.01, respectively.

A vertical slice through the drop (thickness of $0.2D_p$) at different times is shown in Figures 2 and 3 for the two cases. There is good agreement between the simulations using 10^4 and 10^5 particles, indicating that a particle independent solution has been obtained.

Without perturbations (except of the initial random position of the solid particles) the shape of the drop follows the experimental observations, see eg. Figure 7b in [29] and Figure 2 in [14]. The drop deforms from the initial perfect spherical shape into an oblate spheroid due to the resistance exerted by the fluid. An intrusion develops at the rear stagnation point and the drops subsequently forms a spherical cap as shown in Figure 2b-c. The lateral expansion of the drop is approximately 0.2 in good agreement with the value found by Mitts [19] for a similar drop but at a different Reynolds number (Figure 34 of [19]).

The drop ring later becomes unstable due to Rayleigh-Taylor like instability. Bulges develop as shown experimentally in Figure 4b, and similarly in the simulations (Figure 5b). Note however, that the experiment appears to develop four bulges whereas eight are produced in the simulation. This discrepancy can be attributed to the initial conditions of the simulation. The bulges grow in size until a new drops form connected with thin threads. The process is repeated cf. Figure 5, when new vortex rings form.

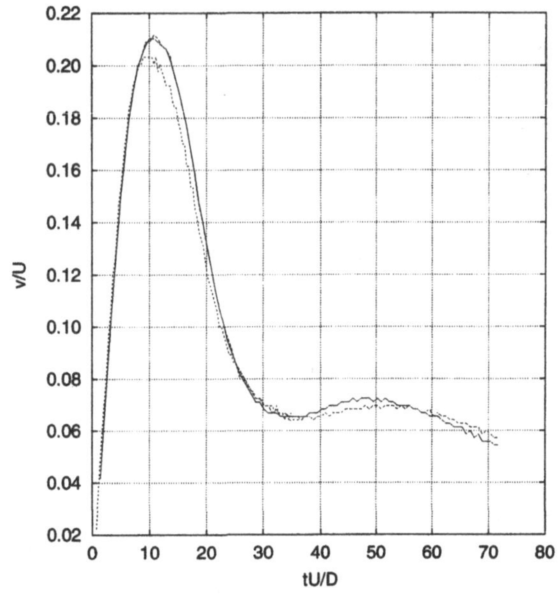

Figure 1: Speed of the drop (centre of gravity) using different number of solid particles and on different mesh resolution. —: Case 1 ($N_p = 10^5$; mesh: 64^3); - - -: Case 3 ($N_p = 10^4$; mesh: 64^3); - - -: Case 4 ($N_p = 10^5$; mesh: 96^3).

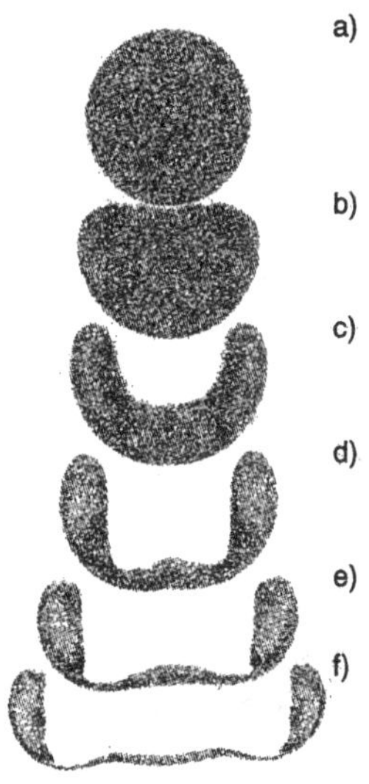

Figure 2: Cross section of the drop at different times. The total number of solid particle is 10^5. a) tU/D_p = 0.00; b) $tU/D_p = 7.18$; c) $tU/D_p = 10.77$; d) tU/D_p = 14.36; e) $tU/D_p = 17.94$; f) $tU/D_p = 21.53$.

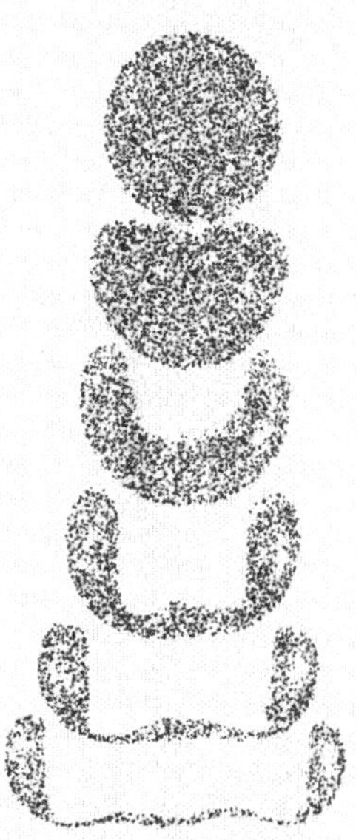

Figure 3: Cross section of the drop at different times. The total number of solid particle is 10^4. a) $tU/D_p = 0.00$; b) $tU/D_p = 7.18$; c) $tU/D_p = 10.77$; d) $tU/D_p = 14.36$; e) $tU/D_p = 17.94$; f) $tU/D_p = 21.53$.

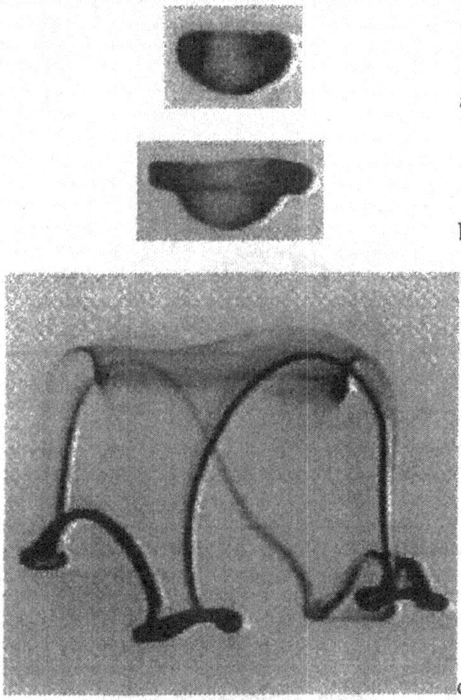

Figure 4: Experimental visualisation of a glycol drop falling in 2-propanol at different times: a) t = 0.018 s; b) t = 0.054 s; c) t = 0.154 s. From Mitts [19].

Since most experiments involve an initial vorticity distribution of the drop as it is 'introduced' into the bath, we conducted a simulation (Case 2) which allowed a temporary one-way coupling of the phases. The drop was allowed to descent through the bath without changing the particle velocity but creating viscous drag and hence fluid vorticity, a model similar to the mechanism proposed by Thomson and Newall [16]. The two-way coupling was activated after 200 time steps ($tU/D_p = 0.72$) and the drop allowed to deform. The cross section of the drop at a late stage ($tU/D_p = 71.78$) in shown in Figure 6. A this time, the drop still forms a stable ring, indicating the importance of the initial vorticity field. A tail appears similar to the vertical tail reported by Nitsche and Batchelor [15] for a similar problem.

5 CONCLUSION

A three-dimensional vortex-in-cell algorithm has been developed for two-phase particular flows with two-way coupling. The two-way coupling is achieved through a vorticity source term, hence allowing treatment of arbitrary body forces. A remeshing strategy is used to secure a regular flow map and to allow proper discretisation of the vorticity source term.

The method is applied to a viscous drop falling in a miscible fluid subject to gravity and viscous drag. The expansion rate of the developed drop ring is found in good agreement with experiments and the method is capable of producing the 'vortex ring cascade' demonstrated in the pioneering study of Thomson and Newall [16].

REFERENCES

[1] Gregory R. Baker, Daniel I. Meiron, and Steven A. Orszag. Vortex simulations of the Rayleigh-Taylor instability. *Phys. Fluids*, 8:1485–1490, 23.

[2] Reiyu Chein and J. N. Chung. Simulation of particle dispersion in a two-dimensional mixing layer. *AIChe J.*, 34(6):946–954, 1988.

[3] Grétar Tryggvason. Numerical simulations of the Rayleigh-Taylor instability. *J. Comp. Phys.*, 75:253–282, 1988.

[4] Juan A. Zufiria. Vortex-in-cell simulation of bubble competition in Rayleigh-Taylor instability. *Phys. Fluids*, 31(11):3199–3212, 1988.

[5] Robert M. Kerr. Simulation of Rayleigh-Taylor flows using vortex blobs. *J. Comp. Phys.*, 76:48–84, 1988.

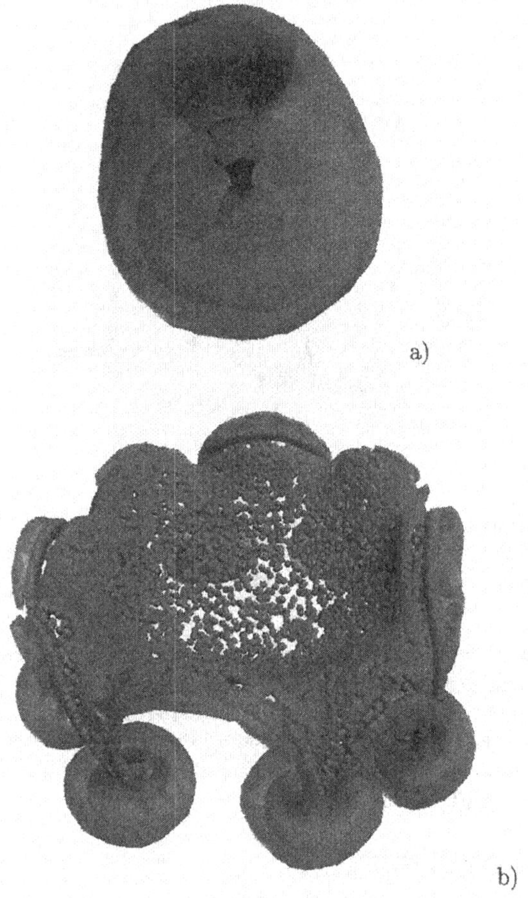

Figure 5: Snap shots of the fluid vorticity and the solid particles at a) $tU/D_p = 1.79$ and b) $tU/D_p = 71.78$ (Case 1). The solid particles are not to scale.

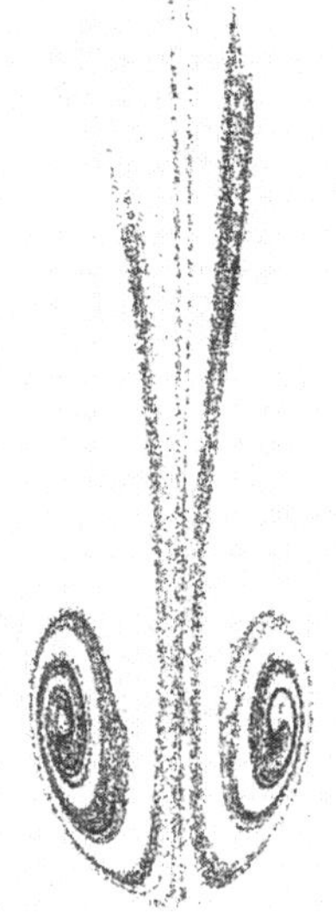

Figure 6: Cross section of the drop at a late stage ($tU/D_d = 71.78$) for the case of a drop with an initial vorticity distribution.

[6] L. Tang, F. Wen, Y. Yang, C. T. Crowe, J. N. Chung, and T. R. Troutt. Self-organizing particle dispersion mechanisms in a plane wake. *Phys. Fluids*, A4:2244–2251, 1992.

[7] H. Chen and J. S. Marshall. A Lagrangian vorticity method for two-phase particulate flows with two-way phase coupling. *J. Comp. Phys.*, 148:169–198, 1999.

[8] Stephen H. Brecht and John R. Ferrante. Vortex-in-cell simulation of buoyant bubbles in three dimensions. *Phys. Fluids A*, 1(7):1166–1191, 1989.

[9] Grétar Tryggvason and Hassan Aref. Numerical experiments on Hele Shaw flow with a sharp interface. *J. Fluid Mech.*, 136:1–30, 1983.

[10] S. Elghobashi. On predicting particle-laden turbulent flows. *Appl. Sci. Res.*, 52:309–329, 1994.

[11] David E. Stock. Particle dispersion in flowing gases — 1994 Freeman scholar lecture. *J. Fluids Engng.*, 118:4–17, 1996.

Figure 7: Snap shot of the fluid vorticity and the solid particles at the late stage ($tU/D_p = 71.78$) for the case of a drop with an initial vorticity distribution (Case 2). The solid particles are not to scale.

[12] C. T. Crowe, T. R. Troutt, and J. N. Chung. Numerical models for two-phase turbulent flows. *Annu. Rev. Fluid Mech.*, 28:11–43, 1996.

[13] R. L. Powell and S. G. Mason. Dispersion by laminar flow. *AIChe J.*, 28(2):286–293, 1982.

[14] Masami Kojima, E. J. Hinch, and Andreas Acrivos. The formation and expansion of a toroidal drop moving in a viscous fluid. *Phys. Fluids*, 27(1):19–32, 1984.

[15] J. M. Nitsche and G. K. Batchelor. Break-up of a falling drop containing dispersed particles. *J. Fluid Mech.*, 340:161–175, 1997.

[16] J. J Thomson and H. F. Newall. On the formation of vortex rings by drops falling into liquids, and some allied phenomena. *Proc. R. Soc.*, 39:417–436, 1885.

[17] David S. Chapman and P. R. Critchlow. Formation of vortex rings from falling drops. *J. Fluid Mech.*, 29(1):177–185, 1967.

[18] Nicholas Baumann, Daniel D. Joseph, Paul Mohr, and Yuriko Renardy. Vortex rings of one fluid in another in free fall. *Phys. Fluids A*, 4(3):567–580, 1992.

[19] Chad J. Mitts. *An Investigation of Transcritical Droplet Dynamics Through the Use of a Miscible Fluid Analog.* Ph.d. thesis, University of Illinois at Chicago, 1996. Unpublished.

[20] Charles K. Birdsall and Dieter Fuss. Clouds-in-clouds, clouds-in-cells physics for many-body plasma simulation. *J. Comp. Phys.*, 3:494–511, 1969.

[21] J. P. Christiansen. Numerical simulation of hydrodynamics by the method of point vortices. *J. Comp. Phys.*, 13:363–379, 1973.

[22] G. H. Cottet and P. Koumoutsakos. *Vortex Methods: Theory and Applications.* Cambridge University Press, 1999.

[23] J. J. Monaghan. Extrapolating B splines for interpolation. *J. Comp. Phys.*, 200:253–262, 1985.

[24] J. C. Wu. Theory for aerodynamis force and moments in viscous flows. *AIAA J.*, 19(4):432–441, 1981.

[25] P. Degond and S. Mas-Gallic. The weighted particle method for convection-diffusion equations. part 1: The case of an isotropic viscosity. *Math. Comput.*, 53(188):485–507, 1989. Oct.

[26] G.-H. Cottet. Private communication, 1999.

[27] G. K. Batchelor. *An Introduction To Fluid Dynamics*. Cambridge University Press, 1 edition, 1967.

[28] T. S. Lundgren. Slow flow through stationary random beads and suspensions of spheres. *J. Fluid Mech.*, 51(2):273–299, 1972.

[29] C. Pozrikidis. The instability of a moving viscous drop. *J. Fluid Mech.*, 210:1–21, 1990.

Numerical Prediction of Rotor Tip-Vortex Roll-Up in Axial Flights by Using a Time-Marching Free-Wake Method

Duck Joo Lee

Department of Aerospace Engineering, Korea Advanced Institute of Science and Technology

373-1 Kusung-dong, Yusong-gu, Taejon 305-701, Korea / E-mail:djlee@hanbit.kaist.ac.kr

ABSTRACT

The wake geometries of a two-bladed rotor in axial flights using a time-marching free-wake method without a non-physical model of the far wake are calculated. The computed free-wake geometries of AH-1G model rotor in climb flight are compared with the experimental visualization results. The time-marching free-wake method can predict the behavior of the tip vortex and the wake roll-up phenomena with remarkable agreements. Tip vortices shed from the two-bladed rotor can interact with each other significantly. The interaction consists of a turn of the tip vortex from one blade rolling around the tip vortex from the other. Wake expansion of wake geometries in radial direction after the contraction is a result of adjacent tip vortices begging to pair together and spiral about each other. Detailed numerical results show regular pairing phenomenon in the climb flights, the hover at high angle of attack and slow descent flight too. On the contrary, unstable motions of wake are observed numerically in the hover at low angle of attack and fast descent flight. It is because of the inherent wake instability and blade-vortex-interaction rather then the effect of recirculation due to the experimental equipment.

1. INTRODUCTION

The tip vortices are one of the most significant but poorly understood aerodynamic features of a helicopter rotor wake. In contrast to fixed-wing aircraft where the tip vortices trail downstream, rotor tip vortices can remain in close proximity of the rotor for a significant amount of time. They are key factors in determining the rotor performance and local blade loads. Under many flight conditions, especially during maneuvers and descent flights, the blades interact closely with the tip vortices resulting in a phenomenon known as Blade-Vortex-Interaction (BVI).

Currently available methods of wake analysis of helicopter flow fields range from relatively simple momentum theory to lifting-surface methods with wake modeling (1,2). The difficulties of lifting-surface methods, in describing the stall and shock for retreating and advancing blade respectively, has led to recent efforts on using computational fluid dynamics (CFD) codes (3,4). Basically the CFD codes can describe the generation and the movement of the vorticity in the wake. However, the inherent numerical dissipation causes a rapid decay of the vortical structures. Another problem occurring in the use of CFD codes is the far wake boundary condition especially for hovering flight, which condition should be carefully implemented also in the free-wake analysis.

Traditionally, the prescribed wake model (5,6), an iterative free-wake model (2,7,8,9), and a time-marching, free-wake model (1,10) have been successfully used to calculate the blade loadings and wake structures in certain conditions. However, more realistic wake geometry, which describes the overall regions of the wake including the far wake, has not been attempted properly. Whether iterative or time-marching free-wake methods, the far-wake model and the artificial initial wake condition should be employed to obtain a converged steady solution in hover using the previously developed methods.

Typically, the far wake is modeled by a vortex ring (7) or a semi-infinite cylinder (8). The far wake may also be truncated after several spirals of the wake (11). The initial state of wake is critical, especially for the time-marching method, because the instability of the wake exists due to the strong starting vortex generated with the assumption of an impulsively rotating blade. Therefore, a helicoidally spiral wake is used initially (11) or a uniform axial velocity (12) is superposed for the impulsively rotating condition. These methods enable the movement of the initial vortex wake downward from the rotor disk to avoid problems of instability during the initial stage. However, the true transient solution and the wake evolution mechanism cannot be predicted.

Recent experiments (13) shows that the tip vortices shed from a two-bladed rotor could interact significantly. The interaction consists of a turn of the tip vortex from a blade rolling around the tip vortex from the other. The wake does not contract monotonically but expands as one of the vortices moves radially outward due to the roll-up. The hovering rotor wake can be represented by three regions based on previous experimental visualization: a well-defined tip vortex contraction region, an intermediate wandering region, and an initially generated far wake bundle. The well-defined 3~4 tip vortices have been properly correlated from experimental data and reasonably predicted by numerous methods (4,5,6,8). Recently, computations of the intermediate wandering region have been attempted to simulate the motion of the tip vortices using Adams-Moulton method to advance the vortices and modified Biot-Savart law to describe the vortex structure (14). But the computation have only achieved the primary goal of demonstrating that the roll-up phenomenon is intrinsic to rotor wake and should not be ignored. The accurate prediction of the rotor-wake geometry is not satisfied yet.

In this paper, a real time-marching free-wake method is described, which does not require the non-physical initial condition and the far-wake model. And the mechanism of the intermediate unsteady region can be explained, which might be very important in predictions of unsteady loads and noise in certain flight conditions. The wake geometries generated by model AH-1G rotor with 41-inch radius are predicted and the results are compared with experimental wake roll-up visualization data (13). These objectives can be fulfilled, not only by using the accurate numerical scheme, but also by observing the physical phenomena carefully. One of the key points is that the rotation speed must be increased slowly from zero to the required speed (15,16,17), a factor that has been overlooked by most previous researchers. The curved vortex filament should be integrated carefully to obtain the induced velocity. This is well proved for the elliptic vortex roll-up (18). The far wake is automatically produced as the blades are rotating slowly without the non-physical far wake model.

Three conditions of flight; climb, hover and descent; are calculated with the time-marching free-wake method, which show stable or inherent unstable conditions of wake roll-up.

2. TIP VORTEX ROLL-UP

Figure 1 shows a sequence of images of the tip vortex development at a rotor collective angle of 9° and climb rate of 3.5 fps (13). The vortices from the two blades are identified by their numbers, i.e. an odd numbered vortex belongs to one blade and even numbered to the other.

The time spacing between each successive frame is not constant, but it was chosen to show the progress of the pairing phenomena. The first frame shows two clear vortices, marked '3' and '4'. In the next frame, vortex '3' has begun to roll up with '4'. This process continues in the next two frames, until '3' and '4' have essentially interchanged their positions while moving downstream (left direction in Figure 1).

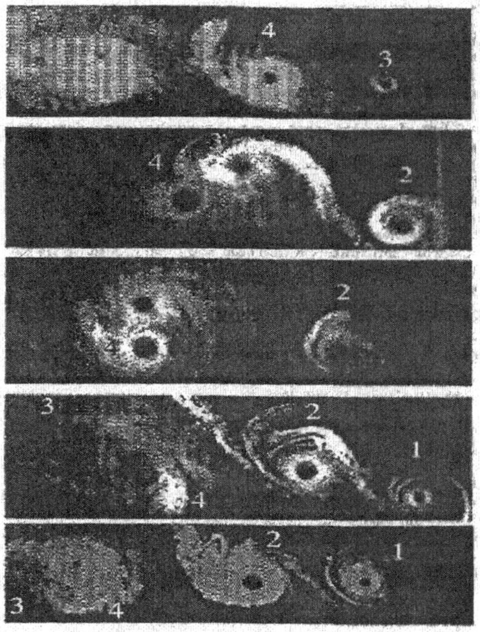

Figure 1. Sequence of vortex roll-up from the individual blade, 9° collective, 3.5 fps climb rate (Caradonna et al and Komerath et al (13))

The notable aspect of this visualization is that the vortex trajectories do not appear to follow a path whose contraction increases monotonically with time. Beyond a certain age, the vortex trajectory tends to expand radially (downward direction in Figure 1). Moreover, the spacing between the tip vortices seems to change. Such vortex expansion have been noted in many previous tests and visualizations, and frequently attributed to being some manifestation of an unstable process. However, the vortex expansion was seen to be a result of adjacent tip vortices beginning to pair together and spiral about each other (13).

3. FORMULATION

The fluid surrounding the body is assumed to be inviscid, irrotational, and incompressible over the entire flow field, excluding the body's solid boundaries and its wakes as shown in Figure 2.

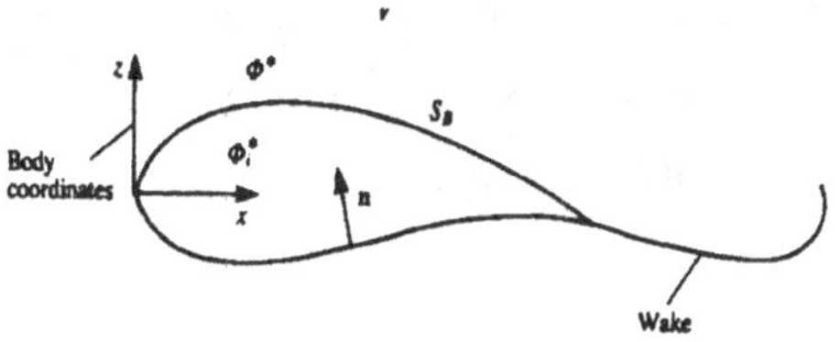

Figure 2. Potential flow over a closed body

Therefore, a velocity potential $\Phi(\vec{x},t)$ can be defined and the continuity equation in the inertial frame becomes:

$$\nabla^2\Phi = 0 \qquad (1)$$

The boundary condition requiring zero normal velocity across the body's solid boundaries is:

$$(\nabla\Phi + \vec{V}_{wake} - \vec{V})\cdot\vec{n} = 0 \qquad (2)$$

Where $\vec{V}_{wake}(\vec{x},t)$ is the induced velocity due to the vorticity field in the wake, $\vec{V}(\vec{x},t)$ is the body surface's velocity, and $\vec{n}(\vec{x},t)$ is the vector normal to the moving surface, as viewed from the blade.

Using Green's second identity, the general solution of equation (1) can be constructed by integrating the contribution of the basic solution of source (σ) and doublet (μ) distributions over the body's surface:

$$\Phi(\vec{x},t) = \frac{1}{4\pi}\int_{body+wake} \mu\vec{n}\cdot\nabla(\frac{1}{r})ds - \frac{1}{4\pi}\int_{body} \sigma(\frac{1}{r})ds \qquad (3)$$

Inserting equation (3) into equation (2) becomes:

$$\{\frac{1}{4\pi}\int_{body+wake} \mu\nabla[\frac{\partial}{\partial n}(\frac{1}{r})]ds - \frac{1}{4\pi}\int_{body} \sigma\nabla(\frac{1}{r})ds - \vec{V}\}\cdot\vec{n} = 0 \qquad (4)$$

The source term is neglected in the case of the thin blade. Thus, only the first part of equation (3) is used to represent the lifting surface. The constant-strength doublet panel is equivalent to a closed vortex lattice with the same strength of circulation, ($\Gamma = \mu$). Then the induced velocity of the vortex lattice in equation (4), representing the blade, can be obtained by using Biot-Savart's law:

$$\vec{V} = -\frac{1}{4\pi}\int_c \frac{\vec{r} \times \Gamma d\vec{l}}{|\vec{r}|^3} \qquad (5)$$

The collocation point is at the mid-span and three-quarter chord of each lattice as shown in Figure 3. The boundary condition of no-flow penetration is satisfied at the collocation point of each lattice.

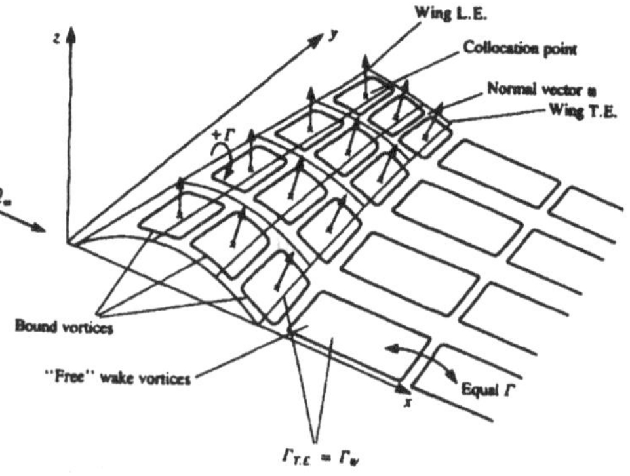

Figure 3. Vortex model for a thin lifting surface

The application of the flow tangency condition (equation 4) to the vortex lattice distribution yields the following linear matrix equation that is to be solved:

$$A_{ij}\,\Gamma_j = R_i \quad (i,j = 1,n) \qquad (6)$$

where A_{ij} is the coefficient matrix of normal induced velocity on the i-th element of the blade due to the j-th vortex lattice with the unit circulation, and Γ_j is the unknown circulation value of the blade vortex lattice. R_i is the normal induced velocity at each control point due to the free stream velocity, the blade-moving velocity, and the wake-induced velocity.

4. TIME-MARCHING FREE-WAKE METHOD

A three-dimensional wing trails the bound circulation (Γ) into the wake. Radial variation of bound circulation produces trailed vorticity in the wake, which direction is parallel to the local free stream direction at each instant it leaves the blade. Azimuthal variation of bound circulation produces shed vorticity, oriented radially in the wake. The strengths of the trailed and shed vorticity are determined by the radial and azimuthal derivatives of bound circulation at the time the wake element leaves the blade. The bound circulation has a peak near the tip, and quickly drops to zero. The trailed sheet therefore has a high strength (proportional to the radial derivative of Γ) at the outer wake, and quickly rolls up into a concentrated tip vortex. The strength of the trailed shed wake vortex at this time step is set equal to the one of the vortex lattice elements, which is located at the trailing edge of the blade ($\Gamma_{T.E,t} = \Gamma_{wake,t}$). This condition is forced to satisfy the Kutta condition ($\gamma_{T.E} = 0$).

Since the wake surface is force-free, each vortex wake element moves with the local stream velocity, which is induced by the other wake element and the blade. The convection velocity of the wake is calculated in the inertial frame. The vortex wakes are generated at each time step. Therefore, the number of wake-elements increases as the blade is rotating. It is clear that a large number of line elements for highly curved and distorted wake region like the tip vortex are necessary to describe the vortex filament distortions accurately. In general, computational time for the calculation of the wake distortion is proportional to the square of the vortex element number. Therefore, the curved element is used for a small number of elements.

There are many mathematical expressions to represent the three dimensional curves. Generally, cubic spline curves are used to describe the curves. However, the cubic spline curves have certain disadvantages; the cubic spline curves require a large tri-diagonal matrix inversion, and the numerical disturbance of position in any one segment affects all the global curve segments. Therefore, the curve is not adequate to represent the vortex filament motion in strong interaction problems. The parabolic blending curves, employed here, maintain the continuity of the first derivative in space, which is critical to our problem. The parabolic blending curve, $C(\xi)$, is given by

$$C(\xi) = (1-\xi)p(r) + \xi q(s) \tag{7}$$

The function of *p(r)* and *q(s)* are parametric parabolas through P_1, P_2, P_3 and P_2, P_3, P_4 as shown in Figure 4, respectively.

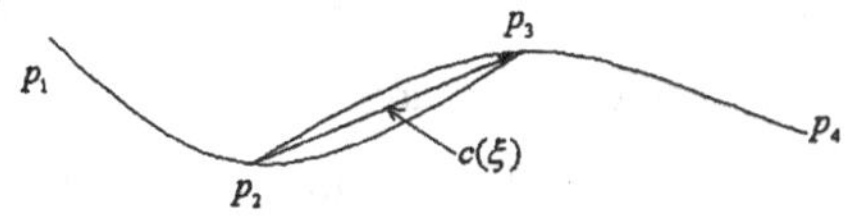

Figure 4. Parabolic blending method

A generalized parametric blending curve is developed from the assumption of normalized chord length approximation for the position parameters, *r* and *s* at P_2 and P_3 which are linearly related with the parameter ξ respectively, *i.e.* $0 \le r, s, \xi \le 1$. Then we apply this blending curve to equation (5).

The induced velocity by a vortex filament with circulation Γ is given by the usual cut-off approach, which is formulated by Moore-Rosenhead. It is defined as

$$\vec{V} = \frac{1}{4\pi}\int \frac{\vec{r}}{\left(|\vec{r}|^2 + \mu^2\right)^{3/2}} \times \Gamma \frac{\partial \mathrm{y}(\xi,t)}{\partial \xi} d\xi \tag{8}$$

Here $\mathrm{y}(\xi,t)$ is the position vector of a material point denoted by Lagrangian variable ξ at an instance, which describes a vortex filament with circulation Γ. The Rosenhead cut-off parameter μ is used to remove the singularity problem in the Biot-Savart's law at the region very closed to the vortex filaments.

The impulsive rotation method of free wake calculation causes non-physical strong instability of the initial wake. So the wake becomes unstable after a few spirals of the wake as shown in Figure 5.

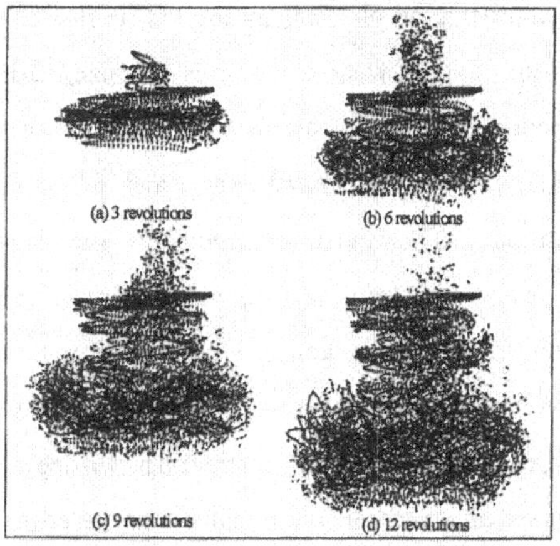

Figure 5. Wake geometry of impulsively rotating case for a one-bladed rotor

One of the key points to overcome these non-physical phenomena is that the rotation speed must be increased slowly from zero to the required speed.

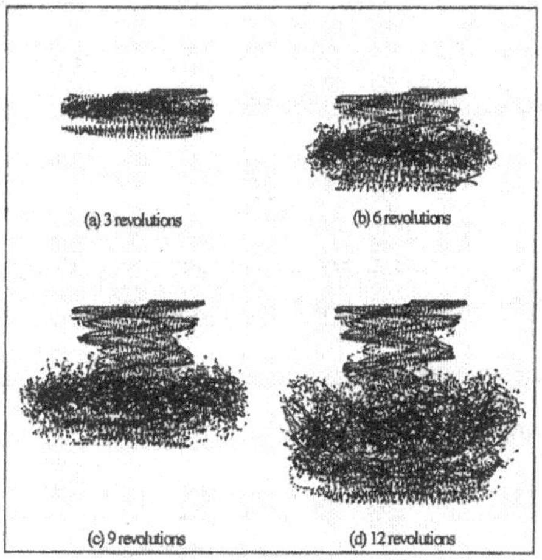

Figure 6. Wake geometry of slowly starting case for a one-bladed rotor

This approach is critical to the curved vortex filament for the simulation of rotor wake (15,16,17). Figure 6 shows that the strength of the initial wake is weak for the slowly starting case, the instability of wake is suppressed, and the wake moves downward slowly where the far wake is automatically produced.

5. TIP VORTEX GEOMETRY CALCULATION

In this approach, no physical assumption is made about the roll-up of inner wake sheet. It is theorized that the computation of the lattice convection is able to correctly predict the roll-up of inner wake sheet and the tip vortex concentration. The sheet influence on the load is introduced by the velocity field induces by all of the filaments of the vortex lattice. No distinction is made between the inner and outer filaments that are thus considered to be potential interacting vortices. The trajectories of tip vortex are re-calculated based on the results of the time-marching free-wake analysis (circulation Γ and location x,y,z of the trailed wake filaments). The circulation Γ_t and location x_t, y_t, z_t of the tip vortex are calculated using the following procedure.

The circulation Γ_t of tip vortex at each time step is obtained by summing up the circulation strength of outside trailed vortices at the maximum lifting point over a spanwise portion along the blade:

$$\Gamma_t = \sum_{i=i_{max}+1}^{i=i_{tip}} \Gamma_i \tag{9}$$

where i_{max} is the index of the maximum lifting point and i_{tip} is that of rotor tip. The radial location r_t of the constructed trip vortex is calculated by using the following barycenter rule:

$$r_t = \left(\sum_{i=i_{max}+1}^{i=i_{tip}} \Gamma_i r_i\right) / \left(\sum_{i=i_{max}+1}^{i=i_{tip}} \Gamma_i\right) \tag{10}$$

The trajectory $z_t(t)$ of the tip vortices is obtained by using the barycenter rule defined previously for the computation of r_t:

$$z_t = \left(\sum_{i=i_{max}+1}^{i=i_{tip}} \Gamma_i z_i\right) / \left(\sum_{i=i_{max}+1}^{i=i_{tip}} \Gamma_i\right) \tag{11}$$

6. RESULTS

The rotor used in this wake calculation is AH-1G's 41-inch radius model rotating at 1800 rpm. The blade is modeled using five chordwise panels and ten spanwise panels. This rotor is the same as that used in the experiments of Caradonna et al and Komerath et al (13). Twenty-four time-steps are taken per blade revolution and the vortex core radius is taken as 10% of the chord length. The tip-vortex pairing process has been quantified by using the trajectory tracking method.

6.1 Tip-vortex roll-up mechanism

The trajectories of the tip-vortices, which are calculated by using time-marching free-wake method, are shown in Figure 7. The left figure shows the three-dimensional view of tip vortex trajectories and the right one shows the cross section view of that.

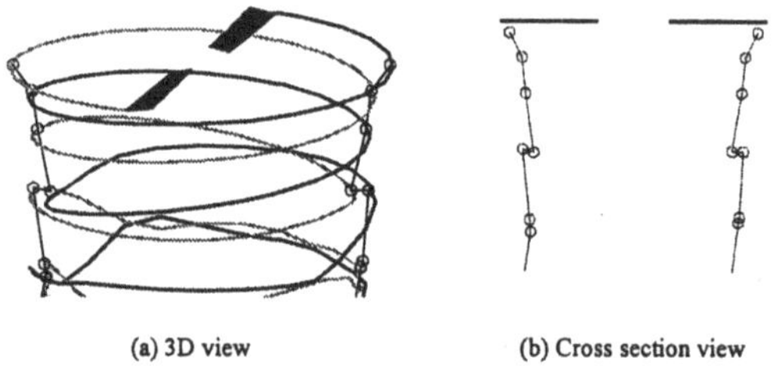

Figure 7. 3D view and cross section view of wake geometries at 11° collective angle, 9.6 fps climb rate

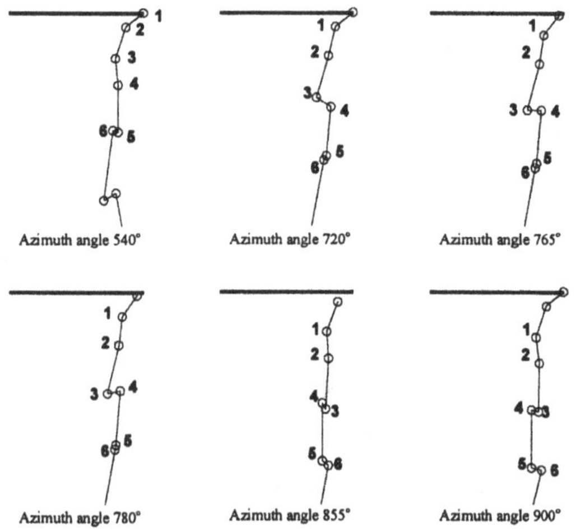

Figure 8. Tip-vortex roll-up process at 11° collective angle, 9.6 fps climb rate

Figure 8 shows the tip-vortex roll-up process for

collective angle of 11° at a climb rate of 9.6 fps. In this figure, vortex '3' begins to roll up with '4'. This process continues until '3' and '4' interchange their positions while moving downstream. We can clearly see that tip-vortex roll-up consists of a turn of the tip-vortex from one blade rolling around the tip-vortex from the other blade as shown in Figure 1 even though different conditions.

6.2 Comparison with experimental data

The trajectories of the tip vortices for collective angle of 6° at a climb rate of 3.5 fps are shown in Fig 9. The tip vortex trajectories from the two blades are offseted by 180° and it is clear that they are influencing each other. The first interaction of the z/R(axial direction) curves of the two vortices indicates the beginning of the vortex pairing process. Subsequent crossings indicate the rotation of the two vortices about each other. In this case the results of time-marching free-wake method has a little error about radial(r) and downstream(z) distance positions but shows that the tip vortex roll-up occurs at 610°, which is the same as the experimental data.

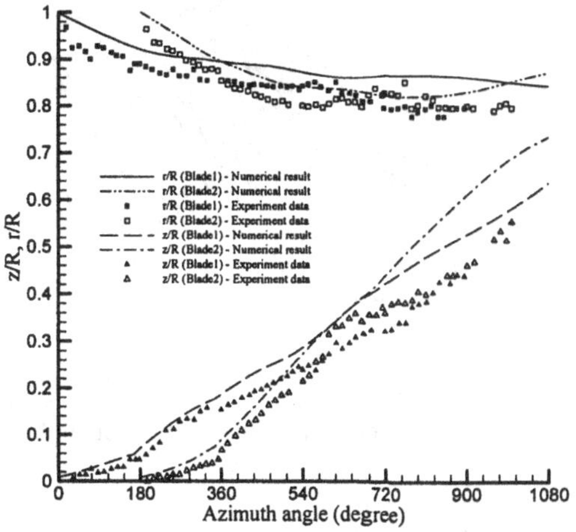

Figure 9. Wake geometries at 6° collective angle, 3.5 fps climb rate

Figure 10 shows the tip-vortices trajectories for higher collective angle of 11° at the same climb rate of 3.5 fps. These trajectories clearly show that the local radial expansion is occurred by the result of adjacent tip vortices begging to pair together and spiral about each other. A comparison of 6° and 11° case clearly shows that the initiation of the wake roll-up is delayed by increasing the collective angle from 610° to 720°.

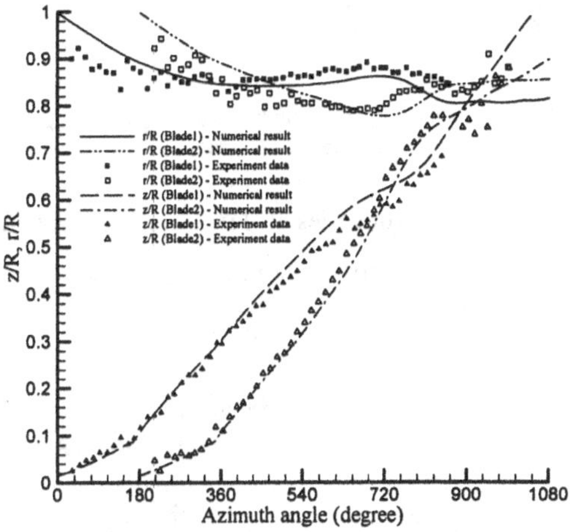

Figure 10. Wake geometries at 11° collective angle, 3.5 fps climb rate

Figure 11 shows the tip vortices trajectories for collective angle of 11° at a faster climb rate of 9.6 fps. These trajectories do not show significant differences compared with those in Figure10 since the effective inflow rates mostly influenced by the collective angle rather than the climb rate in these cases.

The computed wake geometries show excellent agreements with the experimental data as the higher collective angle. The three-dimensional wake trajectories calculated by using time-marching free-wake method at collective angle of 11° and climb rate 9.6 fps are represented in Figure 12. The crossing of the two tip-vortices indicates the commencement of the roll-up process.

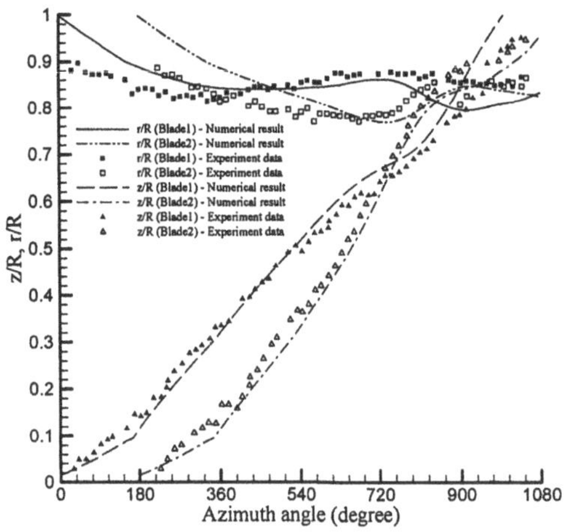

Figure 11. Wake geometries at 11° collective angle, 9.6 fps climb rate

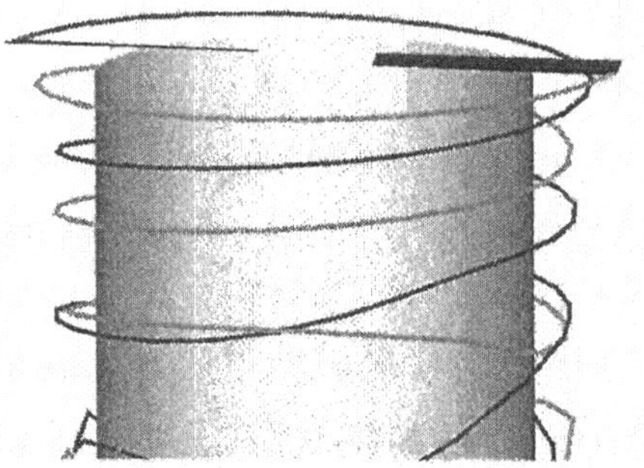

Figure 12. 3-D tip vortex geometry at 11° collective angle, 9.6 fps climb rate

6.3 Instability of wake roll-up

Figure 13 shows detailed view of tip-vortices geometries generated by two blades during 35 revolutions for collective angle of 11° at hover flight. As initial wakes move to far downstream area, the tip-vortex roll-up converges to a certain steady state condition. But Figure 14 shows unstable state of wake roll-up in hover flight for 6° collective angle. It is important to note that this instability is not a manifestation of a numerical instability. This flow unsteadiness seen in hover testing has been reported by Leishman and Bagai (19). This shows that flow unsteadiness is occurred not only by the recirculation and the wind effect of the experimental condition but also by the inherent wake instability.

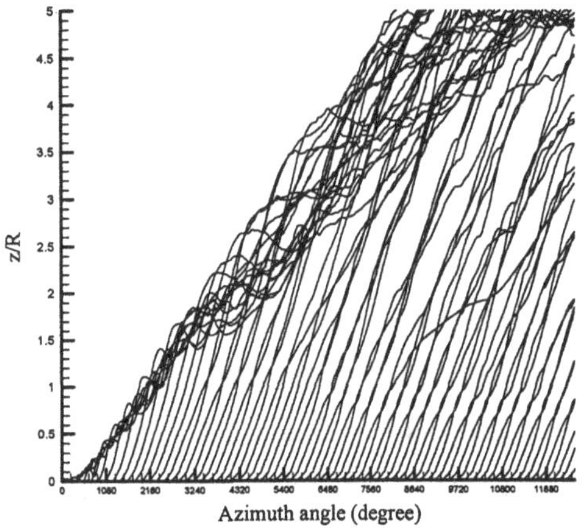

Figure 13. Detailed tip vortex geometry during 35 revolutions at 11° collective angle, at hover flight.

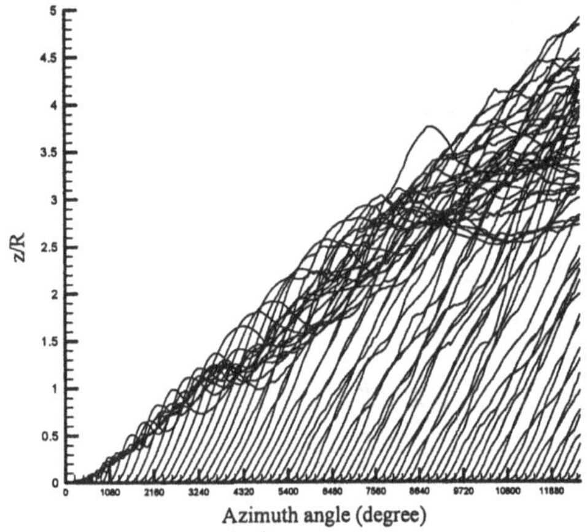

Figure 14. Detailed tip vortex geometry during 35 revolutions at 6° collective angle, at hover flight.

The tip-vortex geometries at 11° collective angle with 3.5 fps climb rate during 35 revolutions also show the steady state of wake roll-up as shown in Figure 15. A comparison of Figure 13 and Figure 15 dose not bring out significant difference but Figure 15 shows faster convergence to a certain steady condition.

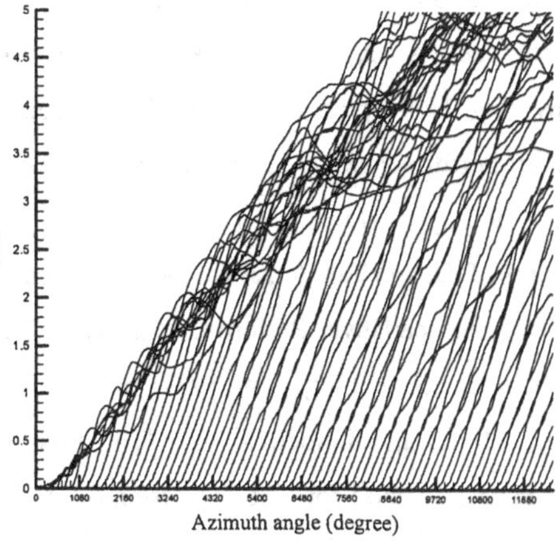

Figure 15. Detailed tip vortex geometry during 35 revolutions at 11° collective angle, 3.5 fps climb rate.

Figure 16 shows the tip-vortex geometries at 11° collective angle with 3.5 fps descent rate during 35 revolutions. In this case, wake roll-up converges to a certain steady state condition and the converged roll-up point is located to more closer to the rotor plane ($z/R=0$) than that of hover and climb case.

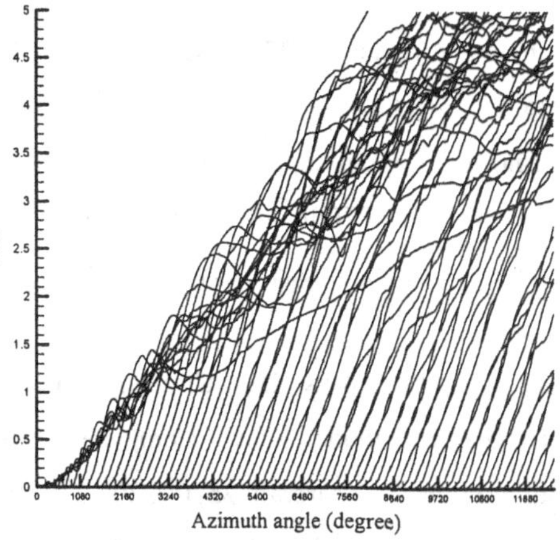

Figure 16. Detailed tip vortex geometry during 35 revolutions at 11° collective angle, 3.5 fps descent rate.

In Figure 17, the tip-vortex geometries at 11° collective angle with faster 9.6 fps descent rate during 35 revolutions show the unstable state of wake roll-up. These random motions of the tip vortices are due to the blade vortex interaction in addition to the instability of the wake.

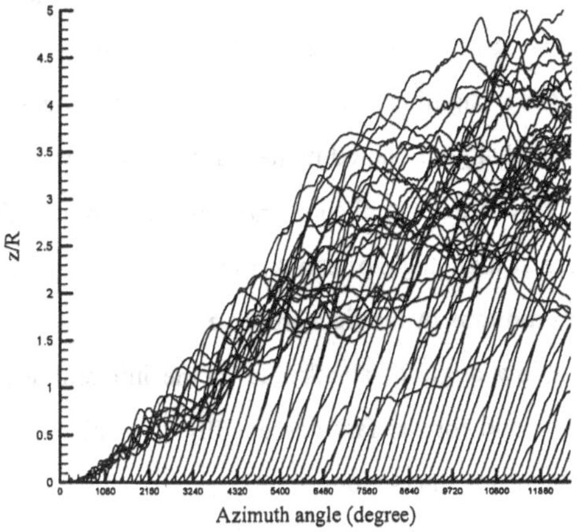

Figure 17. Detailed tip vortex geometry during 35 revolutions at 11° collective angle, 9.6 fps descent rate.

7. CONCLUSION

The ability to predict the performance and aeroacoustics of the entire vehicle is strongly dependent on the ability to predict the highly interactive aerodynamics of the rotor wake. While considerable effort has been invested in the rotor wake problem, computational method, which can handle highly interactive rotor wake problem, has not yet been reported.

In this paper, the radial and axial geometries of the tip vortices are calculated without a far-wake model by using time-marching free-wake method. The time-marching free-wake method can predict the behavior of tip vortex and wake roll-up phenomenon with remarkable agreements. The computed wake geometries show excellent agreements with the experimental data. These excellent agreements can be achieved by using a slow starting rotor based on the physical phenomena.

Generation mechanisms of the wake roll-up are clearly

understood from the free-wake calculations. Tip vortices shed from a two-bladed rotor can interact significantly and the interaction consists of a turn of the tip-vortex from one blade rolling around the tip-vortex from the other blade. The wake expansion is a result of adjacent tip vortices beginning to pair together and spiral about each other. The initiation of the wake roll-up is delayed by increasing the collective angle. The wake trajectories do not show significant differences in these ranges of climb rate change. Detailed numerical results show that a regular vortex pairing phenomenon in the climb flight, the hover at high angle of attack and slow descent flight. Random motions of the tip vortices are observed for the hover at low angle of attack due to the inherent wake instability and for the fast descent flight mainly due to the blade vortex interaction.

ACKNOWLEDGEMENTS

The author would like to thank to Ki-Hoon Chung, Seon-Uk Na and Wan-Ho Jeon in Aeroacoustics Laboratory at KAIST for preparing this material.

REFERENCES

(1) Baron, A., Boffadossi M., "Unsteady Free Wake Analysis of Closely Interacting Helicopter Rotors", Proc. 19th Eur. Rotorcraft Forum, Cernobbio (Como), Italy, Sept. 14-16, 1993.

(2) Felker, F.F., Quackenbush, T.R., Bliss, D.B., and Light, J.L., "Comparisons of Predicted and Measured Rotor Performance Using a New Free Wake Method", Proc. 44th A. Nat. Forum, American Helicopter Society, Washington, D.C., June, 1988.

(3) G.R.Srinivasan, J.D. Baeder, S. Obayashi, and W. J. McCroskey., "Flowfield of a Lifting Rotor in Hover: A Navier-Stokes Simulation", AIAA J., **30**, (10), Oct., 1992, pp.2371-2378.

(4) Strawn, R. C. and Barth, T. J., "A Finite-Volume Euler Solver for Computing Rotary-Wing Aerodynamics on Unstructured Meshes", J. Am. Helicopter Soc., **38**, April, 1993, pp. 61-67.

(5) Landgrebe, A. J., "The Wake Geometry of a Hovering Helicopter Rotor and Its Influence on Rotor Performance", J. Am. Helicopter Soc., **17**, Oct., 1972, pp. 3-15.

(6) Kocurek, J. D. and Tangler, J. L., "A Prescribed Wake Lifting Surface Hover Performance analysis", J. Am. Helicopter Soc., **22**, (1), Jan., 1977, pp. 24-35.

(7) Clark, D. R. and Leiper, A. C., "The Free Wake Analysis a Method for The Prediction of Helicoper Roter Hovering Performance", J. Am. Helicopter Soc., **15**, (1), Jan., 1970, pp. 3-11.

(8) Rosen, A. and Grabe, A., "Free Wake Model of Hovering Rotors Having Straight or Curved Blades", J. Am. Helicopter Soc., **33**, (3), July, 1988, pp. 11-19.

(9) Bagai, A., Leishman, J.G., "Rotor Free-Wake Modeling using a Relaxation Technique - Including Comparisons with Experimental Data", J. Am. Helicopter Soc., **40**, (2), April, 1995, pp. 29-41.

(10) Scully, M.P., "Computation of Helicopter Rotor Wake Geometry and Its Influence on Rotor Harmonic Airloads", MIT ASRL TR 178-1,March, 1975.

(11) Morino, L., Kaprielian, Z. and Sipcic, S. R., "Free Wake Analysis of Helicopter Rotors", Pro. 9th Eur. Rotorcraft Forum, Stresa, Italy, Sept., 1983, Paper No. 3.

(12) Katz, J. and Maskew, B., "Unsteady Low-Speed Aerodynamic model for Complete Aircraft Configurations", J. Aircraft, **25**, (4), 302-310, 1987.

(13) Caradonna, F., Hendley, E., Silva, M., Huang, S., Komerath, N., Reddy, U., Mahalingam, R., Funk, R., Ames R., Darden, L., Villareal, L., Gregory, and Wong, O.,"An Experimental Study OF a Rotor In Axial Flight", AHS Specialists' Meeting on

Aerodynamics and Aeroacoustics, Williamsburg, VA, Oct. 1997.

(14) Rohit, J., A.T. Conlisk, Raghav, M. and N.M. Komerath, "Interaction of Tip-Vortices in the Wake of a Two-Bladed Rotor", AHS 54th Annual Forum, Washington, DC, May 20-22, 1998, pp.182-196.

(15) D. J. Lee and S. U. Na, "Predictions of Helicopter Wake Geometry and Air Loadings by using a Time Marching Free Wake Method", Proc. 1st Forum Russian Helicopter Soc., Moscow, Russia, 1994, pp. 69-85.

(16) D. J. Lee and S. U. Na, "High Resolution Free Vortex Blob Method for Highly Distorted Vortex Wake Generated from a Slowly Starting Rotor Blade in Hover", Pro. 21th Eur. Rotorcraft Forum, Paper No. II-5, Saint-Petersburg, Russia, 1995.

(17) S. U. Na and D. J. Lee, "Numerical Simulations of Wake Structure Generated by Rotating Blades Using a Time Marching Free Vortex Blob Method", European Journal of Mechanics, vol.17, 1998.

(18) K. W. Ryu and D. J. Lee, "Sound Radiation from Elliptic Vortex Rings:Evolution and Interaction", Journal of Sound and Vibration, Vol. 200, No. 3, 1997, pp. 281-301.

(19) Leishman, J.G. and Bagai, A., "Challenges in Understanding the Vortex Dynamics of Helicopter Rotor Wake", AIAA-96-1957, 27th AIAA Fluid Dynamics Coference, June 17-20, 1996, New Orleans, LA.

Experiments and 2D Linear Stability Analysis of the Behavior of Flexible Thin Sheets Cantilevered at the Trailing Edge in a Uniform Flow

Kazuhiko YOKOTA

Mech.Sci., Graduate School of Eng.Sci., Osaka University

1-3, Machikaneyama, Toyonaka, 560-8531, Japan / yokota@me.es.osaka-u.ac.jp

Masatoshi YAMABAYASHI

Aerospace & Electronics, Sumitomo Corporation

1-1-6, Higashisakura, Higashi-ku, Nagoya, 461-8729 Japan / yamabayashi@sumitomocorp.co.jp

Yoshinobu TSUJIMOTO

Mech.Sci., Graduate School of Eng.Sci., Osaka University

1-3, Machikaneyama, Toyonaka, 560-8531, Japan / yokota@me.es.osaka-u.ac.jp

Nobuyuki YAMAGUCHI

Mech. Eng., Meisei University

2-1-1, Hodokubo, Hino, Tokyo, 191-8506 Japan

ABSTRACT

Experiments and 2D linear stability analysis of the behavior of flexible thin sheets cantilevered at the trailing edge in a uniform flow are presented. In the experiment, the sheet displacement is measured temporally with a laser displacement sensor. In the stability analysis, the problem is reduced to an eigenvalue problem by representing the sheet flexural vibration equation for the sheet motion by the finite difference method and the flow field by the vortex method. Reasonable agreement is obtained between experimental and analytical results. Flutter occurs at a lower flow speed. As we increase the flow speed, the flutter ceases once and divergence occurs at a higher flow speed. The reduced frequency increases as the mass ratio increases. The first order bending mode is observed for flutter and divergence. The frictional force makes the flutter onset region smaller and the divergence onset region larger.

1. INTRODUCTION

When a flexible sheet is moved in the air at a high speed, the self-excited oscillation often occurs, called sheet flutter. Flag fluttering is one typical example of sheet flutter. Flexible sheets such as paper and film are treated in copying machines and film manufacturing machines, respectively. When these machines are operated at a higher speed, flutter occurs and this causes various problems such as wrinkling, crumpling and so on. It is, therefore, important to clarify the characteristics of sheet behavior to realize the operations of those machines at higher speeds.

Flutter is one of the aeroelastic phenomena which are caused by the interaction between the motion of the elastic body and the flow. In most of the past flutter studies, the work done by the fluid on the structure is estimated to determine the possibility of flutter. In these analysis, the rigid body motions of the structures are assumed depending on the vibrational modes to be examined. For flutter of flexible materials, however, the mode of flutter depends largely on the fluid forces and cannot be assumed a priori. Therefore, it is indispensable to solve the elastic sheet deformation equation and the fluid motion equation simultaneously. This makes the problem very difficult and very few studies can be found for sheet flutter.

Suzuki and Kaneko(1), and Nagakura and Kaneko(2) analyzed flutter of sheets fixed at both ends as an eigenvalue problems by expanding the oscillation modes of sheet flutter to the eigenmodes of sheet oscillation in vacuum. Huang(3) investigated flutter of sheets fixed at the leading edge by the time marching method. For these treatment, the precision of the analyses depends on the number of the series. Consequently, these analyses are effective for the relatively large mass ratio where the eigenmodes in vacuum become good approximations for the oscillation modes of sheet flutter.

On the other hand, Yamaguchi et al. (4, 5) presented the new treatment in which sheet flutter is reduced to an eigenvalue problem using the finite difference method for the elastic sheet deformation equation and the vortex method for the fluid motion equation. They investigated the flutter characteristics of sheets with a wide mass ratio range fixed at the leading edge in the wide mass ratio range.

In the present paper, we investigated the flutter characteristics of sheets fixed at the trailing edge by using the treatment of Yamaguchi et al. The results are examined based on the comparison with the experiments.

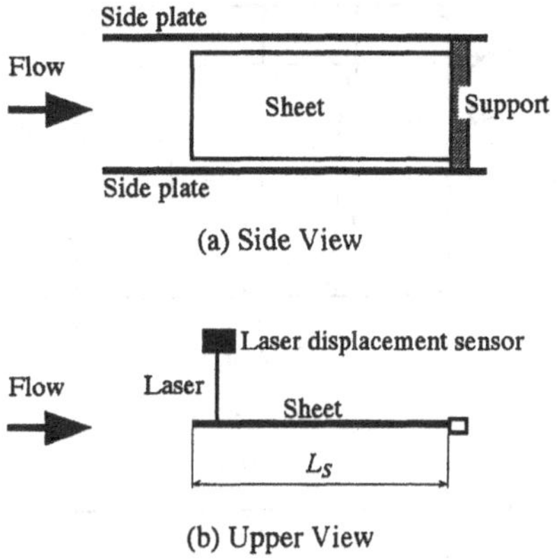

Fig.1 Experimental equipments

2. NOMENCLATURE

c_f : coefficient of skin friction on sheet surface

E_s : Yang's modulus of sheet material

h_s : sheet thickness

i : number of discrete points for discretization

I_s : polar moment of inertia

j : imaginary unit, $j=\sqrt{-1}$

L_s : sheet length

m : number of discretized unit

Δp : pressure difference between upper and lower surfaces of sheet

t : non-dimensional time normalized by L_s / U_a

T : non-dimensional tension due to skin friction defined by Eq.(2)

U_a : flow speed

v : y component of disturbance velocity

x, y: Cartesian coordinates normalized by sheet length

y_s : complex sheet displacement normalized by sheet length

β : relative stiffness defined by Eq.(1)

γ_s, γ_w : complex vortex distributions representing sheet and wake, respectively

μ : mass ratio defined by Eq.(1)

ρ_a, ρ_s : densities of fluid and sheet material, respectively

τ : skin friction stress

ω : reduced angular frequency normalized by U_a / L_s

Superscript

^ : spacial functions

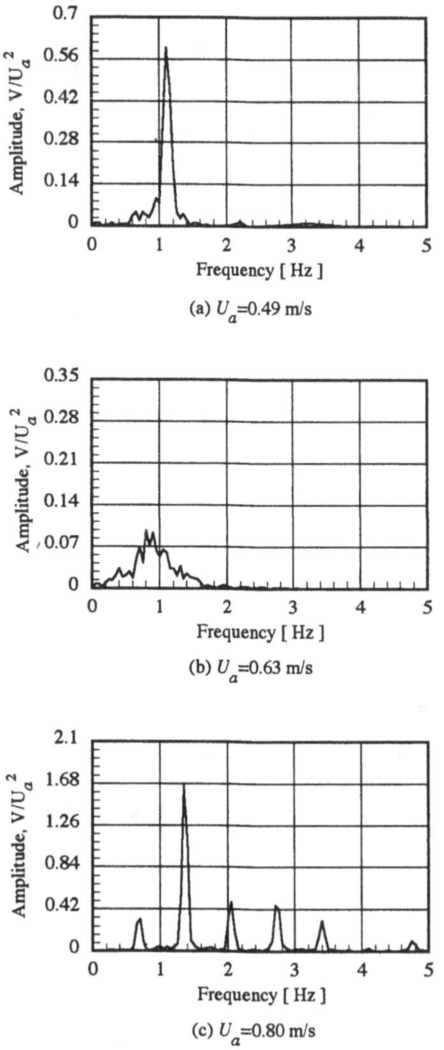

Fig. 2 Spectra of the displacement of the sheet with $E_S I_S = 3.03 \times 10^{-4}$ (μ=0.97)

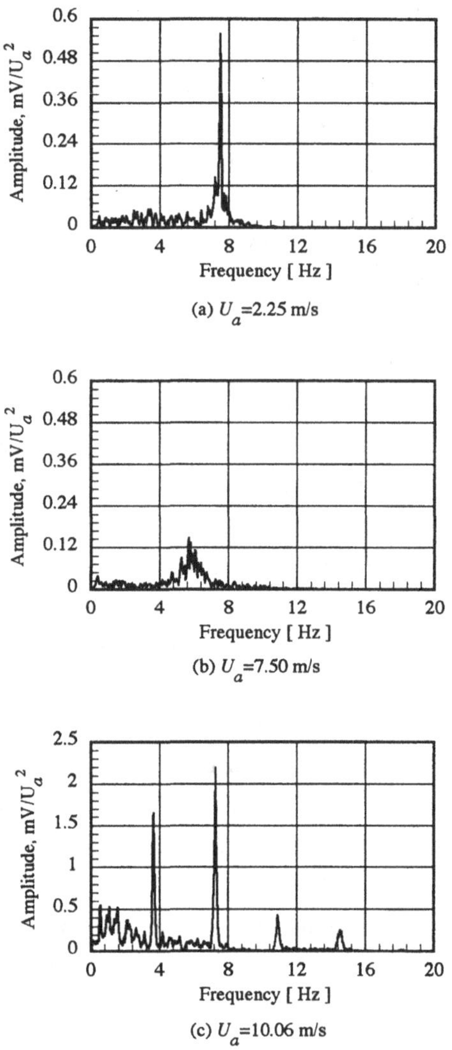

Fig. 3 Spectra of the displacement of the sheet with $E_S I_S = 8.60 \times 10^{-2}$ (μ=5.57)

3. EXPERIMENT

Figure 1 shows a schematic of the experimental apparatus. Sheet is fixed at the trailing edge by an aluminum plate and placed in the uniform flow. To realize the two dimensionality of the flow, side plates were used with a small gap between the plates and the sheet, as shown in Fig.1(a). The displacement of the sheet was measured near the leading edge by using a laser displacement sensor as shown in Fig.1(b). Two kinds of sheets with flexural rigidities $E_s I_s$ of 3.03×10^{-4} and 8.60×10^{-2}[Nm^2] are used for experiments. The mass ratio μ and the relative stiffness β are varied by changing sheet length L_s and the uniform flow speed U_a.

Figures 2 and 3 show the typical spectra of the temporal displacement of the two kinds of sheets under various uniform flow speed. Here the ordinate represents the output voltages from displacement sensor divided by U_a^2. For both sheets, the following common tendencies are seen.

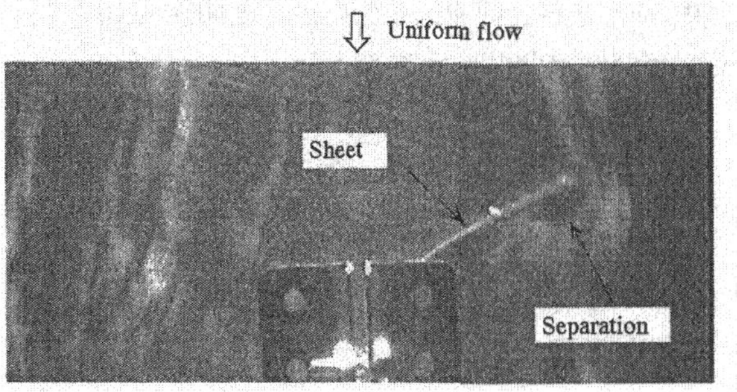

Fig. 4 Leading edge separation in divergence mode

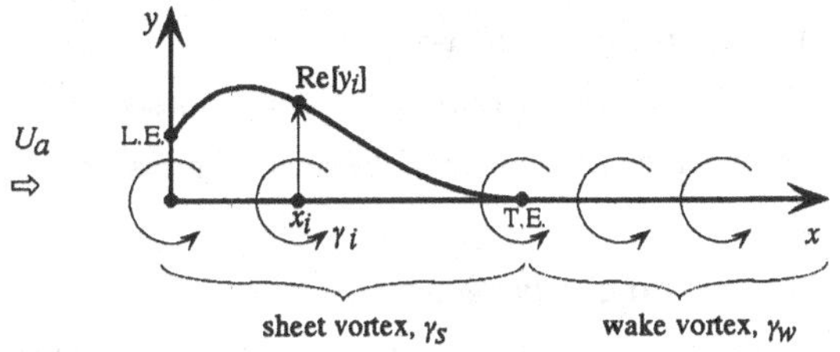

Fig.5 Schematic of sheet displacement, $y_S(x)$ and vortex distribution, $\gamma(x)$

(1) When the uniform flow speed is relatively low as shown in Figs.2(a) and 3(a), each spectrum shows one peak at a typical frequency.

(2) When the uniform flow speed becomes a little higher as shown in Figs.2(b) and 3(b), each spectrum shows no typical peak.

(3) When the uniform flow speed becomes much higher as shown in Figs.2(c) and 3(c), each spectrum shows several peaks at different frequencies from those in Figs.2(a) and 3(a).

In the observation based on a high speed camera, relatively regular oscillation was often seen for each sheet for Figs.2(a) and 3(a). On the other hand, at a little higher speed, irregular oscillations were observed. Therefore, it can be said that the oscillations are regular for Figs.2(a) and 3(a), and irregular for Figs.2(b) and 3(b). Here irregular oscillations are considered to be caused by the irregular disturbances in the uniform flow. For Figs.2(a) and 3(a), the oscillation is regular. The frequency is close to the natural one of the sheet in vacuum. From these results, we judged that this oscillation represents sheet flutter.

For Figs. 2(c) and 3(c), a larger amplitude oscillation was seen by the video observation. The frequency of this oscillation is different from either natural one of the sheet in vacuum. When the displacement is large, the flow separation is observed near the leading edge of the sheet as shown in Fig. 4. To these facts, the analytical results in the following section suggest that this oscillation was caused by the combination of divergence and flow separation: when the displacement due to divergence exceeds certain limit, the flow separation occurs causing the decrease of the fluid force, and then the displacement becomes small - repetition of this cycle is the cause of the oscillation. If the displacement was restricted to be small so as not to cause the flow separation, only divergence without oscillation is observed.

From all experimental results mentioned above, we considered in the present study that the oscillation with the spectra of Figs. 2(a) and 3(a) is flutter, the irregular oscillation with the spectra of Figs. 2(b) and 3(b) is caused by the flow disturbances, and the oscillation with the spectra of Figs. 2(c) and 3(c) is caused by the combination of divergence and the flow separation. As the flow speed increases, the sheet behavior changes from flutter to divergence with some stable region between them.

4. STABILITY ANALYSIS

4.1 Assumption

The following assumptions are made for the analysis.

- The phenomena are two-dimensional. Therefore, the flow disturbances due to the sheet deformation are also two-dimensional.
- The sheet exists in the range of $0 \le x \le L_s$, as shown in Fig.5. The displacement y_s of the sheet deformation is sufficiently small.
- The flow disturbances due to the sheet deformation are irrotational and sufficiently small corresponding to the small displacement y_s.

- The sheet and its vortex wake are represented by the vortices distributed on the x-axis.
- The kinetic boundary conditions are imposed on the x-axis under the assumption of the small disturbances.

4.2 Governing equations

Based on the assumptions in section 4.1, the non-dimensional equation of the sheet flexural vibration can be represented as follows.

$$\mu\frac{\partial^2 y_s}{\partial t^2}+\beta\frac{\partial^4 y_s}{\partial x^4}-\frac{\partial}{\partial x}\left(T\frac{\partial y_s}{\partial x}\right)=\Delta p. \tag{1}$$

$$\mu=\frac{\rho_s h_s}{\rho_a L_s},\quad \beta=\frac{E_s I_s}{\rho_a U_a^2 L_s^3}.$$

Here, the coordinate x and the sheet displacement y_s are reduced by the sheet length L_s, and time t by L_s/U_a. The mass ratio μ and relative stiffness β represent the ratio of the inertia of the sheet to the inertia of the fluid and the ratio of the flexural rigidity of the sheet to the fluid force, respectively. The non-dimensional tension T caused by the skin friction is also included. The coefficient of the skin friction c_f on the sheet surface is assumed to be constant over the sheet. Under the boundary condition of $T=0$ at $x=0$ for a sheet fixed at the leading edge, the non-dimensional tension is obtained as follows.

$$T=-c_f x \quad , \quad \left(c_f=\frac{\tau}{\frac{1}{2}\rho U^2}\right). \tag{2}$$

Here, τ is the skin friction stress. The negative tension means a compression force.

From the linearized 1-D Euler equation, the non-dimensional pressure difference is represented by the sheet vortex as follows.

$$\Delta p=-\left(\int_0^x\frac{\partial\gamma_s}{\partial t}+\gamma_s\right). \tag{3}$$

The kinetic boundary condition on the sheet surface is expressed by

$$\frac{\partial y_s}{\partial t}+\frac{\partial y_s}{\partial x}=v. \tag{4}$$

Here, the normal velocity induced on the sheet surface by the vortex distribution is given by

$$v=\frac{1}{2\pi}\left[\int_0^1\frac{\gamma_s(\xi,t)}{x-\xi}d\xi+\int_1^\infty\frac{\gamma_w(\xi,t)}{x-\xi}d\xi\right]. \tag{5}$$

The free vortex shed from the sheet is assumed to be transported on the free stream. Then, the wake vortex distribution γ_w can be related with the sheet vortex distribution γ_s at the trailing edge as follows.

$$\gamma_w(\xi,t)=\gamma_s(1,t-\xi+1)\quad(\xi>1). \tag{6}$$

The following Kutta's condition is applied at the trailing edge of the sheet.

$$\Delta p(1)=\int_0^1\frac{\partial\gamma_s}{\partial t}+\gamma_s(1)=0. \tag{7}$$

From the equations mentioned above, the governing equations can be represented by the sheet displacement y_s and the sheet vortex distribution γ_s as follows.

- The first is the flexural vibration equation for the displacement of the sheet:

$$\mu\frac{\partial^2 y_s}{\partial t^2}+\beta\frac{\partial^4 y_s}{\partial x^4}+c_f\frac{\partial y_s}{\partial x}+c_f x\frac{\partial^2 y_s}{\partial x^2}+\int_0^x\frac{\partial\gamma_s}{\partial t}+\gamma_s=0. \tag{8}$$

- The second is the kinetic boundary condition:

$$\frac{\partial y_s}{\partial t}+\frac{\partial y_s}{\partial x}-\frac{1}{2\pi}\left[\int_0^1\frac{\gamma_s(\xi,t)}{x-\xi}d\xi+\int_1^\infty\frac{\gamma_s(1,t-\xi+1)}{x-\xi}d\xi\right]=0. \tag{9}$$

- The third is the boundary conditins for γ_s and y_s. The boundary condition for γ_s is the Kutta's condition:

$$\int_0^1\frac{\partial\gamma_s}{\partial t}+\gamma_s=0. \tag{10}$$

The boundary conditions for y_s are the conditions that sheet is cantilevered at the trailing edge:

$$\frac{\partial^2 y_s}{\partial x^2}=\frac{\partial^3 y_s}{\partial x^3}=0\quad\text{at } x=0, \tag{11}$$

and

$$y_s=\frac{\partial y_s}{\partial x}=0\quad\text{at } x=1. \tag{12}$$

4.3 Method of analysis

In the equations from (8) to (12), the displacement and the vortex distribution are assumed to be expressed by

$$y_s = \hat{y}_s(x)e^{j\omega t}, \quad \gamma_s = \hat{\gamma}_s(x)e^{j\omega t} \quad (0 \le x \le 1). \tag{13}$$

Here, ω is the reduced angular frequency (simply, the reduced frequency) of the harmonic oscillation for the sheet-fluid system, and $\hat{y}_s(x)$ and $\hat{\gamma}_s(x)$ are unknown complex amplitudes.

We consider $y_k = \hat{y}_s(x_k)$ and $\gamma_k = \hat{\gamma}_s(x_k)$ at a discrete point $x = x_k$ $(1 \le k \le m+1)$ as unknowns. Here, m is the number of discretized unit. The integral and differential terms in the governing equations are represented in terms of y_s and γ_s using the trapezoidal rule and the fourth order finite difference methods, respectively.

Then, the governing equations are represented by a set of homogeneous simultaneous equations with respect to y_k and γ_k:

$$[A(\omega)]\begin{bmatrix} y_k \\ \gamma_k \end{bmatrix} = \begin{bmatrix} 0 \\ 0 \end{bmatrix}. \tag{14}$$

Here, $A(\omega)$ is a $2(m+1) \times 2(m+1)$ complex matrix. For the existence of non-trivial solution, the determinant should equal to zero:

$$\det[A(\omega)] = 0. \tag{15}$$

This is the characteristic equation of the harmonic oscillation for the sheet-fluid system. The solution ω (or $\omega / 2\pi$) gives the complex eigenfrequency, whose real and imaginary parts represent eigenfrequency and temporal damping ratio, respectively. The corresponding eigenvector (y_k, γ_k) represents eigenmode of the harmonic oscillation. When the imaginary part of ω is positive, zero and negative, the sheet-fluid system is stable, neutrally stable and unstable, respectively. In addition, when the real part of ω is zero and non-zero, the solution represents sheet divergence and sheet flutter, respectively.

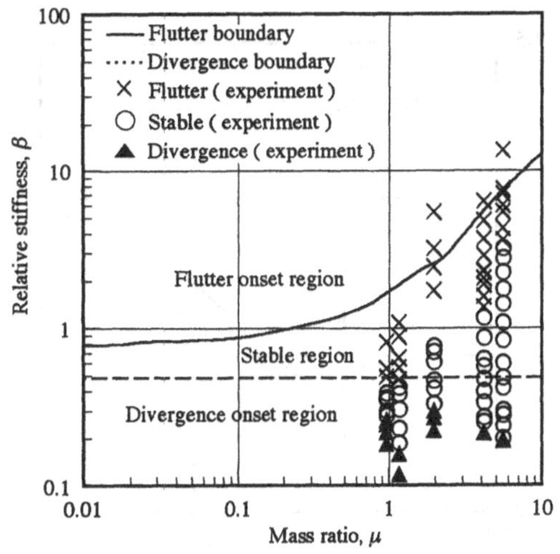

Fig.6 Stability map

5. RESULT AND DISCUSSION

Figure 6 shows the experimental and the analytical stability maps with the skin friction coefficient set to be zero. The abscissa and the ordinate represent the mass ratio μ and the relative stiffness β, respectively.

The solid line represents the neutrally stable condition for flutter and the regions upper/lower than the line are unstable/stable for flutter. The broken line represents the neutrally stable condition for divergence and the regions upper/lower than the line are stable /unstable for divergence. Note that we have a region stable for both flutter and divergence.

For the experimental results, the symbols of ×, ○ and ▲ are plotted when the spectra such as Fig.2(a), Fig.2(b) and Fig.2(c) are seen, respectively. In Fig.6, the space between the symbols of × and ○ for the mass ratio of 1.97 is caused by the difficulty in distinguishing the spectra.

As we increase the flow speed or decrease the value of β keeping the value of μ constant, the sheet behavior shifts from flutter to stable and then to divergence. The experimental and analytical results agree qualitatively in this respect.

The value of relative stiffness β for flutter boundary is larger for the larger mass ratio μ, while the value of β

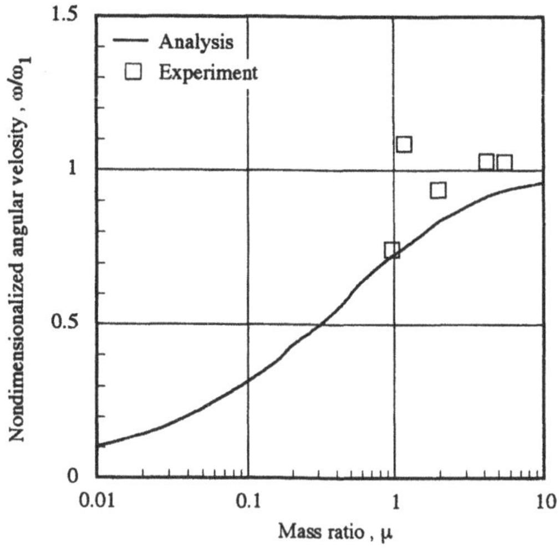

Fig.7 Angular frequency

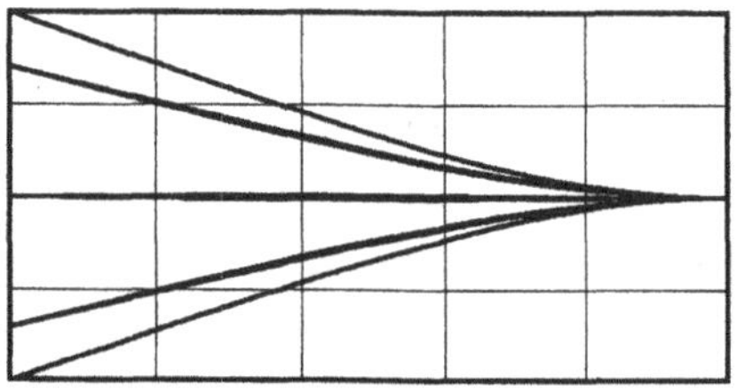

(a) Flutter mode (μ =1.0 , β =1.8)

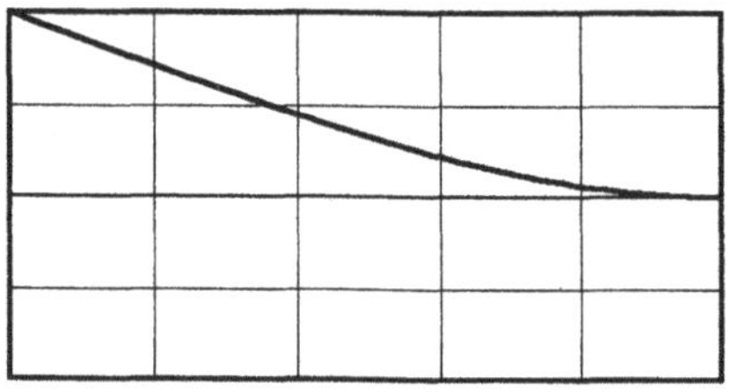

(b) Divergence mode (μ =1.0 , β =0.47)

Fig.8 Modes of flutter and divergence

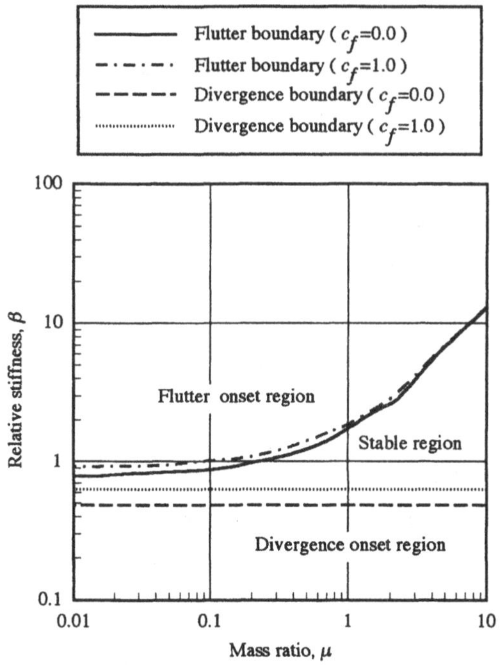

Fig.9 Stability map for c_f=0.0 and 1.0

for divergence boundary is independent of the value of μ. The latter result is reasonable since divergence is a static process. Agreement between analysis and experiments is also found in this respect.

Figure 7 shows the ratio of the reduced frequency of flutter on the flutter boundary to the natural one of the sheet in vacuum. Except the datum for the mass ratio of 1.18, the experimental and the analytical results agree well. Figure 7 also indicates that the ratio of the reduced frequency increases as the mass ratio increases.

Figure 8 shows the modes of flutter and divergence boundaries (β=1.8 for flutter and β=0.47 for divergence) for the sheet with μ= 1.0. The first order mode is obtained for both flutter and divergence. This agrees with the observation using a high speed video.

Figure 9 shows the analytical stability maps for the skin friction coefficient of zero and unity. Here the skin friction coefficient is chosen to be relatively large value of unity so as to make clear the effects of the coefficient. This result indicates that the skin friction has the effect to enlarge the flutter stable region and the effect becomes larger as the mass ratio decreases. On the contrary, the skin friction has the effect to enlarge the divergence onset region and the effect is independent of the mass ratio. For the present case with the sheet fixed at the trailing edge, the skin friction causes a compression stress over the sheet. Because divergence means buckling due to the fluid force buckling is caused by a compression force, it is reasonable that the compressional skin friction enhances divergence. In addition, since divergence is a static process as mentioned before, the result that the effect of the skin friction is independent of the mass ratio is also reasonable.

6. CONCLUSIONS

From the results mentioned above, the following conclusions are obtained for the behavior of the sheet fixed at the trailing edge.

(1) The sheet behavior transforms from flutter to divergence through a stable region as the flow speed is increased.

(2) As the mass ratio increases, the stable region for flutter becomes larger. On the contrary, the stability for divergence is not affected by the mass ratio.

(3) The skin friction enlarges the stable region for flutter, especially for the sheet with the smaller mass ratio.

(4) The skin friction enlarges the divergence onset region, and its effect is independent of the mass ratio.

(5) The linear stability analysis can predict the sheet behavior fairly well.

REFERENCE

(1) Suzuki, Y. And Kaneko, S. (1997),"Flutter of Paper Sheet with Tensile Force," Proceedings of Japan Sciety of Mechanical Engineers, No.97-19(B), p.115-116, in Japanese.

(2) Nagakura, H. And Kaneko, S. (1992), "The Stability of a Cantilever Beam Subjected to One-Dimensional Leakage Flow," Transactions of Japan Sciety of Mechanical Engineers, Series C, Vol. 58, No.546, p.352-359, in Japanese.

(3) Huang, L. H. (1995), "Flutter of Cantilevered Plates in Axial Flow," Journal of Fluids and Structures, No.9, p.127-147.

(4) Yamaguchi, N., Yokota, K. and Tsujimoto, Y. (1999), "Fluttering Behavior of a Flexible Thin Sheet in High-Speed Flow (1st Report, Theoretical Method for Prediction of the Sheet Behavior for Small Perturbed Motion)," Transactions of Japan Sciety of Mechanical Engineers, Series B, Vol. 65, No.632, p.1224-1231, in Japanese.

(5) Yamaguchi, N., Sekiguchi, T., Yokota, K. and Tsujimoto, Y. (1999), "Fluttering Behavior of a Flexible Thin Sheet in High-Speed Flow (2nd Report, Test Results and Predicted Behavior for Low Mass Ratios)," Transactions of Japan Sciety of Mechanical Engineers, Series B, Vol. 65, No.632, p.1232-1239, in Japanese.

VORTEX ELEMENT MODELLING FOR THE DESIGN AND FLOW ANALYSIS OF AXIAL, RADIAL AND MIXED-FLOW TURBOMACHINES.

R. Ivan Lewis
Department of M.M.M.Eng,, Newcastle University,
Room 2-16 Bruce Building, Newcastle upon Tyne , NE1 7RU, UK/ Email R.I.Lewis@ncl.ac.uk

ABSTRACT

The full three-dimensional flow through a turbomachine may be represented by the superposition of a series of blade-to-blade flows upon a circumferentially averaged meridional flow through the annulus. The paper shows how surface vorticity analysis may then be used to simulate the complete design/analysis problem including the effects of interference between the blade-to-blade and meridional flows. These include the shedding of vorticity from the blades into the meridional flow and the inverse effect of meridional streamtube thickness variation upon the blade-to-blade flows. The effects of relative eddy or slip may also be imposed upon the blade-to-blade surface vorticity analysis with precision. The essential underlying equations will be presented together with examples of use of the author's software for design and analysis of an axial fan and a mixed flow fan.

1. INTRODUCTION

In general there has been a tendency recently for commercial CFD codes to aim directly for simulation of the fully three-dimensional flow through duct systems. In the case of turbomachines on the other hand there is considerable design advantage to continue with the traditional approach of superimposing *blade-to-blade* flows located on surfaces of revolution upon an assumed circumferentially averaged *meridional* flow. The two prime functions of a turbomachine are (a) to move fluid at a prescribed mass flow rate between two locations and (b) to arrange a transfer of energy between the fluid and the rotor by rotodynamic action. The major energy exchange process between rotor and fluid, is in fact largely controlled by such blade-to-blade flows, item (b), while the meridional flow deals with requirement (a). This breakdown of the fully three-dimensional flow into a series of superimposed but interacting two-dimensional flows is thus not only a means for simplifying the flow analysis problem but is extremely important both for the designer to understand and keep control of his design problem and to express the detailed fluid-dynamic performance of any turbomachine. However it is essential for such modelling to be based on accurate codes for the two separate problems of the blade-to-blade simulation and meridional analysis. Furthermore it is crucial to include in such a scheme the interference effects between the blade-to-blade and meridional flows which are often of major proportions.

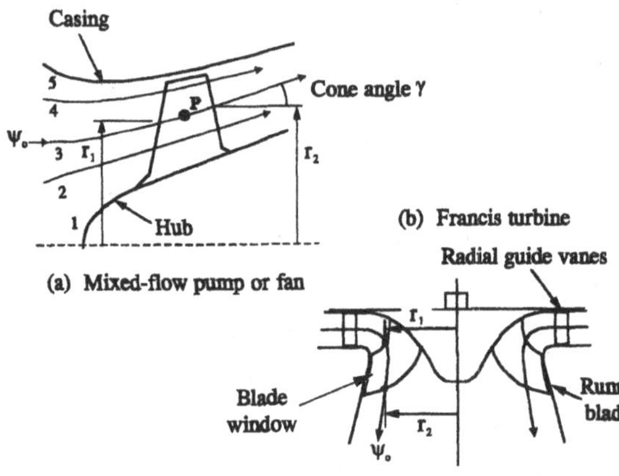

Figure 1 Meridional flow through a turbomachine.

Views of the meridional flow through a mixed-flow fan and a Francis turbine are illustrated in Figure 1, exhibiting the following characteristics[1] :

1. The meridional flow is assumed for numerical convenience to be axisymmetric, based on the circumferentially averaged blade-to-blade flow.
2. The blades form a so-called blade window defined by the leading and trailing edges and the hub and casing.
3. The local cone angle γ may vary considerably from hub to casing or from inlet to outlet.
4. Consequently the inlet and exit radii r_1 and r_2 of meridional streamlines crossing the blade window may vary significantly. This will result in the presence of Coriolis accelerations within a rotor and associated

blade loadings and also in the phenomenon known relative eddy and its associated slip factor.

5. The gap between neighbouring meridional streamlines may also vary resulting in meridional interference with the blade-to-blade flow. (This is equivalent to the effects of overall axial velocity ratio in axial cascades[2] and will be referred to later as AVR). *Blade-to-blade analysis must be able to take both Coriolis/slip and AVR effects into account.*

6. Due to variations in loading, vorticity will be shed from the blades and convected downstream by the meridional flow. The effects of this upon the meridional velocity c_s can be enormous and may even cause instability or reverse flow. *Meridional flow analysis must be able to take these effects into account.*

It is clear from the above that the full design/analysis process must involve successive iterations between the meridional and blade-to-blade flows as follows:

Meridional analysis	↔	Blade-to-blade analysis
↓		↓
Estimation of Meridional velocity c_s, r_2/r_1, AVR	→ ←	Estimation of loading, Vortex shedding, Merid. vorticity ω_θ

It has been the author's long term aim to solve both analysis problems by vortex element methods and to implement them into such an iterative scheme suited both for design and analysis of a wide range of turbomachines and this has now been finally achieved and implemented into his software suite MIXFLO[3]. The aim here will be to introduce these two analyses in brief in sections 2 and 3 and then to illustrate their application to turbomachine design in section 4. Before proceeding with this however it will be helpful to refer to the overall technique adopted for handling blade-to-blade analysis, and this is illustrated in Figure 2. We observe that in view of its two-dimensional nature, the blade-to-blade flow on a meridional surface of revolution intersecting the blade of a mixed-flow fan may be transformed conformally into an equivalent infinite straight cascade in the ζ plane. Further details will be left until section 4 but the main advantages of this technique are worthy of comment at this point, namely:

1. The well established vortex element numerical procedures[4][5][6] for rectilinear cascades may be adapted and implemented with minimal modification.
2. The standard cascade geometrical parameters such as stagger λ, camber θ and pitch/chord ratio t/l become applicable even for mixed-flow blade rows.
3. In view of this all turbomachines become members of the same family and may share the same treatment and classification for geometrical design.

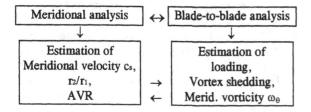

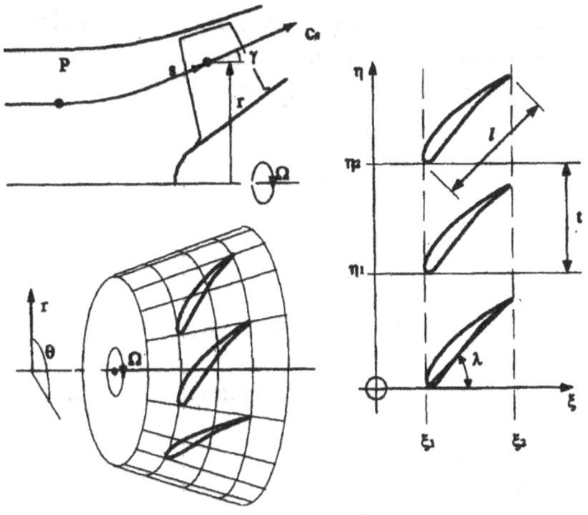

(a) Mixed-flow cascade in the z plane. (b) Straight cascade in the ζ plane.

Figure 2 Transformation of a mixed-flow cascade into an equivalent infinite straight cascade.

2. MERIDIONAL FLOW ANALYSIS

The plan in this section will be twofold. First the governing equations for meridional flow will be developed in section 2.1. Following this a vortex element numerical model will be presented for analysis of these equations in section 2.2.

2.1 Governing Equations for Meridional Flow

The governing equations comprise the continuity equation and the momentum equations, which may be expressed in cylindrical polar coordinates x,r,θ and reduced to axisymmetric incompressible flow, whereupon the continuity equation becomes

$$\frac{\partial c_x}{\partial x}=-\frac{1}{r}\frac{\partial (rc_r)}{\partial r} \qquad (1)$$

from which the Stokes stream function $\psi(x,r)$ may be derived as follows.

$$c_x=\frac{1}{r}\frac{\partial \psi}{\partial r},\quad c_r=-\frac{1}{r}\frac{\partial \psi}{\partial x} \qquad (2)$$

If the meridional flow is to be rotational then we must take into consideration the presence of distributed tangential vorticity ω_θ which will induce velocity components c_x and c_r in the x,r meridional plane where ω_θ is defined through

$$\omega_\theta=\frac{\partial c_r}{\partial x}-\frac{\partial c_x}{\partial r} \qquad (3)$$

If the stream function is now introduced into this through equations(2), the governing equation for axisymmetric meridional flow is obtained, namely

$$\frac{\partial^2 \psi}{\partial x^2}-\frac{1}{r}\frac{\partial \psi}{\partial r}+\frac{\partial^2 \psi}{\partial x^2}=-\omega_\theta r \qquad (4)$$

A second equation is now needed to enable us to express the tangential vorticity distribution $\omega_\theta(x,r)$ in terms of useful turbomachinery design parameters so that equation(4) is then fully specified, and this follows from the momentum equations expressed in the form $-\frac{1}{\rho}\, grad\, p_o = \hat{q} \times curl\hat{q}$, namely

$$\left.\begin{aligned} &\frac{1}{\rho}\frac{\partial p_o}{\partial x} = c_r\omega_\theta - c_\theta\omega_r \\ &\frac{1}{\rho}\frac{\partial p_o}{\partial r} = c_\theta\omega_x - c_x\omega_\theta \\ &\frac{1}{\rho r}\frac{\partial p_o}{\partial \theta} = c_x\omega_r - c_r\omega_x = 0 \quad (for\ axisymmetry) \end{aligned}\right\} \qquad (5)$$

where the two remaining vorticity components are defined

$$\begin{aligned} \omega_x &= \frac{1}{r}\frac{\partial c_\theta r}{\partial r} - \frac{1}{r}\frac{\partial c_r}{\partial \theta} \rightarrow \frac{1}{r}\frac{\partial c_\theta r}{\partial r} \\ \omega_r &= \frac{1}{r}\frac{\partial c_x}{\partial \theta} - \frac{\partial c_\theta}{\partial x} \rightarrow -\frac{\partial c_\theta}{\partial x} \end{aligned} \qquad (6)$$

As proved in full by Lewis[1], it follows that the angular momentum $c_\theta r$ and stagnation pressure p_o obey the relationships

$$\left.\begin{aligned} c_\theta r &= f_1(\psi) \\ p_o &= f_2(\psi) \end{aligned}\right\} \qquad (7)$$

Taking the total derivative of the stagnation pressure

$$\frac{1}{\rho}\frac{dp_o}{d\psi} = \frac{1}{\rho}\frac{\partial p_o}{\partial x}\frac{dx}{d\psi} + \frac{1}{\rho}\frac{\partial p_o}{\partial r}\frac{dr}{d\psi} \qquad (8)$$

Substituting from the momentum equations(5), further manipulation given in full in ref (1) results finally in the following equation for tangential vorticity ω_θ auxiliary to the main governing equation(4).

$$\omega_\theta = c_\theta\frac{d(c_\theta r)}{d\psi} - \frac{r}{\rho}\frac{dp_o}{d\psi} \qquad (9)$$

One further clarification should be made, namely that this analysis has been undertaken in the absence of distributed body forces and thus the foregoing equations apply strictly speaking only to empty annulus locations upstream or downstream of blade rows. The equations thus apply to vorticity convection within annular spaces free of body forces. More advanced formulations have been made by the author elsewhere[6], but for practical numerical schemes it is possible simply to apply the above simplified equations with good accuracy by assuming that the tangential vorticity ω_θ is generated at and shed from planes (e.g. grid lines) specified within the blade rows. Some interpretation of equation(9) would be helpful at this point as follows:

1. We note that the tangential vorticity ω_θ comes from two sources, namely gradients of angular momentum $c_\theta r$ and stagnation pressure p_o normal to the meridional streamlines defined by ψ = constant.

2. We note also from equations(7) that $c_\theta r$ and p_o are conserved along the meridional streamlines (as we would expect from physical considerations).

3. All that remains is for the designer to provide the distributions in the annulus of angular velocity and stagnation pressure $c_\theta r = f_1(\psi)$ and $p_o = f_2(\psi)$ in between each blade row (or equivalent vorticity shedding plane), and the numerical problem is fully specified. For flow though a rotor this is simply accommodated by making use of the Euler pump equation. Thus for the mixed-flow fan illustrated in Figures 1 and 2, the Euler rise in stagnation pressure across the rotor in ideal flow or specific work input $\overline{W}$ may be expressed

$$\frac{1}{\rho}(p_{o2} - p_{o1}) = \Omega(r_2c_{\theta 2} - r_1c_{\theta 1}) = \overline{W} \qquad (10)$$

The two well known problems of turbomachines may now be stated namely:

i) ***The design problem*** - in which the designer specifies the specific work input, stagnation pressure rise or angular momentum distribution which he requires of his turbomachine and seeks to design blade geometry to achieve this.

ii) ***The analysis problem*** - in which the blade geometry is already known and the designer wishes to predict its fluid dynamic performance, including finally the specific work input and the distribution of stagnation pressure and swirl velocity throughout the annulus.

We observe from the Euler pump equation(10) that the specific work input or stagnation pressure rise are completely determined by the fluid angular momentum change and vice versa. We note also that only the Euler or frictionless ideal design has been stated above. However losses and their impact upon the stagnation pressure distribution and hence upon the meridional flow variations may be introduced into the formulations if need be by means of the total-to-total efficiency η_{tt}, defined for a pumping machine through

$$\eta_{tt} = \frac{Actual\ stagnation\ press.rise}{Frictionless\ stagnation\ press.rise} \qquad (11)$$

or the inverse of this for a turbine.

2.2 Numerical model for Meridional Analysis making use of Vortex Element Analysis.

Although solution of the meridional flow is perfectly possible by expressing the governing equation(4) in some suitable form such as that for finite difference analysis, an efficient alternative is to replace the equation by a boundary integral equation based on the surface vorticity model originally proposed by Martensen[4] for plane flows but developed by Lewis[6] for axisymmetric duct flows. The relevant numerical model for this is illustrated in Figure 3 which shows the annulus grid structure for a mixed-flow fan with suitable annotations to bring out various features of the model.

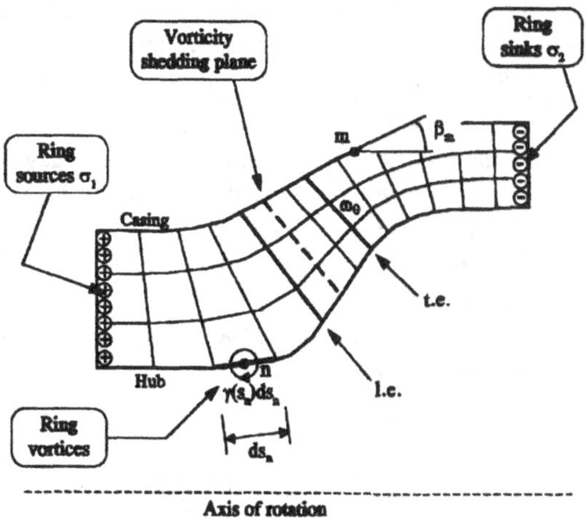

Figure 3 Grid for Vortex Modelling of meridional flow through a mixed-flow fan.

Attention should be drawn to the following:

1. The annulus is filled with a grid defined by surfaces of revolution upon which the blade-to-blade flows will be undertaken. Grid lines (quasi-normals) are draw from hub to casing to include the rotor leading and trailing edges l.e. and t.e..
2. The duct walls are covered with surface ring vorticity elements $\gamma(s)ds$ including both the inlet and exit planes. Thus the duct is totally enclosed with the standard boundary condition of zero efflux through the entire bounding surface.
3. The mass flow through the duct is provided by ring source and sink distributions just inside the duct at the inlet and exit planes.
4. Induced velocities due to the tangential vorticity are evaluated from the ω_θ values given at the grid centres by the above equations assuming the ring vorticity to be smearing uniformly over the cells.
5. This permits solution of the Martensen equation for flow through the annulus including the influence of both the source strength and tangential vorticity.
6. As part of an iterative procedure it is then necessary to calculate the meridional velocity at all grid centres and from this to evaluate the stream function ψ at the grid nodes. From this data equations(9) and (7) may then be used to re-estimate the distribution of tangential vorticity ω_θ.

The relevant boundary integral equation to achieve this may be stated as follows[6]

$$\oint K(s_m,s_n)\gamma(s_n)ds_n - \tfrac{1}{2}\gamma(s_m) + \textstyle\sum_{p=1}^{S_1} S_{pm}\sigma_1 + \sum_{q=1}^{S_2} S_{qm}\sigma_2 + \oiint \overline{K}_{km}\,\omega_\theta(k)ds_k dn_k = 0 \quad (12)$$

Although at first sight it does not seem very obvious, equation(12) is in fact exactly equivalent to the previously derived governing equation(4) but expressed in boundary integral form for surface vorticity modelling. It states that the velocity just inside the surface vorticity sheet adjacent to the annulus boundary at element m, parallel to the wall, is zero. This is equivalent to the standard Martensen equation including the induced velocity at m due to all the wall ring vortex elements $\gamma(s_n)ds_n$ but also including the velocities induced by the inlet ring sources σ_1 shown⊕ and exit ring sinks σ_2 shown⊖ and the distributed meridional ring vorticity ω_θ. Thus at a typical grid cell k the total vortex strength within the cell will approximate to $\omega_\theta(k)ds_k dn_k$ where ds_k and dn_k are the cell dimensions along and normal to the meridional grid lines. In reality this is only true for a curvilinear grid but the author's practice is to introduce up to ten sub-elements into the grid cells close to the wall element m in order to improve resolution. The grid illustrated in Figure 3 is of course diagrammatic and sparse in order to simplify presentation. In practice there would be say ten cells from hub to casing and twenty from inlet to outlet. The program MIXFLO[3] has been designed to interpolate additional grid cells from hub to casing if need be.

Now the velocity components at x_m, r_m induced by a ring vortex of unit strength $\Gamma=1$ located at x_n, r_n are as follows[6],

$$u_{\Gamma_{mn}} = -\frac{1}{2\pi r_n\sqrt{\bar{x}^2+(\bar{r}+1)^2}}\left\{K(k)-\left[1+\frac{2(\bar{r}-1)}{\bar{x}^2+(\bar{r}-1)^2}\right]E(k)\right\}$$

$$v_{\Gamma_{mn}} = -\frac{\bar{x}/\bar{r}}{2\pi r_n\sqrt{\bar{x}^2+(\bar{r}+1)^2}}\left\{K(k)-\left[1+\frac{2\bar{r}}{\bar{x}^2+(\bar{r}-1)^2}\right]E(k)\right\} \quad (13)$$

where K(k) and E(k) are complete elliptic integrals of the first and second kind and the dimensionless coordinates $\bar{x}$ and $\bar{r}$ are defined by

$$\bar{x} = (x_m - x_n)/r_n \quad \text{and} \quad \bar{r} = r_m/r_n \quad (14)$$

and the parameter k is given by

$$k = \sqrt{4\bar{r}/\left(\bar{x}^2 + (\bar{r}+1)^2\right)} \quad (15)$$

As shown elsewhere[6][7], the velocity components induced at x_m, r_m by a ring source of strength $\sigma=1$ at x_n and of radius r_n are given by

$$u_{\sigma_{mn}} = \frac{1}{2\pi r_n\sqrt{\bar{x}^2+(1+\bar{r})^2}}\left\{\frac{2\bar{x}E(k)}{\bar{x}^2+(\bar{r}+1)^2}\right\}$$

$$v_{\sigma_{mn}} = \frac{1}{2\pi r_n\bar{r}\sqrt{\bar{x}^2+(1+\bar{r})^2}}\left\{K(k)-\left[1-\frac{2\bar{r}(\bar{r}-1)}{\bar{x}^2+(\bar{r}-1)^2}\right]E(k)\right\} \quad (16)$$

The coupling coefficients $K(s_m,s_n)$ and S_{pm} in the main governing equation(12) represent the velocity parallel to the annulus surface at body point m due unit ring vortex and source singularities and are thus given by

$$K(s_m,s_n) = u_{\Gamma_{mn}}\cos\beta_m + v_{\Gamma_{mn}}\sin\beta_m$$
$$S_{pm} = u_{\sigma 1_{mn}}\cos\beta_m + v_{\sigma 1_{mn}}\sin\beta_m \quad (17)$$

and where $\bar{K}_{km}$ and S_{qm} take on the same form.

To conclude the numerical model the annulus surface is discretised into M elements determined by the grid. The related discretised governing boundary integral equation then becomes the set of M linear equations for the unknown surface vorticity values $\gamma(s)$,

$$\sum_{n=1}^{M} K'(s_m,s_n)\gamma(s_n) = -\sum_{p=1}^{S_1} S_{pm}\sigma_1 - \sum_{q=1}^{S_2} S_{qm}\sigma_2 - \oint \bar{K}_{km}\,\omega_\theta(k)\,ds_k dn_k \tag{18}$$

and the revised coupling coefficient is given by $K'(s_m,s_n)= K(s_m,s_n)ds_n$ where Δs_n is the length of wall element n. The only remaining items to be specified are the ring source and sink strengths and these are chosen to generate or remove the specified volume flow $\dot{Q}$, i.e.

$$\dot{Q} = S_1 \times \sigma_1 = S_2 \times \sigma_2 \tag{19}$$

where it is best to make S_1 and S_2 equal, with the same number of wall elements at the inlet and exit sections.

3. VORTEX ELEMENT MODELLING OF A BLADE-TO-BLADE FLOW

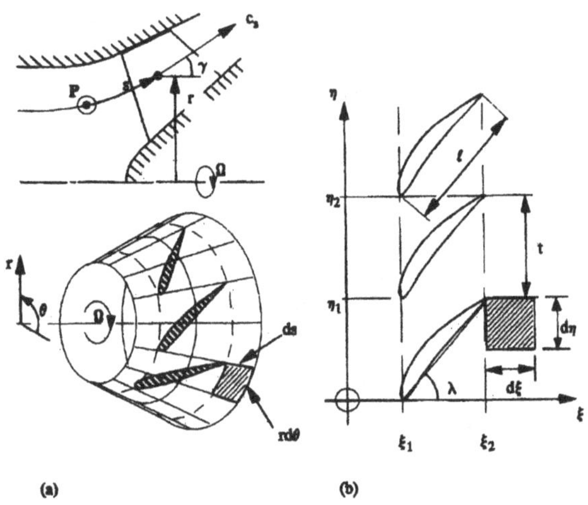

Figure 4 Transformation of a blade meridional flow intersection into a straight infinite cascade.

Since this subject is being dealt with in a second paper[8], only a brief overview will be given here. In particular we will consider the technique employed in the author's codes MIX and MIXFLO for the geometrical transformation of blade-to-blade geometry from a conical surface of revolution into an equivalent straight cascade in the cartesian ξ,η plane as illustrated in Figure 4. For conformality the equivalent elements $ds{\cdot}rd\theta$ and $d\xi{\cdot}d\eta$ must obey the relationship

$$\frac{d\xi}{d\eta} = \frac{ds}{rd\theta} = \frac{1}{r\sin\gamma}\frac{dr}{d\theta} \tag{20}$$

where γ is the local cone angle and s is measured along the meridional surface. Following L.Young[9] this may be achieved by the coordinate transformations

$$d\xi = \frac{ds}{r} = \frac{1}{r\sin\gamma}dr, \quad d\eta = d\theta \tag{21}$$

For numerical purposes these equations may be integrated to yield the direct coordinate relationships

$$\left.\begin{array}{l} \xi - \xi_1 = \int_{s_1}^{s} \frac{1}{r} = \int_{s_1}^{s} \frac{1}{r}\sin\gamma\, d\gamma \\ \eta - \eta_1 = \theta - \theta_1 \end{array}\right\} \tag{22}$$

The modified Martensen boundary integral equation when discretised becomes

$$\sum_{n=1}^{M} K_{mn}\gamma(s_n) = -U_\infty \cos\beta_m - \left(V_\infty + \Omega\left\{r^2 - \tfrac{1}{2}(r_1^2 + r_2^2)\right\}\right)\sin\beta_m \tag{23}$$

where K_{mn} is the cascade coupling coefficient

$$K_{mn} = \frac{\Delta s_n}{2t}\left\{\frac{\sin\frac{2\pi}{t}(y_m-y_n)\cos\beta_m - \sinh\frac{2\pi}{t}(x_m-x_n)\cos\beta_m}{\cosh\frac{2\pi}{t}(x_m-x_n) - \cos\frac{2\pi}{t}(y_m-y_n)}\right\} \tag{24}$$

and the right hand side of equation(23) contains the extra term involving the speed of rotation Ω which accounts for the relative eddy effects. In fact the numerical application of this analysis is fairly complex involving procedures to ensure zero implied vorticity within the blade profile interior region. A full exposition is given in ref.(6), including the introduction of further components to the r.h.s. to correct for change in thickness of the meridional stream surface.

4. SAMPLE CALCULATIONS.

All of the foregoing analyses for vortex element modelling of both the meridional and cascade flows have been implemented in the author's code MIXFLO and the remainder of this paper will be committed to examples of the application of this to several test cases. The first two will provide bench mark tests of the two analyses against exact solutions, sections 4.1 and 4.2 and the third will provide illustrative output for the flow through a mixed-flow fan in section 4.3.

4.1 Meridional analysis for an axial fan.

The first example is that of a single stage fan for which the swirl velocity downstream of the rotor is to be a mixture if free-vortex and forced vortex as follows.

$$c_\theta = \frac{a}{r} + b\,r \tag{25}$$

The chosen grid geometry for this is shown in Figure (5) which is taken from a screen presentation from the author's code MIXFLO[3] for the Test case MVFAN.

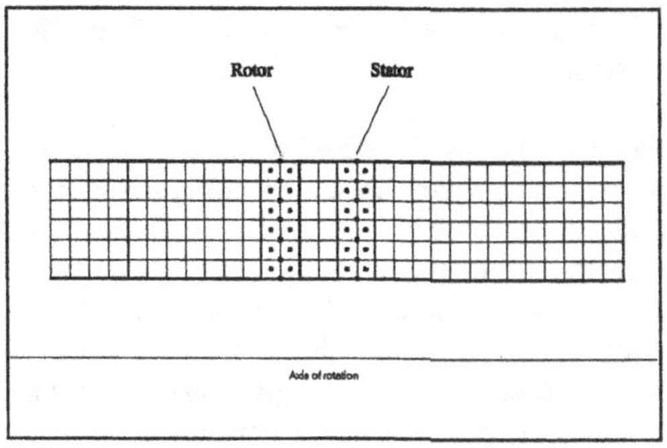

Figure 5 Grid specification for an axial fan.

The design data are as follows.

1 The mean meridional velocity is to be $c_s = 1.0$m/s corresponding to a flow rate of 19.63495 m^3/s.
2 The rotor speed is to be 25.078 revs/min.
3 The rotor is to be designed to deliver a swirl velocity distribution c_θ conforming to equation(25) and as specified in Table 1 below.
4 The stator is to deliver modest residual swirl as prescribed in Table 1 below.
5 The prescribed rotor exit swirl was chosen to deliver an acceptable stage duty[1] of $\phi=0.5$, $\psi=0.3$ at the rms radius $r_{ms}=0.762$m.

Table 1 - Specification of exit conditions for rotor and stator of the mixed-vortex fan.

Radius r	Rotor exit swirl $c_{\theta 2}$ m/s	Stator exit angle β_2°
0.4	-0.63027	-10.0
0.5	-0.59200	-9.0
0.6	-0.58275	-8.0
0.7	-0.59008	-7.0
0.8	-0.60776	-6.0
0.9	-0.63235	-5.0
1.0	-0.66178	-4.0

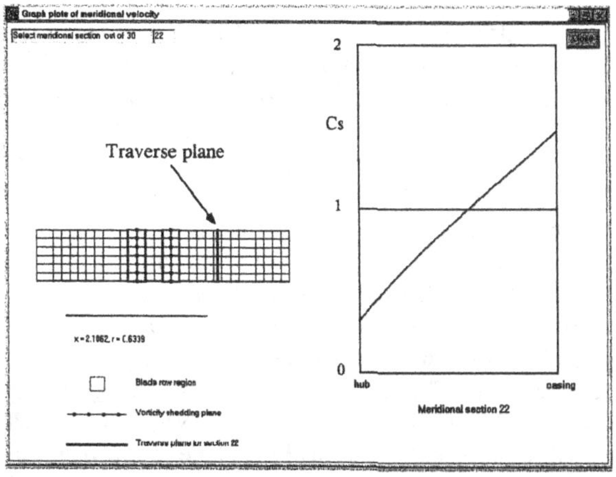

Figure 6 MIXFLO screen presentation of predicted meridional velocity downstream of the axial fan.

As shown in Figure 6, which is taken directly from the screen presentation of the program MIXFLO, the predicted meridional velocity C_s some distance downstream varies considerably from hub to casing, indicating the strong influence of the distributed vorticity ω_θ shed from the rotor in particular. Although no exact solution is available to check this numerical result it is possible to modify the statement of the problem to fit the condition of radial equilibrium for which the exact solution may be obtained from the well known radial equilibrium equation,

$$\frac{1}{r}\frac{dp_o}{dr} = c_x\frac{dc_x}{dr} + \frac{c_\theta}{r}\frac{d(rc_\theta)}{dr} \tag{26}$$

Since c_θ is prescribed through equation(25) and Table 1, and p_o is given in terms of c_θ by the Euler pump equation(10), the radial equilibrium equation may be integrated directly to obtain the axial velocity profile c_x which, for radial equilibrium, is equal to the meridional velocity C_s. The implication here is that the meridional flow is purely cylindrical and has no radial velocity components c_r. Even though we have chosen a cylindrical annulus for this fan there will of course have to be radial velocity components c_r present close to the blade rows to accommodate the meridional streamline shifts associated with the meridional velocity variations which we see are quite strong. However we can create a slightly artificial situation as illustrated in Figure 7 below in which the axial chord of the rotor is stretched out axially in order to spread the vortex shedding and minimise the related radial velocities. This is enhanced if we also assume that the tangential vorticity ω_θ is assumed to be shed at the rotor leading edge and the stator is placed downstream as far as possible. Radial equilibrium is then approximated by the meridional flow at the rotor trailing edge.

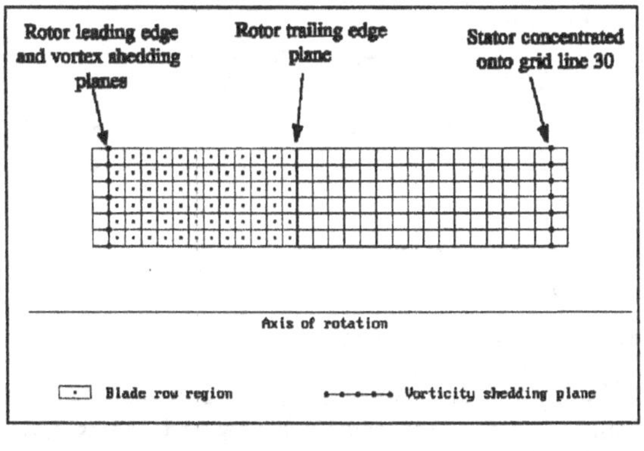

Figure 7 Grid model to simulate radial equilibrium.

To conclude this section the meridional and velocity and swirl angle at the rotor trailing edge predicted by MIXFLO using the above vortex element numerical analysis are compared with those predicted by radial equilibrium theory in Table 2 below. The results are most encouraging bearing in mind that only six grid cells were used between hub and casing to discretise the distributed vorticity ω_θ.

Table 2 - Comparison of MIXFLO output with solution of the radial equilibrium equation for flow downstream of axial fan rotor.

	Output from MIXFLO for rotor t.e. plane		Solution of radial equilibrium equation	
r	c_x	α_2	c_x	α_2
0.4	0.60910	-45.906	0.60917	-45.975
0.5	0.69310	-40.430	0.69491	-40.428
0.6	0.80597	-35.799	0.80642	-35.853
0.7	0.93111	-32.298	0.93224	-32.332
0.8	1.06130	-29.735	1.06611	-29.686
0.9	1.19774	-27.772	1.20465	-27.696
1.0	1.34482	-26.144	1.34601	-26.182

4.2 Slip Factors For Radial Machines

For the second example let apply the mixed-flow cascade analysis given in section 3 to centrifugal radially bladed compressors in order to check its ability to predict the slip factor in comparison with the well known Busemann exact analysis[10]. Making use of the blade-to-blade analysis part of the code MIXFLO[3], predicted slip factors for a wide range of geometries are as illustrated in Figure 8. Also shown are two results predicted by Stanitz[11] using a finite difference method. The present numerical analysis is thus in extremely good agreement with Busemann's important bench-mark results.

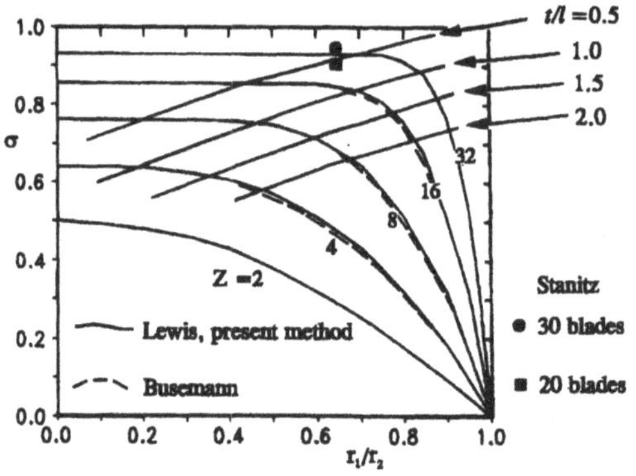

Figure 8 Slip factor for radially bladed rotors.

Slip factors are shown here for a wide range of radially bladed rotors with Z blades and radius ratio r_1/r_2. The numerical results were calculated with a prewhirl velocity $c_\theta = r_1\Omega$ equal to inlet blade speed to provide smooth inflow. Also superimposed on the curves are contours of constant pitch/chord ratio t/l in the transformed ς plane, revealing one advantage of the present transformation method. Thus it is revealed by these results that for t/l values<1.0 the slip factors level off to a constant value, a very significant design observation.

4.3 Design/Analysis Of Mixed-Flow Fans

The mixed-flow fan, illustrated in Figure(9) from an actual design test case[3], involves a complex meridional flow due to both the swan neck annulus shape and also the presence of Coriolis body forces within the rotor due to the radial streamline shift. This is thus the most complex of turbomachinery design problems but one to which vortex element analysis is well suited.

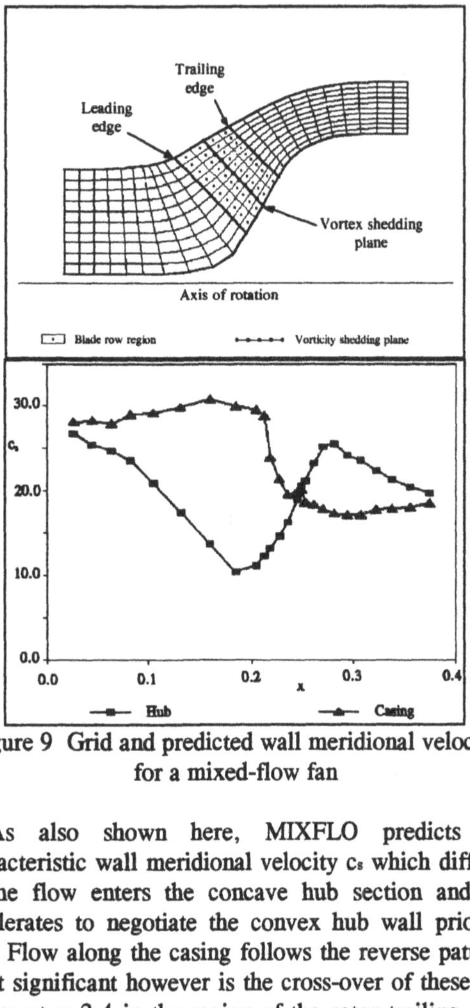

Figure 9 Grid and predicted wall meridional velocity for a mixed-flow fan

As also shown here, MIXFLO predicts the characteristic wall meridional velocity c_s which diffuses as the flow enters the concave hub section and the accelerates to negotiate the convex hub wall prior to exit. Flow along the casing follows the reverse pattern. Most significant however is the cross-over of these two curves at $x \approx 2.4$ in the region of the rotor trailing edge where it is helpful to have fairly uniform meridional velocity. It is also clear from the predicted data that there will be strong acceleration of c_s through the rotor hub blade section which will help to stabilise the blade-to-blade flow. This also results in a large change in the meridional streamline thickness or AVR value which strongly influences the blade-to-blade aerodynamics. Conversely however there are strong meridional diffusions through the casing section which will add to the blade-to-blade duties as a diffuser. This particular fan was designed with a blade loading close to a free-vortex and from the evidence of the axial fan considered in section 4.1, it is clear that a shift here to mixed-vortex loading, equation(25), would offer some advantage in using the deliberate shedding of tangential vorticity ω_θ in order to reduce the casing diffusion.

Table 3 Blade cascade design parameters.

Section	Stagger $\lambda°$	Camber $\theta°$	η_{tt}
1. Hub	40.20	25.0	0.968
2. Mean	56.35	10.0	0.939
3. Casing	58.75	0.0	0.903

The cascade stagger and camber values, as recorded in Table 3, were chosen to produce shock-free inflow and the evidence that this was achieved is provided by the predicted surface pressure distribution for the mid section shown in Figure 10.

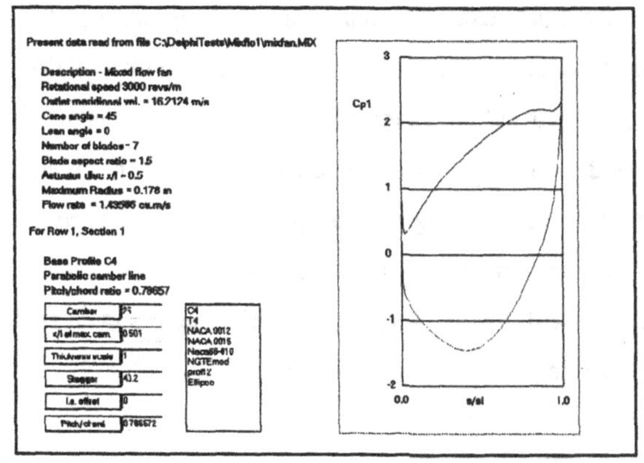

Figure 10 Predicted blade surface pressure distribution for the mid section.

In addition the computer code MIXFLO can be used to view the blade geometry, views of which are given in Figures 11 and 12 below.

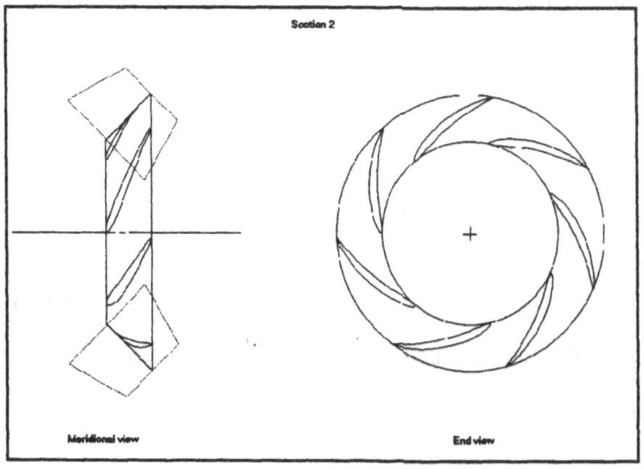

Figure 11 View of the mid blade-to-blade section.

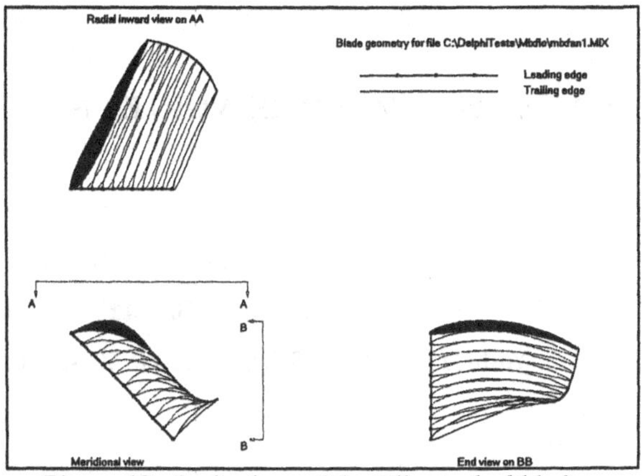

Figure 12 Projections of the fully stacked blade.

Although the original design comprised three sections only, a total of eleven sections were then obtained by interpolation resulting in the blade geometry shown in Figure 12 which is stacked on the leading edge. MIXFLO provides some freedom to adjust the blade stacking if needs so dictate.

(1)Lewis, R.I. (1996), Turbomachinery Performance Analysis. W.Arnold.

(2)Horlock, J.H. (1958), Axial Flow Compressors. Fluid Mechanics and Thermodynamics. Butterworth.

(3)Lewis, R.I. (1999), User instructions for MIXFLO, a computer program for the design and analysis of axial, radial and mixed-flow turbomachines, cascades and blade rows.

(4)Martensen,E. (1959), Berechnung der Druckverteilung an Gitterprofilen in ebener Potentialstromung mit einer Fredholmschen Integralgleichung. Arch. Rat. Mech., Anal., **3**, pp. 235-270.

(5)Wilkinson, D. (1967), A numerical solution of the analysis and design problems for the flow past one or more aerofoils or cascades. ARC R&M, No. 3545.

(6)Lewis, R.I. (1991), Vortex Element Methods for Fluid Dynamic analysis of Engineering Systems. Engine Technology Series, No. 1, Cambridge University Press.

(7)Ryall, D.L. and Collins, I.F. (1967), Design and test of a series of annular aerofoils. Min. Tech., A.R.C. R&M. No. 3492.

(8)Lewis, R.I. (1999) Vortex Element Methods, the most natural approach to flow simulation. Proc. of the First Int. Conf. On Vortex Methods, Kobe, Japan.

(9)Young, L. (1958) Runners of experimental turbo-machines. Engineering, London, **185**, p.376.

(10)Busemann, A. (1928), Das Förderhöhenverhältnis radialer Kreiselpumpen mit logarithmischspiraligen Schaufeln. Z, Angew. Math. Mech., **8**(5), 372-384.

(11)Stanitz, J.D. (1952) Some theoretical aerodynamic investigations of impellers in radial and mixed-flow centrifugal compressors. Trans. ASME, **74**, No. 4.

SOME APPLICATION OF VORTEX METHOD TO WIND ENGINEERING PROBLEM

Hiromichi Shirato and Masaru Matsumoto
Department of Global Environment Engineering, Kyoto University
Yoshida Hon-machi, Sakyo-ku, Kyoto 606-8501, Japan / E-mail:shirato@brdgeng.gee.kyoto-u.ac.jp

ABSTRACT

Some improvement is applied to the conventional 2-D discrete vortex method in order to observe its performance in reproducing viscous flow around bluff body, and to investigate the applicability to the pressure field simulation. Furthermore, the vortex method will be used as one of the tools for visualization technique based on surface pressure information measured in wind tunnel experiments.

1. INTRODUCTION

Robustness, pressure-independence, mesh-free handling are the advantage of the vortex method. On the contrary, the treatment of viscosity, overestimation of induce velocity between adjacent vortex elements should be improved. In this study, some trials are attempted to the conventional 2-D discrete vortex method in order to discuss its applicability to wind engineering problems.

2. REYNOLDS NUMBER DEPENDENCE OF FLOW PATTERN AROUND BLUFF BODY [1]

2.1 Computational Procedure

The numerical viscosity induced by the VIC method is quantified and some modification to reduce it is developed. Its effectiveness is confirmed by applying this modified method to the simulation of flow around 2-D rectangular cross section for various Reynolds number.

Vortices are placed on the body surface, whose strength is determined by the condition of zero-normal velocity together with the Kelvin's theorem.[2] All surface vortices are released simultaneously to the surrounding fluid in every time step and convecting with the flow velocity at each location of a vortex. The procedure is based on the "time-splitting method"[3], which decomposes the original transport equation of vorticity into convection and diffusion sub-step, respectively. Once the diffusion process of vorticity field is evaluated by solving the following equation,

$$\frac{\partial \omega}{\partial t} = \nu \nabla^2 \omega \tag{1}$$

where, ω : vorticity, ν : kinematic viscosity of fluid

then, the velocity components at each mesh point are calculated by solving the Poisson's equation:

$$\nabla^2 \psi = -\omega \tag{2}$$

where, ψ : stream function

The local convecion velocity of a vortex is evaluated by the conventional area-weightening manner (see Fig.1):

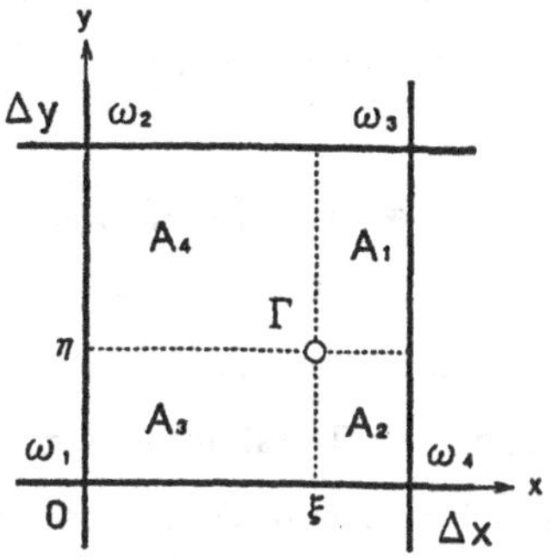

Fig. 1 Area-weightening for a vortex element in a cell

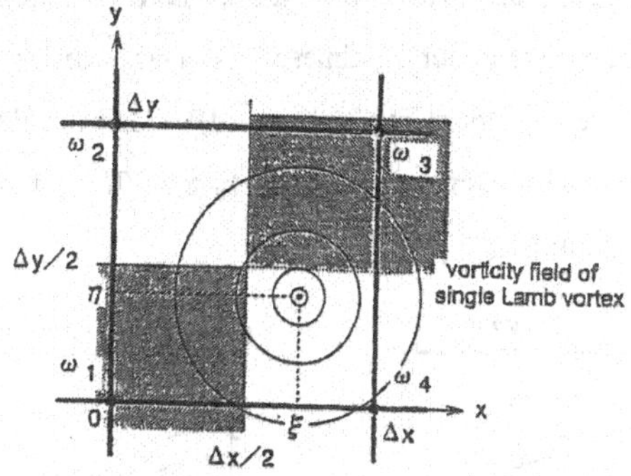

Fig.2 Lamb vortex in a cell

$$\vec{u}_k = \sum_{i=1}^{4} \vec{u}_i A_i / \Delta x / \Delta y \tag{3}$$

where, $\vec{u}_k$: local convection velocity at the center of vortex k, $\vec{u}_k=(u_k, v_k)$, $\vec{u}_i$: velocity at mesh point i, $\vec{u}_i=(u_i, v_i)$

Location of each released vortex in the next time step is elavulated by taking the rate of velocity change into account:

$$\vec{x}_k^{(n+1)} = \vec{x}_k^{(n)} + \left\{ \frac{3}{2}\vec{u}_k^{(n)} - \frac{1}{2}\vec{u}_k^{(n-1)} \right\} \Delta t \tag{4}$$

where, $\vec{x}_k^{(n)}$: position of vortex k at time $t=n\Delta t$, $\vec{x}_k^{(n)}=(x_k^{(n)}, y_k^{(n)})$

2.2 Numerical Viscosity Evaluation and Modified VIC Method

The conventional VIC method is to convert the circulation of one point vortex Γ to the vorticity of the surrounding four-grid points, ω_1, ω_2, ω_3, ω_4 by the area-weightening technique. The assigned vorticity, ω_1, in Fig.1 by VIC method is expressed as:

$$\omega_1 = \Gamma(1 - \frac{\xi}{\Delta x})(1 - \frac{\eta}{\Delta y}) / \Delta x / \Delta y \tag{5}$$

It is pointed out that this method induces numerical viscosity to some extent. This can be approximately quantified by comparing with the diffusive expansion of single Lamb vortex. Assume one Lamb vortex at (ξ , η) in a cell (see Fig.2), whose vorticity distribution is:

$$\omega(r,t) = \frac{\Gamma}{4\pi\nu t} \exp\left\{ \frac{r^2}{4\pi\nu t} \right\} \tag{6}$$

where, t: time after its origin

The vorticity origined at t=0 diffuses acccording to the spatial growth of the standard deviation:

$$\sigma_\omega(t) = \sqrt{2\nu t} \tag{7}$$

At $t=\Delta t$, the most appropriate way to convert $\omega(r, \Delta t)$ field to the vorticity at the surronding four grid points would be assigning each fraction of vorticity field divided by the line $x=\Delta x/2$, $y=\Delta y/2$ to the nearest grid point (see Fig.2). Namely, for ω_1:

$$\omega_1 = \int_{-\infty}^{\Delta x/2} \int_{-\infty}^{\Delta y/2} \omega(r, \Delta t) dA / \Delta x / \Delta y \tag{8}$$

here,

$$\sum_{i=1}^{4} \omega_i = \iint \omega dA / \Delta x / \Delta y = \Gamma / \Delta x / \Delta y \tag{9}$$

Fig.3 plots the isoline of the assigning rate for $\omega_1/(\Gamma/\Delta x/\Delta y)$, with various $\sigma_\omega(\Delta t)/\Delta x$ together with that of VIC method (eqs.(5),(8)). By comparing the assigning rate, it concluded that the numerical viscosity by the VIC method per Δt is approximately equal to the diffusion of single Lamb vortex during the standard deviation σ_ω growing up to about 0.3 times of the mesh size Δx, when $\Delta x = \Delta y$ (Compare Fig.3(b) and (d)). Consequently, the numerical viscosity ν_{num} per Δt by the VIC method can be evaluated as:

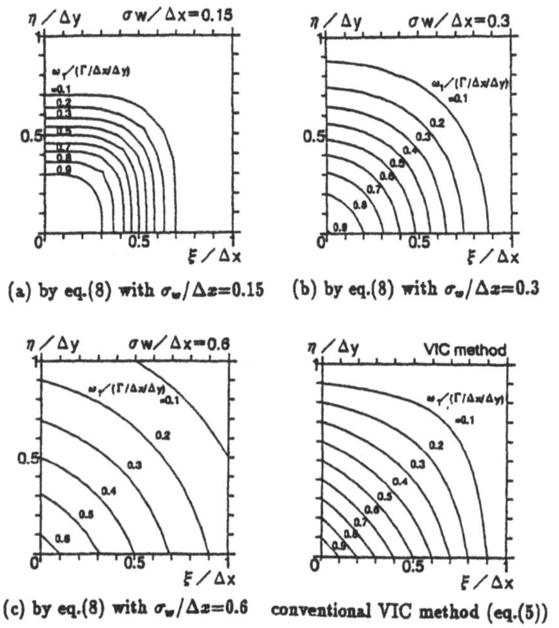

(a) by eq.(8) with $\sigma_\omega/\Delta x$=0.15 (b) by eq.(8) with $\sigma_\omega/\Delta x$=0.3

(c) by eq.(8) with $\sigma_\omega/\Delta x$=0.6 conventional VIC method (eq.(5))

Fig.3 Isoline of assigning rate for ω_1

(assigning rate = $\omega_1/(\Gamma/\Delta x/\Delta y)$)

$$\nu_{num} = \frac{\sigma_w^2}{2\Delta t} = \frac{0.09(\Delta x)^2}{2\Delta t} \tag{10}$$

Following the same approach as before, the proposed modified VIC method is derived by use of an inviscid Lamb vortex, in which the standard deviation is kept to zero. That is, the vorticity field is expressed by a delta function, $\omega(x,y) = \Gamma\,\delta(x-\xi)\,\delta(y-\eta)/\Delta x/\Delta y$. Using eq.(8), the assigned vorticity, ω_1, can be formulated as follows:

$$\omega_1 = \Gamma\{1 - h(\xi - \Delta x/2)\}\{1 - h(\eta - \Delta y/2)\}/\Delta x/\Delta y \tag{11}$$

where, h(x)=0(x<0), 0.5(x=0), 1.0(x>0)

Eq.(11) represents the assigned vorticity ω_1 for an inviscid vortex, which gives the minimum numerical viscosity.

2.3 Application to Flow Around Bluff Body

The modified procedure in previous section is applied to the flow around a 2-D rectangular cross section for various Reynolds number. The slenderness ratio, B/D (B: width, D: depth of cross section), is chosen as 2. The computational domain, 20D in width and 21B in length, is covered be the regular mesh with $\Delta x = \Delta y = 0.125$D. The center of the rectangular section is placed 5.5B downstream from the inlet. Vortices are distributed on the body surface with the pitch Δx. After determined the circulation of these vortices by the zero-normal velocity condition, all the surface vortices are released simultaneously to the surrounding fluid. The convecting velocity of the released vortex is equivalent to the velocity being apart from the body surface with small distance, 0.025D. If a vortex penetration through the body surface is predicted during the next convection sub-step, the motion is modified so that the velocity component normal to the body surface is nullified, while the tangential component is maintained.[4]

To change the Reynolds number, the usage of another value of ν or Δt in the diffusion sub-step may be the alternative, when the time-splitting technique is employed. Computation for various Reynolds number is performed by manipulating Δt_d in the diffusion sub-step in this study. According to eq.(8), Δt gives the same order of effect to σ_ω as ν, then Δt can be variable depending on the target Reynolds number in the diffusion sub-step. That is, $\Delta t_d = (Re_0/Re_{target})\,\Delta t_c$ ($Re_0 = UD/\nu = 300$ in this study). In the convection sub-step, Δt is chosen from the optimal range, $\Delta t_c = 0.1$B/U according to the literature. The target Reynolds number, Re_{target}, is chosen as 100, 300, 800, 3000, and ∞. The diffusion sub-step is by-passed in the last case. Eq.(2) and fully implicit form of eq.(1) are solved by the SOR method. Lift force on the rectangular cross section is evaluated by the extended Blasius formula for non-stationary flow.

$$Lift = -Imag\left\{\frac{i\rho}{2}\oint_C\left(\frac{df}{dz}\right)^2 dz + i\rho\frac{\partial}{\partial t}\oint_C \bar{f}d\bar{z}\right\} \quad (12)$$

where, f: complex potential, f= ϕ +i ϕ , ϕ : velocity potential, $\bar{()}$: conjugate operator

The wake pattern of a rectangular cross section is sensitively modulated by the separated flow reattachment, and, correspondingly, the discontinuity of Strouhal number St (=f_sD/U, f_s: vortex shedding frequency) exists at B/D=2.8 and 6.0. Even for the specified slenderness ratio, B/D=2, the discontinuity of St is also recognized at Re=500. Flow changes from the reattachment type to the fully separated type at this critical Reynolds number (Re=500). Fig.4 plots St for various Re together with the values obtained from wind tunnel experiment[5]. St by the computation is evaluated from the time history of lift force. Power spectra of the lift force for each Re are also shown in the figure. The spectral peak is not so sharp as

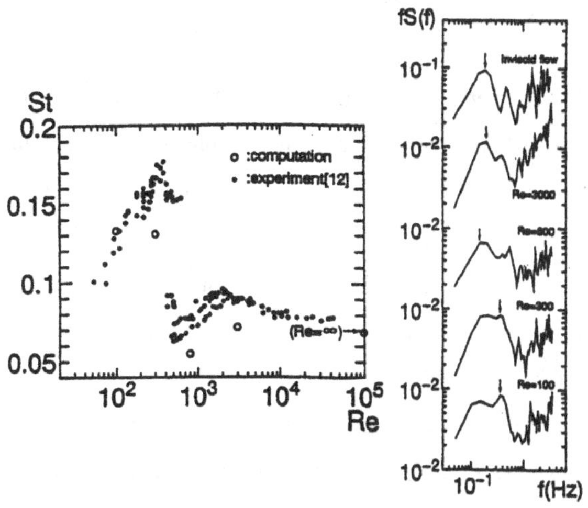

Fig.4 Variation of St with Re

(●: computation, ○: experiment[5])

the experiment. Although most of St by the computation gives a bit of smaller value than that from the experiment, it can be concluded that St jump at Re=500 is reproduced by the computation.

3. UNSTEADY PRESSURE ON MOVING BODY SURFACE[6]

3.1 Unsteady Pressure Properties

For this rectangular section (B/D=2), the movement-induced type vortex-induced oscillation can be observed in the reduced wind velocity Vr=U/fD=5. (f: vibrational frequency of cross section in heaving mode) Furthermore, the pressure convection towards the trailing edge has been confirmed by the experiments. This nature is caused by the generation of the separated vortex due to body movement and its traveling toward downstream with the convection velocity. In the movement-induced type vortex-induced oscillation, the traveling time duration for one separated vortex generated at the leading edge reaching to the trailing edge is exactly one natural period of heaving body oscillation. This property is one of the major condition for the occurrence of movement-induced type vortex-induced oscillation.[7]

3.2 Pressure Evalution by Vortex Method

The unsteady pressure fluctuation on the body surface is evaluated by the unsteady Bernoulli's formula:

$$C_p = \frac{p - p_0}{\frac{1}{2}\rho U^2} = 1 - \frac{u^2 + v^2}{U^2} - \frac{\frac{\partial\phi}{\partial t}}{\frac{1}{2}U^2} \quad (13)$$

where, p: pressure on body surface, p_0: static pressure, u, v: local wind velocity components on body surface

To clarify whether the simulated pressure contains the convection nature, the heaving frequency component is extracted from the "row" pressure signal and the phase lag to the body displacement is evaluated at each point on the side surface. The time delay of the negative peak of pressure with heaving frequency component to the upper-most heaving displacement can be evaluated by the cross-correlation between "row" pressure signal and heaving displacement.

Fig.5 shows the RMS distribution of C_p on the side surface. Observed data in wind tunnel experiments [7][8] and numerical simulation by DNS [9] are plotted together. Present calculation agrees with the other data to a certain extent. It should be noted that there are two distinct peaks in measured RMS distribution of C_p. The upstream side and the second peak correspond respectively to the development of movement-induced vortex, and the periodic expansion of separation bubble. The present calculation gives only monotonous increase of C_p along side surface.

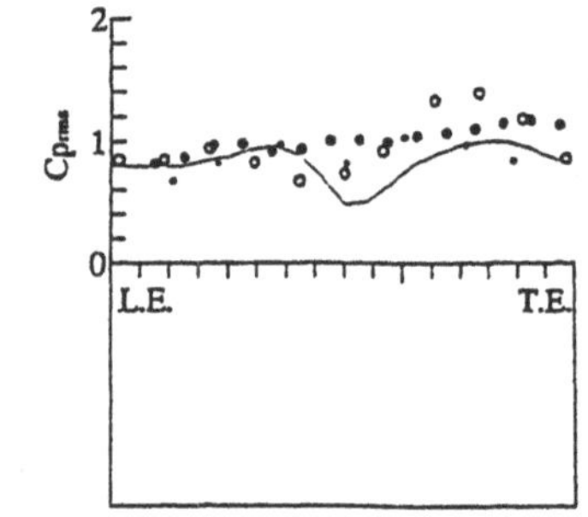

Fig.5 RMS distribution of C_p on body side surface

●: present calculation ($U/fD=5$, $y_0/D=0.1$),

○: experiment ($U/fD=4.95$, $y_0/D=0.05$ [7]),

• : experiment ($U/fD=5.5$, $y_0/D=0.1$ [8])

—: DNS ($U/fD=5$, $y_0/D=0.114$ [9])

Fig.6 plots the phase lag between pressure and body displacement. The simulated result has a sudden drop from nearly 0deg. to 360deg. at the mid point on side surface, which means no tendency of pressure convection, but that the pressure fluctuation of heaving frequency component shows almost in-phase anywhere on side surface.

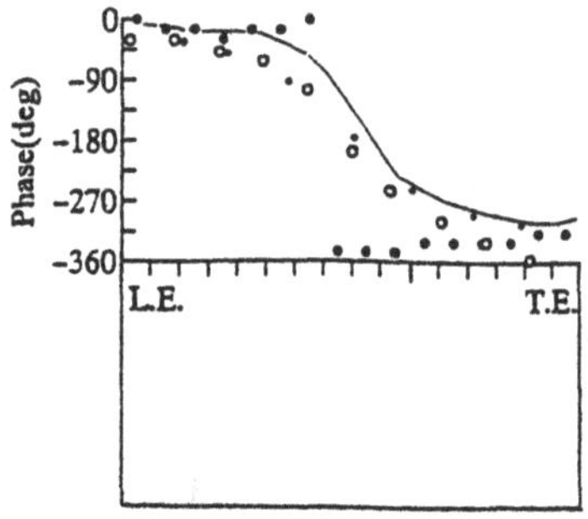

Fig.6 Phase lag distribution of unsteady pressure on body side surface (legend as in Fig.5)

3.3 Discussion on Phase Characteristics Evaluation

In the unsteady Bernoulli's formula, the term on time varying rate of the velocity potential, $\partial\Phi/\partial t$, can be expressed in terms of complex potential function as follows:

$$\frac{\partial\Phi}{\partial t} = Real[\frac{\partial f}{\partial t}] = (\frac{\partial\Phi}{\partial t})_1 + (\frac{\partial\Phi}{\partial t})_2 + (\frac{\partial\Phi}{\partial t})_3 \qquad (14)$$

where,

$$(\frac{\partial\Phi}{\partial t})_1 = -\sum_j \frac{\dot{\Gamma}_j}{2\pi}\arg(z - z_j) - \sum_l \frac{\Gamma_0^l}{2\pi\Delta t}\arg(z - z_0^l) \qquad (14a)$$

$$(\frac{\partial\Phi}{\partial t})_2 = \sum_j \frac{\Gamma_j}{2\pi}\frac{\dot{y}_j(x - x_j) - \dot{x}_j(y - y_j)}{|z - z_j|^2} \qquad (14b)$$

$$(\frac{\partial\Phi}{\partial t})_3 = \sum_l \sum_k \frac{\Gamma_k^l}{2\pi}\frac{\dot{y}_k^l(x - x_k^l) - \dot{x}_k^l(y - y_k^l)}{|z - z_k^l|^2} \qquad (14c)$$

The first term in the equation describes the time varying rate of bound vortex circulation, while the second term reflects the effect of body movement, and the third term corresponds to the shedding vortices convection.

Fig.7 shows the phase distribution on side surface evaluated from eq.(14c), the third term. It gives more realistic result, that is, the phase lag is increased towards the trailing edge due to the convecting nature of separated flow. It is understood that the phase lag is caused by the convection of the movement-induced vortex generated at the leading edge. Hence, the third term plays most important role to reproduce phase characteristics.

In order to improve the prediction of pressure convection, the first term should be reduced, because the magnitude of the first term is strongly related to the time step, Δt, which has no direct relationship with the pressure evaluation.

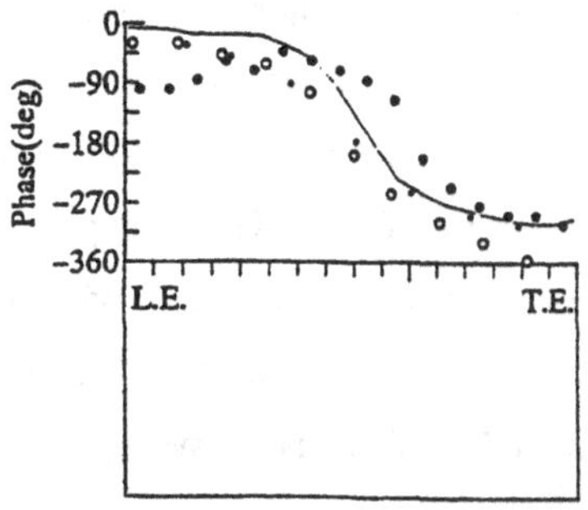

Fig.7 Phase lag distribution by the third term, $(\partial \Phi / \partial t)_3$ (legend as in Fig.5)

4. VORTEX METHOD AS FLOW VISUALIZATION TOOL[10]

There are some technical difficulties when flow visualization in relatively higher wind velocity is required in wind tunnel. Unsteady flow pattern around an oscillating bluff body is simulated by substituting observed surface pressure data in wind tunnel experiments into vortex method.

4.1 Computational Procedure

Based on the idea as shown in Fig.8, the strength of bound vortices is determined by the surface pressure data measured in wind tunnel experiments as follows:

$$\Gamma_j(t) = \bar{U}\sqrt{1 - \alpha Cp_j(t)}\,\Delta s \qquad (15)$$

where, $\Gamma_j(t)$: strength of j-th bound vortex on body surface at time t, Cp_j (t): pressure coefficient at time t in the position of j-th bound vortex, Δs: distance between neighboring bound vortices, α : reduction factor.

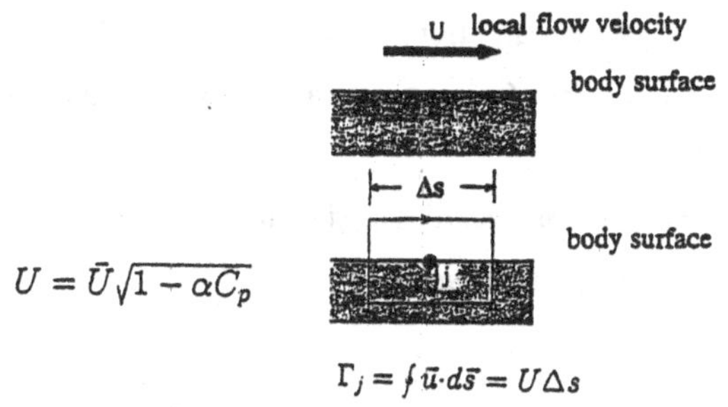

Fig.8 Relationship between surface pressure and bound vortex strength

In exact sense, the relationship in eq.(15) should contain the time derivative of velocity potential according to the unsteady Bernoulli's formula. The previous section concludes that the time derivative of velocity potential plays most important role on surface pressure evaluation. Especially, the term related to the movement of released vortices determines the convective nature of surface pressure on body side surface in state of heaving oscillation. The reduction factor introduced in eq.(15) is to compensate the effect of omitting the time derivative of velocity potential, taking 0.5 in this study. However, the validity should be further investigated in the future.

A 2-D rectangular cross section of B/D = 2 in heaving oscillation is again focused. The cross section keeps its vibrational amplitude y_0/D = 0.1, the reduced wind velocity Vr is chosen as 10. The unsteady pressure on side surface were measured under the above condition in the wind tunnel experiment.[11] All pressure signals were filtered to get the single frequency component whose frequency is f_0. The filtered pressure signals, therefore, fluctuate sinusoidally with a certain phase lag to body displacement as follows:

$$Cp_j(t) = \overline{Cp_j} + \widetilde{Cp_j}\cos(2\pi f_0 t - \theta) \qquad (16)$$

where, $\overline{Cp_j}$: average pressure coefficient at j-th bound vortex on upper side surface, $\widetilde{Cp_j}$: fluctuating pressure coefficient at j-th bound vortex on upper side surface, θ : phase lag from neutral position of heaving displacement (moving upwards) to negative peak pressure.

The pressure quantities on side surface mesured in wind tunnel experiments are shown in Fig.9.

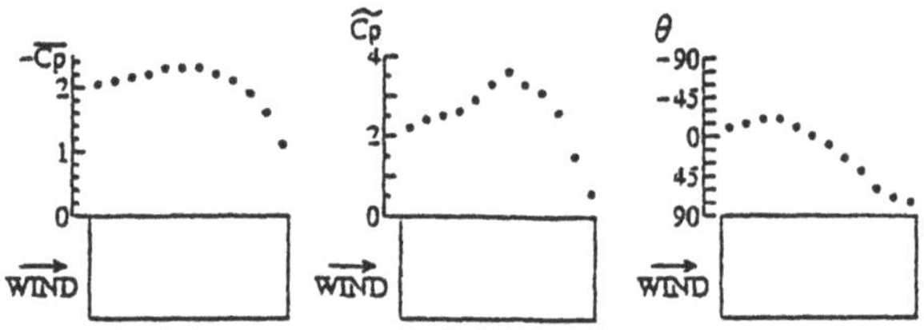

Fig.9 Observed surface pressure for B/D=2 rectangular cross section [11]

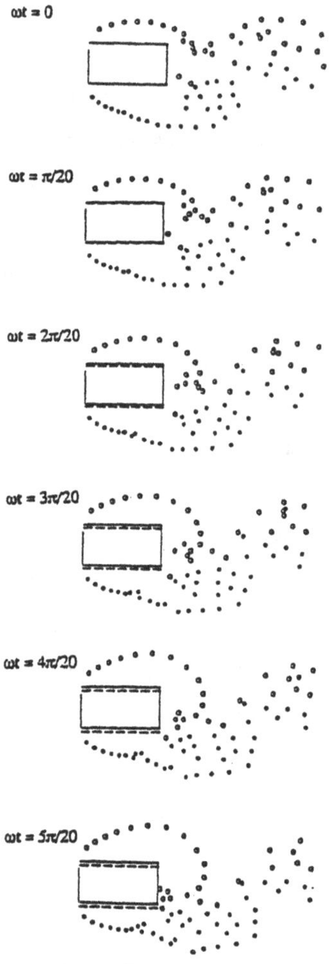

Fig.10 Simulated result

4.2 Simulated Result

Fig.10 shows the simulated result by the proposed technique. Instantaneous position of released vortices and body displacement is plotted. Each small circle represents a released vortex. The open circle indicates clock-wise circulation, whereas the solid one counter clock-wise. The released vortices moving downstream by the separating flow tends to approach to the trailing edge without reattachment on the side surface, then to assemble to form larger scale vortex in the wake region near the rear surface of body. This flow pattern repeats every natural period synchronizing with the body oscillation. These unsteady behavior of flow pattern agrees qualitatively with the phase averaged stream lines around an oscillating 2-D rectangular section with the same condition based on the measured local velocity distribution around the section in the wind tunnel experiment.[12]

5. CONCLUSION

Some concluding remarks are pointed out as follows:

1) The discrete vortex method with the modified vortex-in-cell (VIC) scheme is performed to simulate the flow around a stationary 2-D rectangular cross section (slenderness ratio B/D=2).
2) The numerical viscosity induced by the VIC method is evaluated by comparing with the diffusion of a single Lamb vortex. By use of an inviscid vortex, some modification is proposed to the VIC method, which can reduce the numerical viscosity.
3) Applying the proposed method (modified VIC nethod) to the flow around a 2-D rectangular cross section, the change of flow pattern due to reattachment of separated flow is reproduced. The discontinuity of Strouhal number at Re=500 gives qualitative agreement with the experimental result.
4) The unsteady pressure on an oscillating rectangular 2-D cross section with B/D=2 is simulated by the vortex

method. In case of convection dominating flow, such as focused in this study, the third term of the time varying rate of velocity potential function should be most important.

5) This observation gives the way to tune-up the unsteady pressure prediction by introducing artificial reduction factor to the first term of time varying rate of velocity potential in the unsteady Bernoulli's formula.

6) The unsteady flow pattern was predicted by the surface pressure data obtained in wind tunnel experiments with the aid of the vortex methods. The predicted flow pattern agrees qualitatively with the visualized pattern based on wind tunnel experiment.

7) The validity of conversion to the strength of bound vorticity from surface pressure should be verified by applying this proposed method to other conditions and other cross sections.

References

[1] H.Shirato, M.Matsumoto, N.Shiraishi; Viscous flow sumilation around bluff body by vortex method, Journal of Wind Engineering and Industrial Aerodynamics, vol.46&47, 1993, pp.371-379

[2] H.Sakata, T.Adachi, R.Inamuro; Simulation method for unsteady flow with separation by discrete vortex method, Proceedings of the JSME, vol.49, No.440, 1983

[3] A.J.Chorin; Vortex sheet approximation of boundary layers, Journal of Computational Physics, vol.27, 1978

[4] N.Arai, K.Taguchi, T.Tani; Flow simulation around twin rectangular prisms in side-by-side arrangement by discrete vortex method, Proceedings of the JSME, vol.53, No.486, 1987

[5] A.Okajima; Flow around a rectangular cylinder with a section of various width/height ratios, Journal of Wind Engineering, No.17, 1983

[6] H.Shirato,M.Matsumoto; Unsteady Pressure Evaluation on Oscillating Bluff Body by Vortex Method, Proceedings of the ICWE'96, Fort Collins, 1996

[7] N.Shiraishi, M.Matsumoto; Vortex-induced oscillation on bridge box-girders, Proceedings of JSCE, 1982, pp.17-30

[8] H.Yamada, T.Miyata; Unsteady pressure on rectangular cross section in state of vortex-induced oscillation, Journal of Wind Engineering, No.38, 1989

[9] T.Shimada; Flow simulation of 2-D rectangular cross section, Proceedings of the 13th National Syposium on Wind Engineering, 1994

[10] H.Shirato, M.Matsumoto; Flow visualization using CFD tecnique, Proceeding of the Seventh KAIST-NTU-KU Tri-lateral Seminar on Civil Engineering, 1997

[11] M.Matsumoto; Flutter instabilities for rectangular cylinders, Proceedings of JSCE Structural Engineering, vol.41A, 1995

[12] T.Mizota; Experimental study on flow around an oscillating rectanglar prisms and its flow-induced forces in smooth flow, Ph.D Thesis, 1983